A BOOK OF
ABSTRACT ALGEBRA

A BOOK OF ABSTRACT ALGEBRA

Charles C. Pinter

Professor of Mathematics
Bucknell University

McGraw-Hill Book Company

New York   St. Louis   San Francisco   Auckland   Bogotá   Hamburg
Johannesburg   London   Madrid   Mexico   Montreal   New Delhi
Panama   Paris   São Paulo   Singapore   Sydney   Tokyo   Toronto

This book was set in Times Roman by Santype-Byrd.
The editors were John J. Corrigan and James S. Amar;
the production supervisor was Leroy A. Young.
The drawings were done by VIP Graphics.
The cover was designed by Scott Chelius.
R. R. Donnelley & Sons Company was printer and binder.

A BOOK OF ABSTRACT ALGEBRA

1234567890 DODO 8987654321

ISBN 0-07-050130-0

Library of Congress Cataloging in Publication Data

Pinter, Charles C., date
    A book of abstract algebra.

    Includes index.
    1. Algebra, Abstract.   I. Title.
QA162.P56   512′.02   81-8420
ISBN 0-07-050130-0   AACR2

To my colleagues in Brazil, especially
  Newton da Costa,
    Ayda Arruda, and
      Elias Alves,
as well as to others. In appreciation of their loyal and valued friendship.

CONTENTS*

* Italic headings indicate topics discussed in the exercise sections.

PREFACE

Once, when I was a student struggling to understand modern algebra, I was told to view this subject as an intellectual chess game, with conventional moves and prescribed rules of play. I was ill served by this bit of extemporaneous advice, and vowed never to perpetutate the falsehood that mathematics is purely—or primarily—a formalism. My pledge has strongly influenced the shape and style of this book.

While giving due emphasis to the deductive aspect of modern algebra, I have endeavored here to present modern algebra as a lively branch of mathematics, having considerable imaginative appeal and resting on some firm, clear, and familiar intuitions. I have devoted a great deal of attention to bringing out the *meaningfulness* of algebraic concepts, by tracing these concepts to their origins in classical algebra and at the same time exploring their connections with other parts of mathematics, especially geometry, number theory, and aspects of computation and equation-solving.

In an introductory chapter entitiled Why Abstract Algebra?, as well as in numerous historical asides, concepts of abstract algebra are traced to the historic context in which they arose. I have attempted to show that they arose without artifice, as a natural response to particular needs, in the course of a natural process of evolution. Furthermore, I have endeavored to bring to light, explicitly, the *intuitive content* of the algebraic concepts used in this book. Concepts are more meaningful to students when the students are able to represent those concepts in their minds by clear and familiar mental images. Accordingly, the process of concrete concept-formation is developed with care throughout this book.

I have deliberately avoided a rigid conventional format, with its succession of *definition, theorem, proof, corollary, example*. In my experience, that kind of format encourages some students to believe that mathematical concepts have a merely conventional character, and may encourage rote memorization. Instead, each chapter has the form of a discussion with the student, with the accent on explaining and motivating.

In an effort to avoid fragmentation of the subject matter into loosely related definitions and results, each chapter is built around a central theme, and remains anchored to this focal point. In the later chapters, especially, this focal point is a specific application or use. Details of every topic are then woven into the general discussion, so as to keep a natural flow of ideas running through each chapter.

The arrangement of topics is designed to avoid tedious proofs and long-winded explanations. Routine arguments are worked into the discussion whenever this seems natural and appropriate, and proofs to theorems are seldom more than a few lines long. (There are, of course, a few exceptions to this.) Elementary background material is filled in *as it is needed*. For example, a brief chapter on functions precedes the discussion of permutation groups, and a chapter on equivalence relations and partitions paves the way for Lagrange's theorem.

This book addresses itself especially to the *average* student, to enable him or her to learn and *understand* as much algebra as possible. In scope and subject-matter coverage, it is no different from many other standard texts. It begins with the promise of demonstrating the unsolvability of the quintic, and ends with that promise fulfilled. Standard topics are discussed in their usual order, and many advanced and peripheral subjects are introduced in the exercises, accompanied by ample instruction and commentary.

I have included a copious supply of exercises — probably more exercises than in other books at this level. They are designed to offer a wide range of experiences to students at different levels of ability. There is some novelty in the way the exercises are organized: at the end of each chapter, the exercises are grouped into *Exercise Sets*, each Set containing about six to eight exercises and headed by a descriptive title. Each Set touches upon an idea or skill covered in the chapter.

The first few Exercise Sets in each chapter contain problems which are essentially computational or manipulative. Then, there are two or three Sets of simple proof-type questions, which require mainly the ability to put together definitions and results with understanding of their meaning. After that, I have endeavored to make the exercises more interesting by arranging them so that in each Set a new result is proved, or new light is shed on the subject of the chapter.

As a rule, all the exercises have the same weight: very simple exercises are grouped together as parts of a single problem, and conversely, problems which require a complex argument are broken into several subproblems which the student may tackle in turn. I have selected mainly problems which have intrinsic relevance, and are not merely drill, on the premiss that this is much more satisfying to the student.

ACKNOWLEDGMENTS

I would like to express my thanks for the many useful comments and suggestions provided by colleagues who reviewed this text during the course of its development, especially to William P. Berlinghoff, Southern Connecticut State College; John Ewing, Indiana University; Grant A. Fraser, The University of Santa Clara; Eugene Spiegel, University of Connecticut; Sherman K. Stein, University of California at Davis; and William Wickless, University of Connecticut.

My special thanks go to Carol Napier, mathematics editor at McGraw-Hill during the writing of this book. She found merit in the manuscript at an early stage and was a moving spirit in its subsequent development. I am grateful for her steadfast encouragement, perceptiveness, imagination, and advice which was always "on target."

Charles C. Pinter

WHY ABSTRACT ALGEBRA?

When we open a textbook of abstract algebra for the first time and peruse the table of contents, we are struck by the unfamiliarity of almost every topic we see listed. Algebra is a subject we know well, but here it looks surprisingly different. What are these differences, and how fundamental are they?

First, there is a major difference in emphasis. In elementary algebra we learned the basic symbolism and methodology of algebra; we came to see how problems of the real world can be reduced to sets of equations and how these equations can be solved to yield numerical answers. This technique for translating complicated problems into symbols is the basis for all further work in mathematics and the exact sciences, and is one of the triumphs of the human mind. However, algebra is not only a technique, it is also a branch of learning, a *discipline*, like calculus or physics or chemistry. It is a coherent and unified body of knowledge which may be studied systematically, starting from first principles and building up. So the first difference between the elementary and the more advanced course in algebra is that, whereas earlier we concentrated on technique, we will now develop that branch of mathematics called algebra in a systematic way. Ideas and general principles will take precedence over problem solving. (By the way, this does not mean that modern algebra has no applications—quite the opposite is true, as we will see soon.)

Algebra at the more advanced level is often described as *modern* or *abstract* algebra. In fact, both of these descriptions are partly misleading. Some of the great discoveries in the upper reaches of present-day algebra (for example, the so-called Galois theory) were known many years before the American Civil War; and the broad aims of algebra today were clearly stated by Leibniz in the seventeenth century. Thus, "modern" algebra is not so very modern, after all! To what extent is it *abstract*? Well, abstraction is all relative; one person's abstraction is another person's bread and butter. The abstract tendency in mathematics is a little like the situation of changing moral codes, or changing tastes in music: What shocks one generation becomes the norm in the next. This has been true throughout the history of mathematics.

For example, 1000 years ago negative numbers were considered to be an outrageous idea. After all, it was said, numbers are for counting: we may have one orange, or two oranges, or no oranges at all; but how can we have minus an orange? The *logisticians*, or professional calculators, of those days used negative numbers as an aid in their computations; they considered these numbers to be a useful fiction, for if you believe in them then every linear equation $ax + b = 0$ has a solution (namely $x = -b/a$, provided $a \neq 0$). Even the great Diophantus once described the solution of $4x + 6 = 2$ as an *absurd* number. The idea of a system of numeration which included negative numbers was far too abstract for many of the learned heads of the tenth century!

The history of the complex numbers (numbers which involve $\sqrt{-1}$) is very much the same. For hundreds of years, mathematicians refused to accept them because they couldn't find concrete examples or applications. (They are now a basic tool of physics.)

Set theory was considered to be highly abstract a few years ago, and so were other commonplaces of today. Many of the abstractions of modern algebra are already being used by scientists, engineers, and computer specialists in their everyday work. They will soon be common fare, respectably "concrete," and by then there will be new "abstractions."

Later in this chapter we will take a closer look at the particular brand of abstraction used in algebra. We will consider how it came about and why it is useful.

Algebra has evolved considerably, especially during the past 100 years. Its growth has been closely linked with the development of other branches of mathematics, and it has been deeply influenced by philosophical ideas on the nature of mathematics and the role of logic. To help us understand the nature and spirit of modern algebra, we should take a brief look at its origins.

ORIGINS

The order in which subjects follow each other in our mathematical education tends to repeat the historical stages in the evolution of mathematics. In this scheme, elementary algebra corresponds to the great classical age of algebra, which spans about 300 years from the sixteenth through the eighteenth centuries. It was during these years that the art of solving equations became highly developed and modern symbolism was invented.

The word "algebra"—*al jebr* in Arabic— was first used by Mohammed of Kharizm, who taught mathematics in Baghdad during the ninth century. The word may be roughly translated as "reunion," and describes his method for collecting the terms of an equation in order to solve it. It is an amusing fact that the word "algebra" was first used in Europe in quite another context. In Spain barbers were called *algebristas*, or bonesetters (they re*united* broken bones), because medieval barbers did bonesetting and bloodletting as a sideline to their usual business.

The origin of the word clearly reflects the actual content of algebra at that time, for it was mainly concerned with ways of solving equations. In fact, Omar Khayyam, who is best remembered for his brilliant verses on wine, song, love, and friendship which are collected in the *Rubaiyat*—but who was also a great mathematician—explicitly defined algebra as the *science of solving equations.*

Thus, as we enter upon the threshold of the classical age of algebra, its central theme is clearly identified as that of solving equations. Methods of solving the linear equation $ax + b = 0$ and the quadratic $ax^2 + bc + c = 0$ were well known even before the Greeks. But nobody had yet found a general solution for *cubic* equations

$$x^3 + ax^2 + bx = c$$

or *quartic* (fourth-degree) equations

$$x^4 + ax^3 + bx^2 + cx = d$$

This great accomplishment was the triumph of sixteenth century algebra.

The setting is Italy and the time is the Renaissance—an age of high adventure and brilliant achievement, when the wide world was reawakening after the long austerity of the Middle Ages. America had just been discovered, classical knowledge had been brought to light, and prosperity had returned to the great cities of Europe. It was a heady age when nothing seemed impossible and even the old barriers of birth and rank could be overcome. Courageous individuals set out for great adventures in the far corners of the earth, while others, now confident once again of the power of the human mind, were boldly exploring the limits of knowledge in the sciences and the arts. The ideal was to be bold and many-faceted, to "know something of everything, and everything of at least one thing." The great traders were patrons of the arts, the finest minds in science were adepts at political intrigue and high finance. The study of algebra was reborn in this lively milieu.

Those men who brought algebra to a high level of perfection at the beginning of its classical age—all typical products of the Italian Renaissance—were as colorful and extraordinary a lot as have ever appeared in a chapter of history. Arrogant and unscrupulous, brilliant, flamboyant, swaggering, and remarkable, they lived their lives as they did their work: with style and panache, in brilliant dashes and inspired leaps of the imagination.

The spirit of scholarship was not exactly as it is today. These men, instead of publishing their discoveries, kept them as well-guarded secrets to be used against each other in problem-solving competitions. Such contests were a popular attraction: heavy bets were made on the rival parties, and their reputations (as well as a substantial purse) depended on the outcome.

One of the most remarkable of these men was Girolamo Cardan. Cardan was born in 1501 as the illegitimate son of a famous jurist of the city of Pavia. A man of passionate contrasts, he was destined to become famous as a physician, astrologer, and mathematician—and notorious as a compulsive gambler, scoundrel, and heretic. After he graduated in medicine, his efforts to build up a medical practice were so unsuccessful that he and his wife were forced to seek refuge in the poorhouse. With the help of friends he became a lecturer in mathematics, and, after he cured the child of a senator from Milan, his medical career also picked up. He was finally admitted to the college of physicians and soon became its rector. A brilliant doctor, he gave the first clinical description of typhus fever, and as his fame spread he became the personal physician of many of the high and mighty of his day.

Cardan's early interest in mathematics was not without a practical side. As an inveterate gambler he was fascinated by what he recognized to be the laws of chance. He wrote a gamblers' manual entitled *Book on Games of Chance*, which presents the first systematic computations of probabilities. He also needed mathematics as a tool in casting horoscopes, for his fame as an astrologer was great and his predictions were highly regarded and sought after. His most important achievement was the publication of a book called *Ars Magna* (*The Great Art*), in which he presented systematically all the algebraic knowledge of his time. However, as already stated, much of this knowledge was the personal secret of its practitioners, and had to be wheedled out of them by cunning and deceit. The most important accomplishment of the day, the general solution of the cubic equation which had been discovered by Tartaglia, was obtained in that fashion.

Tartaglia's life was as turbulent as any in those days. Born with the name of Niccolò Fontana about 1500, he was present at the occupation of Brescia by the French in 1512. He and his father fled with many others into a cathedral for sanctuary, but in the heat of battle the soldiers massacred the hapless citizens even in that holy place. The father was killed, and the boy, with a split skull and a deep saber cut across his jaws and palate, was left for dead. At night his mother stole into the cathedral and managed to carry him off; miraculously he survived. The horror of what he had witnessed caused him to stammer for the rest of his life, earning him the nickname *Tartaglia*, "the stammerer," which he eventually adopted.

Tartaglia received no formal schooling, for that was a privilege of rank and wealth. However, he taught himself mathematics and became one of the most gifted mathematicians of his day. He translated Euclid and Archimedes and may be said to have originated the science of ballistics, for he wrote a treatise on gunnery which was a pioneering effort on the laws of falling bodies.

In 1535 Tartaglia found a way of solving any cubic equation of the form $x^3 + ax^2 = b$ (that is, without an x term). When he announced his accomplishment (without giving any details, of course), he was challenged to an algebra contest by a certain Antonio Fior, a pupil of the celebrated professor of mathematics Scipio del Ferro. Scipio had already found a method for solving any cubic equation of the form $x^3 + ax = b$ (that is, without an x^2 term), and had confided his secret to his pupil Fior. It was agreed that each contestant was to draw up 30 problems and hand the list to his opponent. Whoever solved the greater number of problems would receive a sum of money deposited with a lawyer. A few days before the contest, Tartaglia found a way of extending his method so as to solve *any* cubic equation. In less than 2 hours he solved all his opponent's problems, while his opponent failed to solve even one of those proposed by Tartaglia.

For some time Tartaglia kept his method for solving cubic equations to himself, but in the end he succumbed to Cardan's accomplished powers of persuasion. Influenced by Cardan's promise to help him become artillery adviser to the Spanish army, he revealed the details of his method to Cardan under the promise of strict secrecy. A few years later, to Tartaglia's unbelieving amazement and indignation, Cardan published Tartaglia's method in his book *Ars Magna*. Even though he gave Tartaglia full credit as the originator of the method, there can be no doubt that he broke his solemn promise. A bitter dispute arose between the mathematicians, from which Tartaglia was perhaps lucky to escape alive. He lost his position as public lecturer at Brescia, and lived out his remaining years in obscurity.

The next great step in the progress of algebra was made by another member of the same circle. It was Ludovico Ferrari who discovered the general method for solving quartic equations—equations of the form

$$x^4 + ax^3 + bx^2 + cx = d$$

Ferrari was Cardan's personal servant. As a boy in Cardan's service he learned Latin, Greek, and mathematics. He won fame after defeating Tartaglia in a contest in 1548, and received an appointment as supervisor of tax assessments in Mantua. This position brought him wealth and influence, but he was not able to dominate his own violent, blasphemous disposition. He quarreled with the regent of Mantua, lost his position, and died at the age of 43. Tradition has it that he was poisoned by his sister.

As for Cardan, after a long career of brilliant and unscrupulous achievement, his luck finally abandoned him. Cardan's son poisoned his unfaithful wife and was executed in 1560. Ten years later, Cardan was arrested for heresy because he published a horoscope of Christ's life. He spent several months in jail and was released after renouncing his heresy privately, but lost his university position and the right to publish books. He was left with a small pension which had been granted to him, for some unaccountable reason, by the Pope.

As this colorful time draws to a close, algebra emerges as a major branch of mathematics. It became clear that methods can be found to solve many different types of equations. In particular, formulas had been discovered which yielded the roots of all cubic and quartic equations. Now the challenge was clearly out to take the next step, namely to find a formula for the roots of equations of degree 5 or higher (in other words, equations with an x^5 term, or an x^6 term, or higher). During the next 200 years, there was hardly a mathematician of distinction who did not try to solve this problem, but none succeeded. Progress was made in new parts of algebra, and algebra was linked to geometry with the invention of analytic geometry. But the problem of solving equations of degree higher than 4 remained unsettled. It was, in the expression of Lagrange, "a challenge to the human mind."

It was therefore a great surprise to all mathematicians when in 1824 the work of a young Norwegian prodigy named Niels Abel came to light. In his work, Abel showed that *there does not exist any formula* (in the conventional sense we have in mind) for the roots of an algebraic equation whose degree is 5 or greater. This sensational discovery brings to a close what is called the classical age of algebra. Throughout this age algebra was conceived essentially as the science of solving equations, and now the outer limits of this quest had apparently been reached. In the years ahead, algebra was to strike out in new directions.

THE MODERN AGE

About the time Niels Abel made his remarkable discovery, several mathematicians, working independently in different parts of Europe, began raising questions about algebra which had never been considered before. Their researches in different branches of mathematics had led them to investigate "algebras" of a very unconventional kind—and in connection with these algebras they had to find answers to questions which had nothing to do with solving equations. Their work had important applications, and was soon to compel mathematicians to greatly enlarge their conception of what algebra is about.

The new varieties of algebra arose as a perfectly natural development in connection with the application of mathematics to practical problems. This is certainly true for the example we are about to look at first.

The Algebra of Matrices

A *matrix* is a rectangular array of numbers such as

$$\begin{pmatrix} 2 & 11 & -3 \\ 9 & 0.5 & 4 \end{pmatrix}$$

Such arrays come up naturally in many situations, for example, in the solution of simultaneous linear equations. The above matrix, for instance, is the *matrix of coefficients* of the pair of equations

$$2x + 11y - 3z = 0$$

$$9x + 0.5y + 4z = 0$$

Since the solution of this pair of equations depends only on the coefficients, we may solve it by working on the matrix of coefficients alone and ignoring everything else.

We may consider the entries of a matrix to be arranged in *rows* and *columns*; the above matrix has two rows which are

$$(2 \quad 11 \quad -3) \quad \text{and} \quad (9 \quad 0.5 \quad 4)$$

and three columns which are

$$\begin{pmatrix} 2 \\ 9 \end{pmatrix} \quad \begin{pmatrix} 11 \\ 0.5 \end{pmatrix} \quad \text{and} \quad \begin{pmatrix} -3 \\ 4 \end{pmatrix}$$

It is a 2×3 matrix.

To simplify our discussion, we will consider only 2×2 matrices in the remainder of this section.

Matrices are added by adding corresponding entries:

$$\begin{pmatrix} a & b \\ c & d \end{pmatrix} + \begin{pmatrix} a' & b' \\ c' & d' \end{pmatrix} = \begin{pmatrix} a + a' & b + b' \\ c + c' & d + d' \end{pmatrix}$$

The matrix

$$\mathbf{O} = \begin{pmatrix} 0 & 0 \\ 0 & 0 \end{pmatrix}$$

is called the *zero matrix* and behaves, under addition, like the number zero.

The multiplication of matrices is a little more difficult. First, let us recall that the *dot product* of two vectors (a, b) and (a', b') is

$$(a, b) \cdot (a', b') = aa' + bb'$$

that is, we multiply corresponding components and add. Now, suppose we want to multiply two matrices $\mathbf{A}$ and $\mathbf{B}$; we obtain the product $\mathbf{AB}$ as follows:

The entry in the first row and first column of $\mathbf{AB}$, that is, in *this* position

$$\begin{pmatrix} x & | \\ \text{---} & | \text{---} \end{pmatrix}$$

is equal to the dot product of the first row of $\mathbf{A}$ by the first column of $\mathbf{B}$. The entry in the first row and *second* column of $\mathbf{AB}$, in other words, *this* position

$$\begin{pmatrix} | & x \\ \text{---} & | \end{pmatrix}$$

is equal to the dot product of the first row of $\mathbf{A}$ by the *second* column of $\mathbf{B}$.

And so on. For example,

$$\begin{pmatrix} 1 & 2 \\ 3 & 0 \end{pmatrix}\begin{pmatrix} 1 & 1 \\ 2 & 0 \end{pmatrix} = \begin{pmatrix} 5 & \\ & \end{pmatrix} \qquad \begin{pmatrix} 1 & 2 \\ 3 & 0 \end{pmatrix}\begin{pmatrix} 1 & 1 \\ 2 & 0 \end{pmatrix} = \begin{pmatrix} & 1 \\ & \end{pmatrix}$$

$$\begin{pmatrix} 1 & 2 \\ 3 & 0 \end{pmatrix}\begin{pmatrix} 1 & 1 \\ 2 & 0 \end{pmatrix} = \begin{pmatrix} & \\ 3 & \end{pmatrix} \qquad \begin{pmatrix} 1 & 2 \\ 3 & 0 \end{pmatrix}\begin{pmatrix} 1 & 1 \\ 2 & 0 \end{pmatrix} = \begin{pmatrix} & \\ & 3 \end{pmatrix}$$

So finally,

$$\begin{pmatrix} 1 & 2 \\ 3 & 0 \end{pmatrix}\begin{pmatrix} 1 & 1 \\ 2 & 0 \end{pmatrix} = \begin{pmatrix} 5 & 1 \\ 3 & 3 \end{pmatrix}$$

The rules of algebra for matrices are very different from the rules of "conventional" algebra. For instance, the commutative law of multiplication, $\mathbf{AB} = \mathbf{BA}$, is not true. Here is a simple example:

$$\underbrace{\begin{pmatrix} 1 & 1 \\ 1 & 1 \end{pmatrix}}_{\mathbf{A}}\underbrace{\begin{pmatrix} 1 & 1 \\ 1 & 0 \end{pmatrix}}_{\mathbf{B}} = \underbrace{\begin{pmatrix} 2 & 1 \\ 2 & 1 \end{pmatrix}}_{\mathbf{AB}} \neq \underbrace{\begin{pmatrix} 2 & 2 \\ 1 & 1 \end{pmatrix}}_{\mathbf{BA}} = \underbrace{\begin{pmatrix} 1 & 1 \\ 1 & 0 \end{pmatrix}}_{\mathbf{B}}\underbrace{\begin{pmatrix} 1 & 1 \\ 1 & 1 \end{pmatrix}}_{\mathbf{A}}$$

If A is a real number and $A^2 = 0$, then necessarily $A = 0$; but this is not true of matrices. For example,

$$\underbrace{\begin{pmatrix} 1 & -1 \\ 1 & -1 \end{pmatrix}}_{\mathbf{A}}\underbrace{\begin{pmatrix} 1 & -1 \\ 1 & -1 \end{pmatrix}}_{\mathbf{A}} = \begin{pmatrix} 0 & 0 \\ 0 & 0 \end{pmatrix}$$

that is, $\mathbf{A}^2 = \mathbf{0}$ although $\mathbf{A} \neq \mathbf{0}$.

In the algebra of numbers, if $AB = AC$ where $A \neq 0$, we may cancel A and conclude that $B = C$. In matrix algebra we cannot. For example,

$$\underbrace{\begin{pmatrix} 0 & 0 \\ 0 & 1 \end{pmatrix}}_{\mathbf{A}}\underbrace{\begin{pmatrix} 1 & 1 \\ 1 & 1 \end{pmatrix}}_{\mathbf{B}} = \begin{pmatrix} 0 & 0 \\ 1 & 1 \end{pmatrix} = \underbrace{\begin{pmatrix} 0 & 0 \\ 0 & 1 \end{pmatrix}}_{\mathbf{A}}\underbrace{\begin{pmatrix} 0 & 0 \\ 1 & 1 \end{pmatrix}}_{\mathbf{C}}$$

that is, $\mathbf{AB} = \mathbf{AC}, \mathbf{A} \neq \mathbf{0}$, yet $\mathbf{B} \neq \mathbf{C}$.

The *identity* matrix

$$\mathbf{I} = \begin{pmatrix} 1 & 0 \\ 0 & 1 \end{pmatrix}$$

corresponds in matrix multiplication to the number 1; for we have

$\mathbf{AI} = \mathbf{IA} = \mathbf{A}$ for every 2×2 matrix $\mathbf{A}$. If A is a number and $A^2 = 1$, we conclude that $A = \pm 1$. Matrices do not obey this rule. For example,

$$\underbrace{\begin{pmatrix} 1 & 0 \\ 1 & -1 \end{pmatrix}}_{\mathbf{A}} \underbrace{\begin{pmatrix} 1 & 0 \\ 1 & -1 \end{pmatrix}}_{\mathbf{A}} = \underbrace{\begin{pmatrix} 1 & 0 \\ 0 & 1 \end{pmatrix}}_{\mathbf{I}}$$

that is, $\mathbf{A}^2 = \mathbf{I}$, and yet $\mathbf{A}$ is neither $\mathbf{I}$ nor $-\mathbf{I}$.

No more will be said about the algebra of matrices at this point, except that we must be aware, once again, that it is a new game whose rules are quite different from those we apply in conventional algebra.

Boolean Algebra

An even more bizarre kind of algebra was developed in the mid-nineteenth century by an Englishman named George Boole. This algebra—subsequently named boolean algebra after its inventor—has a myriad of applications today. It is formally the same as the algebra of sets.

If S is a set, we may consider *union* and *intersection* to be operations on the subsets of S. Let us agree provisionally to write

$$A + B \quad \text{for} \quad A \cup B$$

and

$$A \cdot B \quad \text{for} \quad A \cap B$$

(This convention is not unusual.) Then,

$$A + B = B + A \qquad A \cdot B = B \cdot A$$

$$A \cdot (B + C) = A \cdot B + A \cdot C$$

$$A + \emptyset = A \qquad A \cdot \emptyset = \emptyset$$

and so on.

These identities are analogous to the ones we use in elementary algebra. But the following identities are also true, and they have no counterpart in conventional algebra:

$$A + (B \cdot C) = (A + B) \cdot (A + C)$$

$$A + A = A \qquad A \cdot A = A$$

$$(A + B) \cdot A = A \qquad (A \cdot B) + A = A$$

and so on.

This unusual algebra has become a familiar tool for people who work with electrical networks, computer systems, codes, and so on. It is as different from the algebra of numbers as it is from the algebra of matrices.

Other exotic algebras arose in a variety of contexts, often in connection with scientific problems. There were "complex" and "hypercomplex" algebras, algebras of vectors and tensors, and many others. Today it is estimated that over 200 different kinds of algebraic systems have been studied, each of which arose in connection with some application or specific need.

Algebraic Structures

As legions of new algebras began to occupy the attention of mathematicians, the awareness grew that algebra can no longer be conceived merely as the *science of solving equations*. It had to be viewed much more broadly as a branch of mathematics capable of revealing general principles which apply equally to *all known and all possible algebras*.

What is it that all algebras have in common? What trait do they share which lets us refer to all of them as "algebras"? In the most general sense, every algebra consists of a *set* (a set of numbers, a set of matrices, a set of switching components, or any other kind of set) and certain *operations* on that set. An operation is simply a way of combining any two members of a set to produce a unique third member of the same set.

Thus, we are led to the modern notion of algebraic structure. An *algebraic structure* is understood to be an arbitrary set, with one or more operations defined on it. And algebra, then, is defined to be *the study of algebraic structures*.

It is important that we be awakened to the full generality of the notion of algebraic structure. We must make an effort to discard all our preconceived notions of what an algebra is, and look at this new notion of algebraic structure in its naked simplicity. *Any* set, with a rule (or rules) for combining its elements, is already an algebraic structure. There does not need to be any connection with known mathematics. For example, consider the set of all colors (pure colors as well as color combinations), and the operation of mixing any two colors to produce a new color. This may be conceived as an algebraic structure. It obeys certain rules, such as the commutative law (mixing red and blue is the same as mixing blue and red). In a similar vein, consider the set of all musical sounds with the operation of combining any two sounds to produce a new (harmonious or disharmonious) combination.

As another example, imagine that the guests at a family reunion have made up a rule for picking the *closest common relative* of any two persons present at the reunion (and suppose that, for any two people at the reunion, their closest common relative is also present at the reunion). This, too, is an algebraic structure: we have a set (namely the set of persons at the reunion) and an operation on that set (namely the "closest common relative" operation).

As the general notion of algebraic structure became more familiar (it was not fully accepted until the early part of the twentieth century), it was bound to have a profound influence on what mathematicians perceived algebra to *be*. In the end it became clear that the purpose of algebra is to study algebraic structures, and nothing less than that. Ideally it should aim to be a general science of algebraic structures whose results should have applications to particular cases, thereby making contact with the older parts of algebra. Before we take a closer look at this program, we must briefly examine another aspect of modern mathematics, namely the increasing use of the axiomatic method.

AXIOMS AND MEN

The axiomatic method is beyond doubt the most remarkable invention of antiquity, and in a sense the most puzzling. It appeared suddenly in Greek geometry in a highly developed form—already sophisticated, elegant, and thoroughly modern in style. Nothing seems to have foreshadowed it and it was unknown to ancient mathematicians before the Greeks. It appears for the first time in the light of history in that great textbook of early geometry, Euclid's *Elements*. Its origins—the first tentative experiments in formal deductive reasoning which must have preceded it—remain steeped in mystery.

Euclid's *Elements* embodies the axiomatic method in its purest form. This amazing book contains 465 geometric propositions, some fairly simple, some of astounding complexity. What is really remarkable, though, is that the 465 propositions, forming the largest body of scientific knowledge in the ancient world, are derived logically from only 10 premises which would pass as trivial observations of common sense. Typical of the premises are the following:

> *Things equal to the same thing are equal to each other.*
> *The whole is greater than the part.*
> *A straight line can be drawn through any two points.*
> *All right angles are equal.*

So great was the impression made by Euclid's *Elements* on following generations that it became the model of correct mathematical form and remains so to this day.

It would be wrong to believe there was no notion of demonstrative mathematics before the time of Euclid. There is evidence that the earliest geometers of the ancient Middle East used reasoning to discover geometric principles. They found proofs and must have hit upon many of the same proofs we find in Euclid. The difference is that Egyptian and Babylonian mathematicians considered logical demonstration to be an auxiliary process, like the preliminary sketch made by artists—a private mental process which guided them to a result but did not deserve to be recorded. Such an attitude shows little understanding of the true nature of geometry and does not contain the seeds of the axiomatic method.

It is also known today that many—maybe most—of the geometric theorems in Euclid's *Elements* came from more ancient times, and were probably borrowed by Euclid from Egyptian and Babylonian sources. However, this does not detract from the greatness of his work. Important as are the contents of the *Elements*, what has proved far more important for posterity is the formal manner in which Euclid presented these contents. The heart of the matter was the way he *organized* geometric facts— arranged them into a logical sequence where each theorem builds on preceding theorems and then forms the logical basis for other theorems.

(We must carefully note that the axiomatic method is not a way of discovering facts but of organizing them. New facts in mathematics are found, as often as not, by inspired guesses or experienced intuition. To be accepted, however, they should be supported by proof in an axiomatic system.)

Euclid's *Elements* has stood throughout the ages as the model of organized, rational thought carried to its ultimate perfection. Mathematicians and philosophers in every generation have tried to imitate its lucid perfection and flawless simplicity. Descartes and Leibniz dreamed of organizing all human knowledge into an axiomatic system, and Spinoza created a deductive system of ethics patterned after Euclid's geometry. While many of these dreams have proved to be impractical, the method popularized by Euclid has become the prototype of modern mathematical form. Since the middle of the nineteenth century, the axiomatic method has been accepted as the only correct way of organizing mathematical knowledge.

To perceive why the axiomatic method is truly central to mathematics, we must keep one thing in mind: mathematics by its nature is essentially *abstract*. For example, in geometry straight lines are not stretched threads, but a concept obtained by disregarding all the properties of stretched threads except that of extending in one direction. Similarly, the concept of a geometric figure is the result of idealizing from all the properties of actual objects and retaining only their spatial relationships. Now, since the objects of mathematics are *abstractions*, it stands to reason that we must acquire knowledge about them by logic and not by observation or experiment (for how can one experiment with an abstract thought?).

This remark applies very aptly to modern algebra. The notion of algebraic structure is obtained by idealizing from all particular, concrete systems of algebra. We choose to ignore the properties of the actual objects in a system of algebra (they may be numbers, or matrices, or whatever—we disregard what they *are*), and we turn our attention simply to the way they combine under the given operations. In fact, just as we disregard what the objects in a system *are*, we also disregard what the operations *do* to them. We retain only the equations and inequalities which hold in the system, for only these are relevant to algebra. Everything else may be discarded. Finally, equations and inequalities may be deduced from one another logically, just as spatial relationships are deduced from each other in geometry.

THE AXIOMATICS OF ALGEBRA

Let us remember that in the mid-nineteenth century, when eccentric new algebras seemed to show up at every turn in mathematical research, it was finally understood that sacrosanct laws such as the identities $ab = ba$ and $a(bc) = (ab)c$ are not inviolable—for there are algebras in which they do not hold. By varying or deleting some of these identities, or by replacing them by new ones, an enormous variety of new systems can be created.

Most importantly, mathematicians slowly discovered that all the algebraic laws which hold in any system can be derived from a few simple, basic ones. This is a genuinely remarkable fact, for it parallels the discovery made by Euclid that a few very simple geometric postulates are sufficient to prove all the theorems of geometry. As it turns out, then, we have the same phenomenon in algebra: a few simple algebraic equations offer themselves naturally as axioms, and from them all other facts may be proved.

These basic algebraic laws are familiar to most high school students today. We list them here for reference. We assume that A is any set and there is an operation on A which we designate with the symbol $*$.

$$a * b = b * a \tag{1}$$

If Equation (1) is true for any two elements a and b in A, we say that the operation $*$ is *commutative*. What it means, of course, is that the value of $a * b$ (or $b * a$) is independent of the order in which a and b are taken.

$$a * (b * c) = (a * b) * c \tag{2}$$

If Equation (2) is true for any three elements a, b, and c in A, we say the operation $*$ is *associative*. Remember that an operation is a rule for combining any *two* elements, so if we want to combine *three* elements, we can do so in different ways. If we want to combine a, b, and c *without changing their order*, we may either combine a with the result of combining b and c, which produces $a * (b * c)$; or we may first combine a with b, and then combine the result with c, producing $(a * b) * c$. The associative law asserts that these two possible ways of combining three elements (without changing their order) yield the same result.

$$\text{There exists an element } e \text{ in } A \text{ such that} \tag{3}$$
$$e * a = a \quad \text{and} \quad a * e = a \quad \text{for every } a \text{ in } A$$

If such an element e exists in A, we call it an *identity element* for the operation $*$. An identity element is sometimes called a "neutral" element, for it may be combined with any element a without altering a. For example, 0 is an identity element for addition, and 1 is an identity element for multiplication.

$$\text{For every element } a \text{ in } A \text{, there is an element } a^{-1} \text{ ("a inverse")}$$
$$\text{in } A \text{ such that} \tag{4}$$
$$a * a^{-1} = e \quad \text{and} \quad a^{-1} * a = e$$

If statement (4) is true in a system of algebra, we say that every element has an inverse with respect to the operation $*$. The meaning of the inverse should be clear: the combination of any element with its inverse produces the neutral element (one might roughly say that the inverse of a "neutralizes" a). For example, if A is a set of numbers and the operation is addition, then the inverse of any number a is $(-a)$; if the operation is multiplication, the inverse of any $a \neq 0$ is $1/a$.

Let us assume now that the same set A has a second operation, symbolized by $\perp$, as well as the operation $*$.

$$a * (b \perp c) = (a * b) \perp (a * c) \tag{5}$$

If Equation (5) holds for any three elements a, b, and c in A, we say that $*$ is *distributive* over $\perp$. If there are two operations in a system, they must interact in some way; otherwise there would be no need to consider them together. The distributive law is the most common way (but not the only possible one) for two operations to be related to one another.

There are other "basic" laws besides the five we have just seen, but these are the most common ones. The most important algebraic systems have axioms chosen from among them. For example, when a mathematician nowadays speaks of a *ring*, the mathematician is referring to a set A with two operations, usually symbolized by $+$ and $\cdot$, having the following axioms:

> *Addition is commutative and associative, it has a neutral element commonly symbolized by 0, and every element a has an inverse* $-a$ *with respect to addition. Multiplication is associative, has a neutral element 1, and is distributive over addition.*

Matrix algebra is a particular example of a ring, and all the laws of matrix algebra may be proved from the preceding axioms. However, there are many other examples of rings: rings of numbers, rings of functions, rings of code "words," rings of switching components, and a great many more. Every algebraic law which can be proved in a ring (from the preceding axioms) is true in every *example* of a ring. In other words, instead of proving the same formula repeatedly—once for numbers, once for matrices, once for switching components, and so on—it is sufficient nowadays to prove only that the formula holds in rings, and then of necessity it will be true in all the hundreds of different concrete examples of rings.

By varying the possible choices of axioms, we can keep creating new axiomatic systems of algebra endlessly. We may well ask: is it legitimate to study *any* axiomatic system, with *any* choice of axioms, regardless of usefulness, relevance, or applicability? There are "radicals" in mathematics who claim the freedom for mathematicians to study any system they wish, without the need to justify it. However, the practice in established mathematics is more conservative: particular axiomatic systems are investigated on account of their relevance to new and traditional problems and other parts of mathematics, or because they correspond to particular applications.

In practice, how is a particular choice of algebraic axioms made? Very simply: when mathematicians look at different parts of algebra and notice that a common pattern of proofs keeps recurring, and essentially the same assumptions need to be made each time, they find it natural to single out this choice of assumptions as the axioms for a new system. All the important new systems of algebra were created in this fashion.

ABSTRACTION REVISITED

Another important aspect of axiomatic mathematics is this: when we capture mathematical facts in an axiomatic system, we never try to reproduce the facts in full, but only that side of them which is important or relevant in a particular context. This process of *selecting what is relevant* and disregarding everything else is the very essence of abstraction.

This kind of abstraction is so natural to us as human beings that we practice it all the time without being aware of doing so. Like the Bourgeois Gentleman in Molière's play who was amazed to learn that he spoke in prose, some of us may be surprised to discover how much we think in abstractions. Nature presents us with a myriad of interwoven facts and sensations, and we are challenged at every instant to single out those which are immediately relevant and discard the rest. In order to make our surroundings comprehensible, we must continually pick out certain data and separate them from everything else.

For natural scientists, this process is the very core and essence of what they do. Nature is not made up of forces, velocities, and moments of inertia. Nature is a whole—nature simply *is*! The physicist isolates certain aspects of nature from the rest and finds the laws which govern these *abstractions*.

It is the same with mathematics. For example, the system of the integers (whole numbers), as known by our intuition, is a complex reality with many facets. The mathematician separates these facets from one another and studies them individually. From one point of view the set of the integers, with addition and multiplication, forms a *ring* (that is, it satisfies the axioms stated previously). From another point of view it is an ordered set, and satisfies special axioms of ordering. On a different level, the positive integers form the basis of "recursion theory," which singles out the particular way positive integers may be *constructed*, beginning with 1 and adding 1 each time.

It therefore happens that the traditional subdivision of mathematics into subject matters has been radically altered. No longer are the integers one subject, complex numbers another, matrices another, and so on; instead, particular *aspects* of these systems are isolated, put in axiomatic form, and studied abstractly without reference to any specific objects. The other side of the coin is that each aspect is shared by many of the traditional systems: for example, algebraically the integers form a ring, and so do the complex numbers, matrices, and many other kinds of objects.

There is nothing intrinsically new about this process of divorcing properties from the actual objects *having* the properties; as we have seen, it is precisely what geometry has done for more than 2000 years. Somehow, it took longer for this process to take hold in algebra.

The movement toward axiomatics and abstraction in modern algebra began about the 1830s and was completed 100 years later. The movement was tentative at first, not quite conscious of its aims, but it gained momentum as it converged with similar trends in other parts of mathematics. The thinking of many great mathematicians played a decisive role, but none left a deeper or longer lasting impression than a very young Frenchman by the name of Évariste Galois.

The story of Évariste Galois is probably the most fantastic and tragic in the history of mathematics. A sensitive and prodigiously gifted young man, he was killed in a duel at the age of 20, ending a life which in its brief span had offered him nothing but tragedy and frustration. When he was only a youth his father commited suicide, and Galois was left to fend for himself in the labyrinthine world of French university life and student politics. He was twice refused admittance to the École Polytechnique, the most prestigious scientific establishment of its day, probably because his answers to the entrance examination were too original and unorthodox. When he presented an early version of his important discoveries in algebra to the great academician Cauchy, this gentleman did not read the young student's paper, but lost it. Later, Galois gave his results to Fourier in the hope of winning the mathematics prize of the Academy of Sciences. But Fourier died, and that paper, too, was lost. Another paper submitted to Poisson was eventually returned because Poisson did not have the interest to read it through.

Galois finally gained admittance to the École Normale, another focal point of research in mathematics, but he was soon expelled for writing an essay which attacked the king. He was jailed twice for political agitation in the student world of Paris. In the midst of such a turbulent life, it is hard to believe that Galois found time to create his colossally original theories on algebra.

What Galois did was to tie in the problem of finding the roots of equations with new discoveries on groups of permutations. He explained exactly *which* equations of degree 5 or higher have solutions of the traditional kind—and which others do not. Along the way, he introduced some amazingly original and powerful concepts, which form the framework of much algebraic thinking to this day. Although Galois did not work explicitly in axiomatic algebra (which was unknown in his day), the abstract notion of algebraic structure is clearly prefigured in his work.

In 1832, when Galois was only 20 years old, he was challenged to a duel. What argument led to the challenge is not clear: some say the issue was political, while others maintain the duel was fought over a fickle lady's wavering love. The truth may never be known, but the turbulent, brilliant, and idealistic Galois died of his wounds. Fortunately for mathematics, the night before the duel he wrote down his main mathematical results and entrusted them to a friend. This time, they weren't lost—but they were only published 15 years after his death. The mathematical world was not ready for them before then!

Algebra today is organized axiomatically, and as such it is abstract. Mathematicians study algebraic structures from a general point of view, compare different structures, and find relationships between them. This abstraction and generalization might appear to be hopelessly impractical—but it is not! The general approach in algebra has produced powerful new methods for "algebraizing" different parts of mathematics and science, formulating problems which could never have been formulated before, and finding entirely new kinds of solutions.

Such excursions into pure mathematical fancy have an odd way of running ahead of physical science, providing a theoretical framework to account for facts even before those facts are fully known. This pattern is so characteristic that many mathematicians see themselves as pioneers in a world of *possibilities* rather than facts. Mathematicians study *structure* independently of content, and their science is a voyage of exploration through all the kinds of structure and order which the human mind is capable of discerning.

TWO

OPERATIONS

Addition, subtraction, multiplication, division—these and many others are familiar examples of operations on appropriate sets of numbers.

Intuitively, an operation on a set A is a way of combining any two elements of A to produce another element in the same set A.

Every operation is denoted by a symbol, such as $+$, $\times$, or $\div$. In this book we will look at operations from a lofty perspective; we will discover facts pertaining to operations *generally* rather than to specific operations on specific sets. Accordingly, we will sometimes make up operation symbols such as $*$ and $\odot$ to refer to arbitrary operations on arbitrary sets.

Let us now define formally what we mean by an *operation* on a set A. Let A be *any set*:

An operation $*$ on A is a rule which assigns to each ordered pair (a, b) of elements of A exactly one element $a * b$ in A.

There are three aspects of this definition which need to be stressed:

$a * b$ *is defined for **every** ordered pair* (a, b) *of elements of A.* $\qquad$ (1)

There are many rules which look deceptively like operations but are not, because this condition fails. Often $a * b$ is defined for all the obvious choices of a and b, but remains undefined in a few exceptional cases. For example, division does not qualify as an operation on the set $\mathbb{R}$ of the real numbers, for there are ordered pairs such as $(3, 0)$ whose quotient $3/0$ is undefined. In order to be an operation on $\mathbb{R}$, division would have to associate a real number a/b with *every* ordered pair (a, b) of elements of $\mathbb{R}$. No exceptions allowed!

$a * b$ *must be **uniquely** defined.* $\qquad$ (2)

In other words, the value of $a * b$ must be given unambiguously. For example, one might attempt to define an operation $\square$ on the set $\mathbb{R}$ of the real numbers by letting $a \square b$ be the number whose square is ab. Obviously this is ambiguous because $2 \square 8$, let us say, may be either 4 or -4. Thus, $\square$ does not qualify as an operation on $\mathbb{R}$!

$$\text{If } a \text{ and } b \text{ are in } A, a * b \text{ must be in } A. \tag{3}$$

This condition is often expressed by saying that A is *closed* under the operation $*$. If we propose to define an operation $*$ on a set A, we must take care that $*$, when applied to elements of A, *does not take us out of A*. For example, division cannot be regarded as an operation on the set of the integers, for there are pairs of integers such as (3, 4) whose quotient 3/4 is not an integer.

On the other hand, division does qualify as an operation on the set of all the *positive real numbers*, for the quotient of any two positive real numbers is a uniquely determined positive real number.

An operation is *any* rule which assigns to each ordered pair of elements of A a unique element in A. Therefore it is obvious that there are, in general, many possible operations on a given set A. If, for example, A is a set consisting of just two distinct elements, say a and b, each operation on A may be described by a table such as this one:

(x, y)	$x * y$
(a, a)	
(a, b)	
(b, a)	
(b, b)	

In the left column are listed the four possible ordered pairs of elements of A, and to the right of each pair (x, y) is the value of $x * y$. Here are a few of the possible operations:

(x,y)		(x,y)		(x,y)		(x,y)	
(a, a)	a	(a, a)	a	(a, a)	b	(a, a)	b
(a, b)	a	(a, b)	b	(a, b)	a	(a, b)	b
(b, a)	a	(b, a)	a	(b, a)	b	(b, a)	b
(b, b)	a	(b, b)	b	(b, b)	a	(b, b)	a

Each of these tables describes a *different* operation on A. Each table has four rows, and each row may be filled with either an a or a b; hence there are 16 possible ways of filling the table, corresponding to *16 possible operations* on the set *A*.

We have already seen that any operation on a set *A* comes with certain "options." An operation ∗ may be *commutative*, that is, it may satisfy

$$a * b = b * a \tag{4}$$

for any two elements *a* and *b* in *A*. It may be *associative*, that is, it may satisfy the equation

$$(a * b) * c = a * (b * c) \tag{5}$$

for any three elements *a*, *b*, and *c* in *A*.

To understand the importance of the associative law, we must remember that an operation is a way of combining *two* elements; so if we want to combine *three* elements, we can do so in different ways. If we want to combine *a*, *b*, and *c* *without changing their order*, we may either combine *a* with the result of combining *b* and *c*, which produces *a* ∗ (*b* ∗ *c*); or we may first combine *a* with *b*, and then combine the result with *c*, producing (*a* ∗ *b*) ∗ *c*. The associative law asserts that these two possible ways of combining three elements (without changing their order) produce the same result.

For example, the addition of real numbers is associative because $a + (b + c) = (a + b) + c$. However, division of real numbers is *not* associative: for instance, 3/(4/5) is 15/4, whereas (3/4)/5 is 3/20.

If there is an element *e* in *A* with the property that

$$e * a = a \quad \text{and} \quad a * e = a \quad \text{for every element } a \text{ in } A \tag{6}$$

then *e* is called an *identity* or "neutral" element with respect to the operation ∗. Roughly speaking, Equation (6) tells us that when *e* is combined with any element *a*, it does not change *a*. For example, in the set ℝ of the real numbers, 0 is a neutral element for addition, and 1 is a neutral element for multiplication.

If *a* is any element of *A*, and *x* is an element of *A* such that

$$a * x = e \quad \text{and} \quad x * a = e \tag{7}$$

then *x* is called an *inverse* of *a*. Roughly speaking, Equation (7) tells us that when an element is combined with its inverse it produces the neutral element. For example, in the set ℝ of the real numbers, $-a$ is the inverse of *a* with respect to addition; if $a \neq 0$, then $1/a$ is the inverse of *a* with respect to multiplication.

The inverse of *a* is often denoted by the symbol a^{-1}.

EXERCISES

Throughout this book, the exercises are grouped into Exercise Sets, each Set being identified by a letter A, B, C, etc. and headed by a descriptive title. Each Exercise Set contains six to ten exercises, numbered consecutively. *Generally, the exercises in each Set are independent of each other and may be done separately.* However, when the exercises in a Set are related, with some exercises building on preceding ones, so they must be done in sequence, this is indicated with a symbol † in the margin to the left of the heading.

A. Examples of Operations

Which of the following rules are operations on the indicated set? ($\mathbb{Z}$ designates the set of the integers, $\mathbb{Q}$ the rational numbers, and $\mathbb{R}$ the real numbers.) For each rule which is not an operation, explain why it is not.

Example $a * b = \dfrac{a + b}{ab}$, on the set $\mathbb{Z}$.

SOLUTION This is not an operation on $\mathbb{Z}$. There are integers a and b such that $(a + b)/ab$ is not an integer. (For example,

$$\frac{2 + 3}{2 \cdot 3} = \frac{5}{6}$$

is not an integer.) Thus, $\mathbb{Z}$ is not closed under $*$.

1 $a * b = \sqrt{|ab|}$, on the set $\mathbb{Q}$. *no, may n be rat*

2 $a * b = a \ln b$, on the set $\{x \in \mathbb{R}: x > 0\}$. *no, ln b may be < 0*

3 $a * b$ is a root of the equation $x^2 - a^2 b^2 = 0$, on the set $\mathbb{R}$. *no, not unique*

4 Subtraction, on the set $\mathbb{Z}$. *yes*

5 Subtraction, on the set $\{n \in \mathbb{Z}: n \geqslant 0\}$. *no, may get neg #s*

6 $a * b = |a - b|$, on the set $\{n \in \mathbb{Z}: n \geqslant 0\}$. *yes*

B. Properties of Operations

Each of the following is an operation $*$ on $\mathbb{R}$. Indicate whether or not
  (i) it is commutative,
 (ii) it is associative,
(iii) $\mathbb{R}$ has an identity element with respect to $*$,
 (iv) every $x \in \mathbb{R}$ has an inverse with respect to $*$.

Instructions For (i), compute $x * y$ and $y * x$, and verify whether or not they are equal. For (ii), compute $x * (y * z)$ and $(x * y) * z$, and verify whether or not they are equal. For (iii), first solve the equation $x * e = x$ for e; if the equation cannot be solved, there is no identity element. If it *can* be solved, it is still necessary to check that $e * x = x * e = x$ for any $x \in \mathbb{R}$. If it checks, then e is an identity element. For (iv), first note that if there is no identity element, there can be no inverses. If there *is* an identity element e, first solve the equation $x * x' = e$ for x'; if the equation cannot be solved, x does not have an inverse. If it *can* be solved, check to make sure that $x * x' = x' * x = e$. If this checks, x' is the inverse of x.

Example $x * y = x + y + 1$

Associative		Commutative		Identity		Inverses	
Yes ☒	No ☐	Yes ☒	No ☐	Yes ☒	No ☐	Yes ☒	No ☐

(i) $x * y = x + y + 1$; $y * x = y + x + 1 = x + y + 1$.
  (*Thus, $*$ is commutative.*)
(ii) $x * (y * z) = x * (y + z + 1) = x + (y + z + 1) + 1 = x + y + z + 2$.
  $(x * y) * z = (x + y + 1) * z = (x + y + 1) + z + 1 = x + y + z + 2$.
  ($*$ *is associative.*)
(iii) Solve $x * e = x$ for e: $x * e = x + e + 1 = x$; therefore, $e = -1$.
  Check: $x * (-1) = x + (-1) + 1 = x$; $(-1) * x = (-1) + x + 1 = x$.
  Therefore, -1 is the identity element.
  ($*$ *has an identity element.*)
(iv) Solve $x * x' = -1$ for x': $x * x' = x + x' + 1 = -1$; therefore
  $x' = -x - 2$. Check: $x * (-x - 2) = x + (-x - 2) + 1 = -1$;
  $(-x - 2) * x = (-x - 2) + x + 1 = -1$. Therefore, $-x - 2$ is the inverse of x.
  (*Every element has an inverse.*)

1 $x * y = \sqrt{x^2 + y^2}$

Commutative		Associative		Identity		Inverses	
Yes ☒	No ☐	Yes ☒	No ☐	Yes ☐	No ☒	Yes ☐	No ☒

(i) $x * y = \sqrt{x^2 + y^2}$; $y * x = \sqrt{y^2 + x^2}$
(ii) $x * (y * z) = x * (\sqrt{y^2 + z^2}) = \sqrt{x^2 + (\sqrt{y^2 + z^2})^2} = \sqrt{x^2 + y^2 + z^2}$
  $(x * y) * z = (\sqrt{x^2 + y^2}) * z = \sqrt{(\sqrt{x^2 + y^2})^2 + z^2} = \sqrt{x^2 + y^2 + z^2}$
(iii) Solve $x * e = x$ for e. Check. $\sqrt{x^2 + e^2} = x$ $e = 0$
(iv) Solve $x * x' = e$ for x'. Check. $\sqrt{x^2 + x'^2} = 0$ doesn't work

2 $x * y = |x + y|$

Commutative		Associative		Identity		Inverses	
Yes ☐	No ☐	Yes ☐	No ☐	Yes ☐	No ☐	Yes ☐	No ☐

3 $x * y = |xy|$

Commutative		Associative		Identity		Inverses	
Yes ☒	No ☐	Yes ☒	No ☐	Yes ☐	No ☒	Yes ☐	No ☒

cause if
$x < 0$, $|xe|$
still > 0

$|xe| = x$

4 $x * y = x - y$

$x - e = x$
$e = 0$

$x' = x$

Commutative	Associative	Identity	Inverses
Yes ☐ No ☒	Yes ☐ No ☒	Yes ☒ No ☐	Yes ☒ No ☐

5 $x * y = xy + 1$

Commutative	Associative	Identity	Inverses
Yes ☐ No ☐	Yes ☐ No ☐	Yes ☐ No ☐	Yes ☐ No ☐

6 $x * y = \max \{x, y\} =$ the larger of the two numbers x and y

Commutative	Associative	Identity	Inverses
Yes ☒ No ☐	Yes ☒ No ☐	Yes ☐ No ☒	Yes ☐ No ☒

would have to be inf of ℝ

7 $x * y = \dfrac{xy}{x + y + 1}$

Commutative	Associative	Identity	Inverses
Yes ☒ No ☐	Yes ☐ No ☐	Yes ☐ No ☐	Yes ☐ No ☐

C. Operations on a Two-Element Set

Let A be the two-element set $A = \{a, b\}$.

1 Write the tables of all 16 operations on A. (Use the format explained on page 21.)

Label these operations 0_1 to 0_{16}. Then:

2 Identify which of the operations 0_1 to 0_{16} are commutative.
3 Identify which operations, among 0_1 to 0_{16}, are associative.
4 For which of the operations 0_1 to 0_{16} is there an identity element?
5 For which of the operations 0_1 to 0_{16} does every element have an inverse?

THREE

THE DEFINITION OF GROUPS

One of the simplest and most basic of all algebraic structures is the *group*. A group is defined to be a set with an operation (let us call it *) which is associative, has a neutral element, and for which each element has an inverse. More formally,

> *By a **group** we mean a set G with an operation * which satisfies the axioms:*
>
> (G1) * *is associative.*
>
> (G2) *There is an element e in G such that $a * e = a$ and $e * a = a$ for every element a in G.*
>
> (G3) *For every element a in G, there is an element a^{-1} in G such that $a * a^{-1} = e$ and $a^{-1} * a = e$.*

The group we have just defined may be represented by the symbol $\langle G, * \rangle$. This notation makes it explicit that the group consists of the *set G* and the *operation* *. (Remember that, in general, there are other possible operations on G, so it may not always be clear which is the group's operation unless we indicate it.) If there is no danger of confusion, we shall denote the group simply with the letter G.

The groups which come to mind most readily are found in our familiar number systems. Here are a few examples.

$\mathbb{Z}$ is the symbol customarily used to denote the set

$$\{\ldots, -3, -2, -1, 0, 1, 2, 3, \ldots\}$$

of the integers. The set $\mathbb{Z}$, with the operation of *addition*, is obviously a group. It is called the *additive group of the integers* and is represented by the symbol $\langle \mathbb{Z}, + \rangle$. Mostly, we denote it simply by the symbol $\mathbb{Z}$.

$\mathbb{Q}$ designates the set of the rational numbers (that is, quotients m/n of integers, where $n \neq 0$). This set, with the operation of addition, is called the *additive group of the rational numbers*, $\langle \mathbb{Q}, + \rangle$. Most often we denote it simply by $\mathbb{Q}$.

The symbol $\mathbb{R}$ represents the set of the real numbers. $\mathbb{R}$, with the operation of addition, is called the *additive group of the real numbers*, and is represented by $\langle \mathbb{R}, + \rangle$, or simply $\mathbb{R}$.

The set of all the *nonzero rational numbers* is represented by $\mathbb{Q}^*$. This set, with the operation of *multiplication*, is the group $\langle \mathbb{Q}^*, \cdot \rangle$, or simply $\mathbb{Q}^*$. Similarly, the set of all the *nonzero real numbers* is represented by $\mathbb{R}^*$. The set $\mathbb{R}^*$ with the operation of multiplication, is the group $\langle \mathbb{R}^*, \cdot \rangle$, or simply $\mathbb{R}^*$.

Finally, $\mathbb{Q}^+$ denotes the group of all the positive rational numbers, with multiplication. $\mathbb{R}^+$ denotes the group of all the positive real numbers, with multiplication.

Groups occur abundantly in nature. This statement means that a great many of the algebraic structures which can be discerned in natural phenomena turn out to be groups. Typical examples, which we shall examine later, come up in connection with the structure of crystals, patterns of symmetry, and various kinds of geometric transformations. Groups are also important because they happen to be one of the fundamental building blocks out of which more complex algebraic structures are made.

Especially important in scientific applications are the *finite* groups, that is, groups with a finite number of elements. It is not surprising that such groups occur often in applications, for in most situations of the real world we deal with only a finite number of objects.

The easiest finite groups to study are those called the *groups of integers modulo n* (where n is any positive integer greater than 1). These groups will be described in a casual way here, and a rigorous treatment deferred until later.

Let us begin with a specific example, say, the group of integers modulo 6. This group consists of a set of six elements,

$$\{0, 1, 2, 3, 4, 5\}$$

and an operation called *addition modulo 6*, which may be described as follows: Imagine the numbers 0 through 5 as being evenly distributed on the circumference of a circle. To add two numbers h and k, start with h and

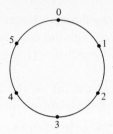

move clockwise k additional units around the circle: $h + k$ is where you end up. For example, $3 + 3 = 0$, $3 + 5 = 2$, and so on. The set $\{0, 1, 2, 3, 4, 5\}$ with this operation is called the *group of integers modulo 6*, and is represented by the symbol $\mathbb{Z}_6$.

In general, the group of integers modulo n consists of the set

$$\{0, 1, 2, \ldots, n - 1\}$$

with the operation of *addition modulo n*, which can be described exactly as previously. Imagine the numbers 0 through $n - 1$ to be points on the unit circle, each one separated from the next by an arc of length $2\pi/n$. To add h

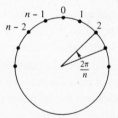

and k, start with h and go clockwise through an arc of k times $2\pi/n$. The sum $h + k$ will, of course, be one of the numbers 0 through $n - 1$. From geometrical considerations it is clear that this kind of addition (by successive rotations on the unit circle) is *associative*. Zero is the neutral element of this group, and $n - h$ is obviously the inverse of h [for $h + (n - h) = n$, which coincides with 0]. This group, the *group of integers modulo n*, is represented by the symbol $\mathbb{Z}_n$.

Often when working with finite groups, it is useful to draw up an "operation table." For example, the operation table of $\mathbb{Z}_6$ is

+	0	1	2	3	4	5
0	0	1	2	3	4	5
1	1	2	3	4	5	0
2	2	3	4	5	0	1
3	3	4	5	0	1	2
4	4	5	0	1	2	3
5	5	0	1	2	3	4

The basic format of this table is as follows,

+	0	1	2	3	4	5
0						
1						
2						
3						
4						
5						

with one *row* for each element of the group, and one *column* for each element of the group. Then $3 + 4$, for example, is located in the row of 3 and the column of 4. In general, any finite group $\langle G, * \rangle$ has a table

$*$	$\cdots\cdots$	y	$\cdots\cdots$
$\vdots$			
x		$x * y$	
$\vdots$			

The entry in the row of x and the column of y is $x * y$.

Let us remember that the commutative law is *not* one of the axioms of group theory, hence the identity $a * b = b * a$ is not true in every group. If the commutative law holds in a group G, such a group is called a *commutative group* or, more commonly, an *abelian group*. Abelian groups are named after the mathematician Niels Abel, who was mentioned in Chapter 1, and who was a pioneer in the study of groups. All the examples of groups mentioned up to now are abelian groups, but here is an example which is not.

Let G be the group which consists of the six matrices

$$I = \begin{pmatrix} 1 & 0 \\ 0 & 1 \end{pmatrix} \qquad A = \begin{pmatrix} 0 & 1 \\ 1 & 0 \end{pmatrix} \qquad B = \begin{pmatrix} 0 & 1 \\ -1 & -1 \end{pmatrix}$$

$$C = \begin{pmatrix} -1 & -1 \\ 0 & 1 \end{pmatrix} \qquad D = \begin{pmatrix} -1 & -1 \\ 1 & 0 \end{pmatrix} \qquad K = \begin{pmatrix} 1 & 0 \\ -1 & -1 \end{pmatrix}$$

with the operation of *matrix multiplication* which was explained on page 8. This group has the following operation table, which should be checked:

	I	A	B	C	D	K
I	I	A	B	C	D	K
A	A	I	C	B	K	D
B	B	K	D	A	I	C
C	C	D	K	I	A	B
D	D	C	I	K	B	A
K	K	B	A	D	C	I

In linear algebra it is shown that the multiplication of matrices is associative. (The details are simple.) It is clear that I is the identity element of this group, and by looking at the table one can see that each of the six matrices in $\{I, A, B, C, D, K\}$ has an inverse in $\{I, A, B, C, D, K\}$. (For example, B is the inverse of D, A is the inverse of A, and so on). Thus, G is a group! Now, we observe that $AB = C$ and $BA = K$, so G is not commutative.

EXERCISES

A. Examples of Abelian Groups

Prove that each of the following sets, with the indicated operation, is an abelian group.

Instructions. Proceed as in Chapter 2, Exercise B.

1 $x * y = x + y + k$ (k a fixed constant), on the set $\mathbb{R}$ of the real numbers.

2 $x * y = \dfrac{xy}{2}$, on the set $\{x \in \mathbb{R} : x \neq 0\}$.

3 $x * y = x + y + xy$, on the set $\{x \in \mathbb{R} : x \neq -1\}$.

4 $x * y = \dfrac{x + y}{xy + 1}$, on the set $\{x \in \mathbb{R}; -1 < x < 1\}$.

B. Groups on the Set $\mathbb{R} \times \mathbb{R}$

The symbol $\mathbb{R} \times \mathbb{R}$ represents the set of all ordered pairs (x, y) of real numbers. $\mathbb{R} \times \mathbb{R}$ may therefore be identified with the set of all the points in the plane. Which of the following subsets of $\mathbb{R} \times \mathbb{R}$, with the indicated operation, is a group? Which is an abelian group?

Instructions. Proceed as in the preceding exercise. To find the identity element, which in these problems is an ordered pair (e_1, e_2) of real numbers, solve the equation $(a, b) * (e_1, e_2) = (a, b)$ for e_1 and e_2. To find the inverse (a', b') of (a, b), solve the equation $(a, b) * (a', b') = (e_1, e_2)$ for a' and b'. [Remember that $(x, y) = (x', y')$ if and only if $x = x'$ and $y = y'$.]

1 $(a, b) * (c, d) = (ad + bc, bd)$, on the set $\{(x, y) \in \mathbb{R} \times \mathbb{R} : y \neq 0\}$.
2 $(a, b) * (c, d) = (ac, bc + d)$, on the set $\{(x, y) \in \mathbb{R} \times \mathbb{R} : x \neq 0\}$.
3 Same operation as in part 2, but on the set $\mathbb{R} \times \mathbb{R}$.
4 $(a, b) * (c, d) = (ac - bd, ad + bc)$, on the set $\mathbb{R} \times \mathbb{R}$ with the origin deleted.
5 Consider the operation of the preceding problem on the set $\mathbb{R} \times \mathbb{R}$. Is this a group? Explain.

C. Groups of Subsets of a Set

If A and B are any two sets, their *symmetric difference* is the set $A + B$ defined as follows:

$$A + B = (A - B) \cup (B - A)$$

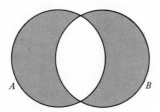

The shaded area is $A + B$

It is perfectly clear that $A + B = B + A$; hence this operation is commutative. It is also associative, as the accompanying pictorial representation suggests: Let the union of A, B, and C be divided into seven regions as illustrated.

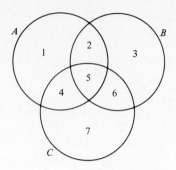

$A + B$ consists of the regions 1, 4, 3, and 6.

$B + C$ consists of the regions 2, 3, 4, and 7.

$A + (B + C)$ consists of the regions 1, 3, 5, and 7.

$(A + B) + C$ consists of the regions 1, 3, 5, and 7.

Thus, $A + (B + C) = (A + B) + C$.

If D is a set, then the *power set* of D is the set P_D of all the subsets of D. That is,

$$P_D = \{A : A \subseteq D\}$$

The operation $+$ is to be regarded as an operation on P_D. *Prove the following:*

1. There is an identity element with respect to the operation $+$, which is: ___∅___ .

2. Every subset A of D has an inverse with respect to $+$, which is: _D—A_ . Thus, $\langle P_D, + \rangle$ is a group!

3. Let D be the three-element set $D = \{a, b, c\}$. List the elements of P_D. (For example, one element if $\{a\}$, another is $\{a, b\}$, and so on. Do not forget the empty set and the whole set D.) Then write the operation table for $\langle P_D, + \rangle$.

D. A Checkerboard Game

1	2
3	4

Our checkerboard has only four squares, numbered 1, 2, 3, and 4. There is a single checker on the board, and it has four possible moves:

V: Move vertically; that is, move from 1 to 3, or from 3 to 1, or from 2 to 4, or from 4 to 2.

H: Move horizontally, that is, move from 1 to 2 or vice versa, or from 3 to 4 or vice versa.

D: Move diagonally, that is, move from 2 to 3 or vice versa, or move from 1 to 4 or vice versa.

I: Stay put.

We may consider an operation on the set of these four moves, which consists of performing moves successively. For example, if we move horizontally and then vertically, we end up with the same result as if we had moved diagonally:

$$H * V = D$$

If we perform two horizontal moves in succession, we end up where we started: $H * H = I$. And so on. If $G = \{V, H, D, I\}$, and $*$ is the operation we have just described, write the table of G.

$*$	I	V	H	D
I				
V				
H				
D				

Granting associativity, explain why $\langle G, * \rangle$ is a group.

E. A Coin Game

different coins

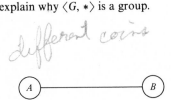

Imagine two coins on a table, at positions A and B. In this game there are eight possible moves:

M_1: Flip over the coin at A.
M_2: Flip over the coin at B.
M_3: Flip over both coins.
M_4: Switch the coins.

M_5: Flip coin at A; then switch.
M_6: Flip coin at B; then switch.
M_7: Flip both coins; then switch.
I: Do not change anything.

We may consider an operation on the set $\{I, M_1, \ldots, M_7\}$, which consists of performing any two moves in succession. For example, if we switch coins, then flip over the coin at A, this is the same as first flipping over the coin at B, then switching:

$$M_4 * M_1 = M_2 * M_4 = M_6$$

If $G = \{I, M_1, \ldots, M_7\}$ and   is the operation we have just described, write the table of $\langle G, * \rangle$.

$*$	I	M_1	M_2	M_3	M_4	M_5	M_6	M_7
I	I	M_1	M_2	M_3	M_4	M_5	M_6	M_7
M_1	M_1							
M_2	M_2				M_6			
M_3	M_3							
M_4	M_4 M_6							
M_5	M_5							
M_6	M_6							
M_7	M_7							

Granting associativity, explain why $\langle G, * \rangle$ is a group. Is it commutative? If not, show why not.

F. Groups in Binary Codes

The most basic way of transmitting information is to code it into strings of 0s and 1s, such as 0010111, 1010011, etc. Such strings are called *binary words*, and the number of 0s and 1s in any binary word is called its *length*. All information may be coded in this fashion.

When information is transmitted, it is sometimes received incorrectly. One of the most important purposes of coding theory is to find ways of *detecting errors*, and *correcting* errors of transmission.

If a word $\mathbf{a} = a_1 a_2 \cdots a_n$ is sent, but a word $\mathbf{b} = b_1 b_2 \cdots b_n$ is received (where the a_i and the b_j are 0s or 1s), then the *error pattern* is the word $\mathbf{e} = e_1 e_2 \cdots e_n$ where

$$e_i = \begin{cases} 0 & \text{if } a_i = b_i \\ 1 & \text{if } a_i \neq b_i \end{cases}$$

With this motivation, we define an operation of *adding* words, as follows: If $\mathbf{a}$ and $\mathbf{b}$ are both of length 1, we add them according to the rules

$$0 + 0 = 0 \qquad 1 + 1 = 0 \qquad 0 + 1 = 1 \qquad 1 + 0 = 1$$

If $\mathbf{a}$ and $\mathbf{b}$ are both of length n, we add them by *adding corresponding digits*. That is (let us introduce commas for convenience),

$$(a_1, a_2, \ldots, a_n) + (b_1, b_2, \ldots, b_n) = (a_1 + b_1, a_2 + b_2, \ldots, a_n + b_n)$$

Thus, the sum of $\mathbf{a}$ and $\mathbf{b}$ is the error pattern $\mathbf{e}$.

For example,

$$
\begin{array}{r}
0010110 \\
+\ 0011010 \\
\hline
=\ 0001100
\end{array}
\qquad
\begin{array}{r}
10100111 \\
+\ 11110111 \\
\hline
=\ 01010000
\end{array}
$$

The symbol $\mathbb{B}^n$ will designate the set of all the binary words of length n. We will prove that the operation of word addition has the following properties on $\mathbb{B}^n$:

1. It is commutative.
2. It is associative.
3. There is an identity element for word addition.
4. Every word has an inverse under word addition.

First, we verify the commutative law for words of length 1:

$$0 + 1 = 1 = 1 + 0$$

1. Show that $(a_1, a_2, \ldots, a_n) + (b_1, b_2, \ldots, b_n) = (b_1, b_2, \ldots, b_n) + (a_1, a_2, \ldots, a_n)$.

2. To verify the associative law, we first verify it for words of length 1:

$$1 + (1 + 1) = 1 + 0 = 1 = 0 + 1 = (1 + 1) + 1$$

$$1 + (1 + 0) = 1 + 1 = 0 = 0 + 0 = (1 + 1) + 0$$

Check the remaining six cases.

3. Show that $(a_1, \ldots, a_n) + [(b_1, \ldots, b_n) + (c_1, \ldots, c_n)] = [(a_1, \ldots, a_n) + (b_1, \ldots, b_n)] + (c_1, \ldots, c_n)$.

4. The identity element of $\mathbb{B}^n$, that is, the identity element for adding words of length n, is:

5. The inverse, with respect to word addition, of any word $(a_1, \ldots, a_n)$ is:

FOUR

ELEMENTARY PROPERTIES OF GROUPS

Is it possible for a group to have *two different* identity elements? Well, suppose e_1 and e_2 are identity elements of some group G. Then

$$e_1 * e_2 = e_2 \qquad \text{because } e_1 \text{ is an identity element, and}$$

$$e_1 * e_2 = e_1 \qquad \text{because } e_2 \text{ is an identity element}$$

Therefore

$$e_1 = e_2$$

This shows that in every group there is *exactly one* identity element.

Can an element a in a group have *two different inverses*? Well, if a_1 and a_2 are both inverses of a, then

$$a_1 * (a * a_2) = a_1 * e = a_1$$

and $\qquad (a_1 * a) * a_2 = e * a_2 = a_2$

By the associative law, $a_1 * (a * a_2) = (a_1 * a) * a_2$, hence $a_1 = a_2$. This shows that in every group, each element has *exactly one* inverse.

Up to now we have used the symbol $*$ to designate the group operation. Other, more commonly used symbols are $+$ and ("plus" and "multiply"). When $+$ is used to denote the group operation, we say we are using *additive notation*, and we refer to $a + b$ as the *sum* of a and b. (Remember that a and b do not have to be numbers and therefore "sum" does not, in general, refer to adding numbers.) When is used to denote the group operation, we say we are using *multiplicative notation*; we usually write *ab* instead of $a \cdot b$, and call *ab* the *product* of *a* and *b*. (Once again, remember that "product" does not, in general, refer to multiplying numbers.) Multiplicative notation is the most popular because it is simple and saves space. In the remainder of this book multiplicative notation will be used except where otherwise indicated. In particular, when we represent a group by a letter such as *G* or *H*, it will be understood that the group's operation is written as multiplication.

There is common agreement that in additive notation the identity element is denoted by 0, and the inverse of *a* is written as $-a$. (It is called the *negative* of *a*.) In multiplicative notation the identity element is *e* and the inverse of *a* is written as a^{-1} ("*a* inverse"). It is also a tradition that $+$ is to be used only for commutative operations.

The most basic rule of calculation in groups is the *cancelation law*, which allows us to cancel the factor *a* in the equations $ab = ac$ and $ba = ca$. This will be our first theorem about groups.

Theorem 1 *If G is a group and a, b, c are elements of G, then*

(i) $ab = ac$ *implies* $b = c$ *and*
(ii) $ba = ca$ *implies* $b = c$

It is easy to see why this is true: if we multiply (on the left) both sides of the equation $ab = ac$ by a^{-1}, we get $b = c$. In the case of $ba = ca$, we multiply on the right by a^{-1}. This is the *idea* of the proof; now here is the proof:

Suppose $$ab = ac$$

Then $$a^{-1}(ab) = a^{-1}(ac)$$

By the associative law, $$(a^{-1}a)b = (a^{-1}a)c$$

that is, $$eb = ec$$

Thus, finally, $$b = c$$

Part (ii) is proved analogously.

In general, we *cannot* cancel *a* in the equation $ab = ca$. (Why not?)

Theorem 2 *If G is a group and a, b are elements of G, then*

$$ab = e \quad \textit{implies} \quad a = b^{-1} \quad \textit{and} \quad b = a^{-1}$$

The proof is very simple: if $ab = e$, then $ab = aa^{-1}$, so by the cancelation law, $b = a^{-1}$. Analogously, $a = b^{-1}$.

This theorem tells us that if the product of two elements is equal to e, these elements are inverses of each other. In particular, if a is the inverse of b, then b is the inverse of a.

The next theorem gives us important information about computing inverses.

Theorem 3 *If G is a group and a, b are elements of G, then*

(i) $(ab)^{-1} = b^{-1}a^{-1}$ *and*

(ii) $(a^{-1})^{-1} = a$

The first formula tells us that the inverse of a product is the product of the inverses in reverse order. The next formula tells us that a is the inverse of the inverse of a. The proof of (i) is as follows:

$$(ab)(b^{-1}a^{-1}) = a[(bb^{-1})a^{-1}] \qquad \text{by the associate law}$$

$$= a[ea^{-1}] \qquad \text{because } bb^{-1} = e$$

$$= aa^{-1}$$

$$= e$$

Since the product of ab and $b^{-1}a^{-1}$ is equal to e, it follows by Theorem 2 that they are each other's inverses. Thus, $(ab)^{-1} = b^{-1}a^{-1}$. The proof of (ii) is analogous but simpler: $aa^{-1} = e$, so by Theorem 2 a is the inverse of a^{-1}, that is, $a = (a^{-1})^{-1}$.

The associative law states that the two products $a(bc)$ and $(ab)c$ are equal; for this reason, no confusion can result if we denote either of these products by writing abc (without any parentheses), and call abc the product of these *three* elements in this order.

We may next define the product of any *four* elements a, b, c, and d in G by

$$abcd = a(bcd)$$

By successive uses of the associative law we find that

$$a(bc)d = ab(cd) = (ab)(cd) = (ab)cd$$

Hence the product $abcd$ (without parentheses, but without changing the order of its factors) is defined without ambiguity.

In general, any two products, each involving the same factors in the same order, are equal. The net effect of the associative law is that *parentheses are redundant.*

Having made this observation, we may feel free to use products of several factors, such as $a_1 a_2 \cdots a_n$, without parentheses, whenever it is convenient. Incidentally, by using the identity $(ab)^{-1} = b^{-1}a^{-1}$ repeatedly, we find that

$$(a_1 a_2 \cdots a_n)^{-1} = a_n^{-1} \cdots a_2^{-1} a_1^{-1}$$

If G is a finite group, the number of elements in G is called the *order* of G. It is customary to denote the order of G by the symbol

$$|G|$$

EXERCISES

Remark on notation. In the exercises below, the exponential notation a^n is used in the following sense: if a is any element of a group G, then a^2 means aa, a^3 means aaa, and, in general, a^n is the product of n factors of a, for any positive integer n.

A. Solving Equations in Groups

Let a, b, c, and x be elements of a group G. In each of the following, solve for x in terms of a, b, and c.

Example $x^2 = b$ *and* $x^5 = e$

From the first equation, $b = x^2$

Squaring, $b^2 = x^4$

Multiplying on the left by x,

$$xb^2 = xx^4 = x^5 = e \qquad \text{(NOTE: } x^5 = e \text{ was given.)}$$

Multiplying on the right side by $(b^2)^{-1}$,

$$xb^2(b^2)^{-1} = e(b^2)^{-1}$$

Therefore: $x = (b^2)^{-1}$

1 $axb = c$

2 $x^2 b = xa^{-1}c$

3 $x^2 a = bxc^{-1}$ and $acx = xac$

4 $ax^2 = b$ and $x^3 = e$

5 $x^2 = a^2$ and $x^5 = e$

6 $(xax)^3 = bx$ and $x^2 a = (xa)^{-1}$

B. Rules of Algebra in Groups

For each of the following rules, either prove that it is true in every group G, or give a counterexample to show that it is false in some groups. (All the counterexamples you need may be found in the group of matrices $\{I, A, B, C, D, K\}$ described on page 30.)

1 If $x^2 = e$, then $x = e$.

2 If $x^2 = a^2$, then $x = a$.

3 $(ab)^2 = a^2b^2$

4 If $x^2 = x$, then $x = e$.

5 For every $x \in G$, there is some $y \in G$ such that $x = y^2$. (This is the same as saying that every element of G has a "square root.")

6 For any two elements x and y in G, there is an element z in G such that $y = xz$.

C. Elements which Commute

If a and b are in G and $ab = ba$, we say that a and b *commute*. Assuming that a and b commute, *prove the following*:

1 a^{-1} and b^{-1} commute.

2 a and b^{-1} commute. (HINT: First show that $a = b^{-1}ab$.)

3 a commutes with ab.

4 a^2 commutes with b^2.

5 xax^{-1} commutes with xbx^{-1}, for any $x \in G$.

6 Prove: $ab = ba$ iff $aba^{-1} = b$.

   (The abbreviation iff stands for "if and only if." Thus, first prove that *if* $ab = ba$, *then* $aba^{-1} = b$. Next, prove that *if* $aba^{-1} = b$, *then* $ab = ba$. Proceed roughly as in Exercise Set A. Thus, assuming $ab = ba$, solve for b. Next, assuming $aba^{-1} = b$, solve for ab.)

7 Prove: $ab = ba$ iff $aba^{-1}b^{-1} = e$.

† D. Group Elements and Their Inverses[1]

Let G be a group. Let a, b, c denote elements of G, and let e be the neutral element of G. *Prove the following*:

---

[1] NOTE: When the exercises in a Set are related, with some exercises building on preceding ones, so they must be done in sequence, this is indicated with a symbol † in the margin to the left of the heading.

1 If $ab = e$, then $ba = e$. (HINT: See Theorem 2.)

2 If $abc = e$, then $cab = e$ and $bca = e$.

3 State a generalization of parts 1 and 2.

4 If $xay = a^{-1}$, then $yax = a^{-1}$.

5 $a = a^{-1}$ iff $aa = e$. (That is, a is its own inverse iff $a^2 = e$.)

6 Let $c = c^{-1}$. Then $ab = c$ iff $abc = e$.

7 Let a, b, and c each be equal to its own inverse. If $ab = c$, then $bc = a$ and $ca = b$.

8 If abc is its own inverse, then bca is its own inverse, and cab is its own inverse.

9 Let a and b each be equal to its own inverse. Then ba is the inverse of ab.

† E. Counting Elements and Their Inverses

Let G be a finite group, and let S be the set of all the elements of G which are *not* equal to their own inverse. That is, $S = \{x \in G : x \neq x^{-1}\}$. The set S can be divided

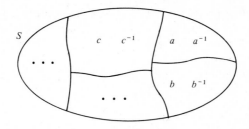

up into pairs so that each element is paired off with its own inverse. *Prove the following*:

1 In any finite group G, the number of elements not equal to their own inverse is an even number.

2 The number of elements of G equal to their own inverse is odd or even, depending on whether the number of elements in G is odd or even.

3 If the order of G is even, there is at least one element x in G such that $x \neq e$ and $x = x^{-1}$.

In parts 4 to 6, let G be a finite *abelian* group, say, $G = \{e, a_1, a_2, \ldots, a_n\}$.
Prove:

4 $(a_1 a_2 \cdots a_n)^2 = e$

5 If there is no element $x \neq e$ in G such that $x = x^{-1}$, then $a_1 a_2 \cdots a_n = e$.

6 If there is exactly one $x \neq e$ in G such that $x = x^{-1}$, then $a_1 a_2 \cdots a_n = x$.

† F. Constructing Small Groups

In each of the following, let G be any group. Let e denote the neutral element of G.

1 If a, b are any elements of G, prove each of the following:
  (i) If $a^2 = a$, then $a = e$.
  (ii) If $ab = a$, then $b = e$.
  (iii) If $ab = b$, then $a = e$.

2 Explain why every row of a group table must contain each element of the group exactly once. (HINT: Suppose x appears twice in the row of a:

		y_1		y_2	
		$\vdots$		$\vdots$	
a	$\cdots$	x	$\cdots$	x	$\cdots$

Now use the cancelation law for groups.)

3 There is *exactly one group* on any set of three distinct elements, say the set $\{e, a, b\}$. Indeed, keeping in mind parts 1 and 2 above, there is only one way of completing the following table. Do so! *You need not prove associativity.*

	e	a	b
e	e	a	b
a	a	b	e
b	b	c	a

4 There is exactly one group G of four elements, say $G = \{e, a, b, c\}$ satisfying the additional property that $xx = e$ for every $x \in G$. Using only part 1, above, complete the following group table of G.

	e	a	b	c
e	e	a	b	c
a	a	e	c	b
b	b	c	e	a
c	c	b	a	e

5 There is exactly one group G of four elements, say $G = \{e, a, b, c\}$, such that $xx = e$ for some $x \neq e$ in G, and $yy \neq e$ for some $y \in G$ (say, $aa = e$ and $bb \neq e$). Complete the group table of G, as in the preceding exercise.

6 Use Exercise E3 to explain why the groups in parts 4 and 5 are the only possible groups of four elements (except for renaming the elements with different symbols).

G. Direct Products of Groups

If G and H are any two groups, their *direct product* is a new group, denoted by $G \times H$, and defined as follows: $G \times H$ consists of all the ordered pairs (x, y) where x is in G and y is in H. That is,

$$G \times H = \{(x, y) : x \in G \quad \text{and} \quad y \in H\}$$

The operation of $G \times H$ consists of multiplying corresponding components:

$$(x, y) \cdot (x', y') = (xx', yy')$$

If G and H are denoted additively, it is customary to also denote $G \times H$ additively:

$$(x, y) + (x', y') = (x + x', y + y')$$

1 Prove that $G \times H$ is a group by checking the three group axioms, $(G1)$ to $(G3)$:

(G1) $(x_1, y_1)[(x_2, y_2)(x_3, y_3)] =$

$\qquad$ $[(x_1, y_1)(x_2, y_2)](x_3, y_3) =$

(G2) Let e_G be the identity element of G, and e_H the identity element of H. The identity element of $G \times H$? is : _____ . Check.

(G3) For each $(a, b) \in G \times H$, the inverse of (a, b) is:_____ . Check.

2 List the elements of $\mathbb{Z}_2 \times \mathbb{Z}_3$, and write its operation table. (NOTE: There are six elements, each of which is an ordered pair. The notation is additive.)

3 If G and H are abelian, prove that $G \times H$ is abelian.

4 Suppose the groups G and H both have the following property:

Every element of the group is its own inverse.

Prove that $G \times H$ also has this property.

H. Powers and Roots of Group Elements

Let G be a group, and $a, b \in G$. For any positive integer n, we define a^n by

$$a^n = \underbrace{aaa \cdots a}_{n \text{ factors}}$$

If there is an element $x \in G$ such that $a = x^2$, we say that a has a square root in G. Similarly, if $a = y^3$ for some $y \in G$, we say a has a cube root in G. In general, a has an nth root in G if $a = z^n$ for some $z \in G$. Prove the following:

1 $(bab^{-1})^n = ba^n b^{-1}$, for every positive integer n. Prove by induction. (Remember that to prove a formula such as this one by induction, you first prove it for $n = 1$; next you prove that *if* it is true for $n = k$, *then* it must be true for $n = k + 1$. You may conclude that it is true for every positive integer n.)

2 If $ab = ba$, then $(ab)^n = a^n b^n$ for every positive integer n. Prove by induction.

3 If $xax = e$, then $(xa)^{2n} = a^n$. Prove by induction.

4 If $a^3 = e$, then a has a square root. (HINT: Try a^2.)

5 If $a^2 = e$, then a has a cube root.

6 If a^{-1} has a cube root, so does a.

7 If $x^2 ax = a^{-1}$, then a has a cube root. (HINT: Show that xax is a cube root of a^{-1}.)

8 If $xax = b$, then ab has a square root.

SUBGROUPS

Let G be a group, and S a nonempty subset of G. It may happen (though it doesn't have to) that the product of every pair of elements of S is in S. If it happens, we say that S is *closed with respect to multiplication*. Then, it may happen that the inverse of every element of S is in S. In that case, we say that S is *closed with respect to inverses*. If both these things happen, we call S a ***subgroup*** of G.

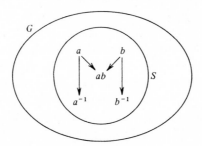

When the operation of G is denoted by the symbol $+$, the wording of these definitions must be adjusted: if the *sum* of every pair of elements of S is in S, we say that S is *closed with respect to addition*. If the negative of every element of S is in S, we say that S is *closed with respect to negatives*. If both these things happen, S is a ***subgroup*** of G.

For example, the *set of all the even integers* is a subgroup of the additive group $\mathbb{Z}$ of the integers. Indeed, the sum of any two even integers is an even integer, and the negative of any even integer is an even integer.

As another example, $\mathbb{Q}^*$ (the group of the nonzero rational numbers, under multiplication) is a subgroup of $\mathbb{R}^*$ (the group of the nonzero real numbers, under multiplication). Indeed, $\mathbb{Q}^* \subseteq \mathbb{R}^*$ because every rational number is a real number. Furthermore, the product of any two rational numbers is rational, and the inverse (that is, the reciprocal) of any rational number is a rational number.

An important point to be noted is this: if S is a subgroup of G, *the operation of S is the same as the operation of G.* In other words, if a and b are elements of S, the product ab computed in S is precisely the product ab computed in G.

For example, it would be meaningless to say that $\langle \mathbb{Q}^*, \cdot \rangle$ is a subgroup of $\langle \mathbb{R}, + \rangle$; for although it is true that $\mathbb{Q}^*$ is a subset of $\mathbb{R}$, the operations on these two groups are different.

The importance of the notion of subgroup stems from the following fact: *if G is a group and S is a subgroup of G, then S itself is a group.*

It is easy to see why this is true. To begin with, the operation of G, restricted to elements of S, is certainly an operation on S. *It is associative:* for if a, b, and c are in S, they are in G (because $S \subseteq G$); but G is a group, so $a(bc) = (ab)c$. Next, *the identity element e of G is in S* (and continues to be an identity element in S): for S is nonempty, so S contains an element a; but S is closed with respect to inverses, so S also contains a^{-1}; thus, S contains $aa^{-1} = e$, because S is closed with respect to multiplication. Finally, *every element of S has an inverse in S* because S is closed with respect to inverses. Thus, S is a group!

One reason why the notion of subgroup is useful is that it provides us with an easy way of showing that certain things are groups. Indeed, if G is already known to be a group, and S is a subgroup of G, we may conclude that S is a group without having to check all the items in the definition of "group." This conclusion is illustrated by the next example.

Many of the groups we use in mathematics are groups whose elements are functions. In fact, historically, the first groups ever studied as such were groups of functions.

$\mathscr{F}(\mathbb{R})$ represents the *set of all functions from $\mathbb{R}$ to $\mathbb{R}$,* that is, the set of all real-valued functions of a real variable. In calculus we learned how to add functions: if f and g are functions from $\mathbb{R}$ to $\mathbb{R}$, their *sum* is the function $f + g$ given by:

$$[f + g](x) = f(x) + g(x) \qquad \text{for every real number } x$$

Clearly, $f + g$ is again a function from $\mathbb{R}$ to $\mathbb{R}$, and is uniquely determined by f and g.

$\mathscr{F}(\mathbb{R})$, with the operation $+$ for adding functions, is the group $\langle \mathscr{F}(\mathbb{R}), + \rangle$, or simply $\mathscr{F}(\mathbb{R})$. The details are simple, but first, let us remember what it means for two functions to be equal. If f and g are functions from $\mathbb{R}$ to $\mathbb{R}$, then f and g are *equal* (that is, $f = g$) if and only if $f(x) = g(x)$ for every real number x. In other words, to be equal f and g must yield the same value when applied to any real number x.

To check that $+$ is associative, we must show that $f + [g + h] = [f + g] + h$, for any three functions f, g, and h in $\mathscr{F}(\mathbb{R})$. This means that for any real number x, $\{f + [g + h]\}(x) = \{[f + g] + h\}(x)$. Well,

$$\{f + [g + h]\}(x) = f(x) + [g + h](x) = f(x) + g(x) + h(x)$$

and $\{[f + g] + h\}(x)$ has the same value.

The neutral element of $\mathscr{F}(\mathbb{R})$ is the function $\mathcal{O}$ given by

$$\mathcal{O}(x) = 0 \qquad \text{for every real number } x$$

To show that $\mathcal{O} + f = f$ one must show that $[\mathcal{O} + f](x) = f(x)$ for every real number x. This is true because $[\mathcal{O} + f](x) = \mathcal{O}(x) + f(x) = 0 + f(x) = f(x)$.

Finally, the inverse of any function f is the function $-f$ given by

$$[-f](x) = -f(x) \qquad \text{for every real number } x$$

One perceives immediately that $f + [-f] = \mathcal{O}$, for every function f.

$\mathscr{C}(\mathbb{R})$ represents the set of all *continuous* functions from $\mathbb{R}$ to $\mathbb{R}$. Now, $\mathscr{C}(\mathbb{R})$, with the operation $+$, is a subgroup of $\mathscr{F}(\mathbb{R})$, because we know from calculus that the sum of any two continuous functions is a continuous function, and the negative $-f$ of any continuous function f is a continuous function. Because any subgroup of a group is itself a group, we may conclude that $\mathscr{C}(\mathbb{R})$, with the operation $+$, is a group. It is denoted by $\langle \mathscr{C}(\mathbb{R}), + \rangle$, or simply $\mathscr{C}(\mathbb{R})$.

$\mathscr{D}(\mathbb{R})$ represents the set of all the *differentiable* functions from $\mathbb{R}$ to $\mathbb{R}$. It is a subgroup of $\mathscr{F}(\mathbb{R})$ because the sum of any two differentiable functions is differentiable, and the negative of any differentiable function is differentiable. Thus, $\mathscr{D}(\mathbb{R})$, with the operation of adding functions, is a group, denoted by $\langle \mathscr{D}(\mathbb{R}), + \rangle$, or simply $\mathscr{D}(\mathbb{R})$.

By the way, in any group G the one-element subset $\{e\}$, containing only the neutral element, is a subgroup. It is closed with respect to multiplication because $ee = e$, and closed with respect to inverses because $e^{-1} = e$. At the other extreme, the whole group G is obviously a subgroup of itself. These two examples are, respectively, the smallest and largest possible subgroups of G. They are called the *trivial* subgroups of G. All the other subgroups of G are called *proper* subgroups.

Suppose G is a group and a, b, and c are elements of G. Define S to be the subset of G which contains *all the possible products of a, b, c, and their inverses*, in any order, with repetition of factors permitted. (Thus, typical elements of S would be $abac^{-1}$, $c^{-1}a^{-1}bbc$, and so on.) It is easy to see that S is a subgroup of G: for if two elements of S are multiplied together, they yield an element of S, and the inverse of any element of S is an element of S. For example, the product of aba and $cb^{-1}ac$ is $abacb^{-1}ac$, and the inverse of $ab^{-1}c^{-1}a$ is $a^{-1}cba^{-1}$. S is called the *subgroup of G generated by a, b, and c*.

If $a_1, \ldots, a_n$ are any finite number of elements of G, we may define the *subgroup generated by $a_1, \ldots, a_n$* in the same way. In particular, if a is a single element of G, we may consider the subgroup generated by a. This subgroup is designated by the symbol $\langle a \rangle$, and is called a *cyclic subgroup* of G; a is called its *generator*. Note that $\langle a \rangle$ consists of all the possible products of a and a^{-1}, for example $a^{-1}aaa^{-1}$ and $aaa^{-1}aa^{-1}$. However, since factors of a^{-1} cancel factors of a, there is no need to consider products involving both a and a^{-1} side by side. Thus, $\langle a \rangle$ contains

$$a, aa, aaa, \ldots,$$

$$a^{-1}, a^{-1}a^{-1}, a^{-1}a^{-1}a^{-1}, \ldots,$$

as well as $aa^{-1} = e$.

If the operation of G is denoted by $+$, the same definitions can be given with "sums" instead of "products."

In the group of matrices whose table appears on page 30, the subgroup generated by D is $\langle D \rangle = \{I, B, D\}$ and the subgroup generated by A is $\langle A \rangle = \{I, A\}$. (The student should check the table to verify this.) In fact, the entire group G of that example is generated by the two elements A and B.

If a group G is generated by a single element a, we call G a *cyclic group*, and write $G = \langle a \rangle$. For example, the additive group $\mathbb{Z}_6$ is cyclic. (What is its generator?)

Every finite group G is generated by one or more of its elements (obviously). A set of equations, involving only the generators and their inverses, is called a set of *defining equations* for G if these equations completely determine the multiplication table of G.

For example, let G be the group $\{e, a, b, b^2, ab, ab^2\}$ whose generators a and b satisfy the equations

$$a^2 = e \qquad b^3 = e \qquad ba = ab^2 \tag{*}$$

These three equations do indeed determine the multiplication table of G. To see this, note first that the equation $ba = ab^2$ allows us to switch powers of a with powers of b, bringing powers of a to the left, and powers of b to the right. For example, to find the product of ab and ab^2, we compute as follows:

$$(ab)(ab^2) = ab\underset{=ab^2}{\underbrace{ab}}b^2 = aab^2b^2 = a^2b^4$$

But by (*), $a^2 = e$ and $b^4 = b^3b = b$, so finally, $(ab)(ab^2) = b$. All the entries in the table of G may be computed in the same fashion.

When a group is determined by a set of generators and defining equations, its structure can be efficiently represented in a diagram called a *Cayley diagram*. These diagrams are explained in Exercise G.

EXERCISES

A. Recognizing Subgroups

In each of the following, determine whether or not H is a subgroup of G. (Assume that the operation of H is the same as that of G.)

Instructions. If H is a subgroup of G, show that both conditions in the definition of "subgroup" are satisfied. If H *is not* a subgroup of G, explain which condition fails.

Example $G = \mathbb{R}^*$, *the multiplicative group of the real numbers.*

$H = \{2^n : n \in \mathbb{Z}\}$ H is ⊠   is not ☐   a subgroup of G.

(i) If $2^n, 2^m \in H$, then $2^n2^m = 2^{n+m}$. But $n + m \in \mathbb{Z}$, so $2^{n+m} \in H$.
(ii) If $2^n \in H$, then $1/2^n = 2^{-n}$. But $-n \in \mathbb{Z}$, so $2^{-n} \in H$.
(Note that in this example the operation of G and H is multiplication. In the next problem, it is addition.)

1 $G = \langle \mathbb{R}, + \rangle$, $H = \{\log a : a \in \mathbb{Q}, a > 0\}$. H is ☐   is not ☐   a subgroup of G.
2 $G = \langle \mathbb{R}, + \rangle$, $H = \{\log n : n \in \mathbb{Z}, n > 0\}$. H is ☐   is not ☐   a subgroup of G.
3 $G = \langle \mathbb{R}, + \rangle$, $H = \{x \in \mathbb{R} : \tan x \in \mathbb{Q}\}$. H is ☐   is not ☐   a subgroup of G.
HINT: Use the following formula from trigonometry:

$$\tan(x + y) = \frac{\tan x + \tan y}{1 - \tan x \tan y}$$

4 $G = \langle \mathbb{R}^*, \cdot \rangle$, $H = \{2^n3^m : m, n \in \mathbb{Z}\}$. H is ☐   is not ☐   a subgroup of G.
5 $G = \langle \mathbb{R} \times \mathbb{R}, + \rangle$, $H = \{(x, y) : y = 2x\}$. H is ☐   is not ☐   a subgroup of G.

6 $G = \langle \mathbb{R} \times \mathbb{R}, + \rangle$, $H = \{(x, y) : x^2 + y^2 > 0\}$. H is □ is not □ a subgroup of G.

7 Let C and D be sets, with $C \subseteq D$. Prove that P_C is a subgroup of P_D. (See Chapter 3, Exercise C.)

B. Subgroups of Groups of Functions

In each of the following, show that H is a subgroup of G.

Example $G = \langle \mathscr{F}(\mathbb{R}), + \rangle$, $H = \{f \in \mathscr{F}(\mathbb{R}) : f(0) = 0\}$

    (i) Suppose f, $g \in H$; then $f(0) = 0$ and $g(0) = 0$, so $[f + g](0) = f(0) + g(0) = 0 + 0 = 0$. Thus, $f + g \in H$.
    (ii) If $f \in H$, then $f(0) = 0$. Thus, $[-f](0) = -f(0) = -0 = 0$, so $-f \in H$.

1 $G = \langle \mathscr{F}(\mathbb{R}), + \rangle$, $H = \{f \in \mathscr{F}(\mathbb{R}) : f(x) = 0 \text{ for every } x \in [0, 1]\}$
2 $G = \langle \mathscr{F}(\mathbb{R}), + \rangle$, $H = \{f \in \mathscr{F}(\mathbb{R}) : f(-x) = -f(x)\}$
3 $G = \langle \mathscr{F}(\mathbb{R}), + \rangle$, $H = \{f \in \mathscr{F}(\mathbb{R}) : f \text{ is periodic of period } \pi\}$

REMARK: A function f is said to be *periodic* of period a if there is a number a, called the period of f, such that $f(x) = f(x + na)$ for every $x \in \mathbb{R}$ and $n \in \mathbb{Z}$.

4 $G = \langle \mathscr{C}(\mathbb{R}), + \rangle$, $H = \{f \in \mathscr{C}(\mathbb{R}) : \int_0^1 f(x)\, dx = 0\}$
5 $G = \langle \mathscr{D}(\mathbb{R}), + \rangle$, $H = \{f \in \mathscr{D}(\mathbb{R}) : df/dx \text{ is constant}\}$
6 $G = \langle \mathscr{F}(\mathbb{R}), + \rangle$, $H = \{f \in \mathscr{F}(\mathbb{R}) : f(x) \in \mathbb{Z} \text{ for every } x \in \mathbb{R}\}$

C. Subgroups of Abelian Groups

In the following exercises, let G be an abelian group.

1 If $H = \{x \in G : x = x^{-1}\}$, that is, H consists of all the elements of G which are their own inverses, prove that H is a subgroup of G.
2 Let n be a fixed integer, and let $H = \{x \in G : x^n = e\}$. Prove that H is a subgroup of G.
3 Let $H = \{x \in G : x = y^2 \text{ for some } y \in G\}$, that is, let H be the set of all the elements of G which have a square root. Prove that H is a subgroup of G.
4 Let H be a subgroup of G, and let $K = \{x \in G : x^2 \in H\}$. Prove that K is a subgroup of G.
5 Let H be a subgroup of G, and let K consist of all the elements x in G such that some power of x is in H. That is, $K = \{x \in G : \text{for some integer } n > 0,\ x^n \in H\}$. Prove that K is a subgroup of G.
6 Suppose H and K are subgroups of G, and define HK as follows:

$$HK = \{xy : x \in H \quad \text{and} \quad y \in K\}$$

Prove that HK is a subgroup of G.

7 Explain why parts 4–6 are not true if G is not abelian.

D. Subgroups of an Arbitrary Group

Let G be a group. *Prove the following*:

1 If H and K are subgroups of a group G, prove that $H \cap K$ is a subgroup of G. (Remember that $x \in H \cap K$ iff $x \in H$ and $x \in K$).

2 Let H and K be subgroups of G. If $H \subseteq K$, then H is a subgroup of K.

3 By the *center* of a group G we mean the set of all the elements of G which commute with every element of G, that is,

$$C = \{a \in G : ax = xa \text{ for every } x \in G\}$$

Prove that C is a subgroup of G. (HINT: If we wish to assume $xy = yx$ and prove $xy^{-1} = y^{-1}x$, it is best to prove first that $yxy^{-1} = x$.)

4 Let $C' = \{a \in G : (ax)^2 = (xa)^2 \text{ for every } x \in G\}$. Prove that C' is a subgroup of G.

5 Let G be a *finite* group, say a group with n elements, and let S be a nonempty subset of G. Suppose $e \in S$, and that S is closed with respect to multiplication. Prove that S is a subgroup of G. (HINT: It remains to prove that G is closed with respect to inverses. Let $G = \{a_1, \ldots, a_n\}$; one of these elements is e. If $a_i \in G$, consider the *distinct* elements $a_i a_1, a_i a_2, \ldots, a_i a_n$.)

6 Let G be a group, and $f : G \to G$ a function. A *period* of f is any element a in G such that $f(x) = f(ax)$ for every $x \in G$. Prove: The set of all the periods of f is a subgroup of G.

7 Let H be a subgroup of G, and let $K = \{x \in G : xax^{-1} \in H \text{ for every } a \in H\}$. Prove:
   (a) K is a subgroup of G;
   (b) H is a subgroup of K.

8 Let G and H be groups, and $G \times H$ their direct product.
   (a) Prove that $\{(x, e) : x \in G\}$ is a subgroup of $G \times H$.
   (b) Prove that $\{(x, x) : x \in G\}$ is a subgroup of $G \times G$.

E. Generators of Groups

1 List all the cyclic subgroups of $\langle \mathbb{Z}_{10}, + \rangle$.

2 Show that $\mathbb{Z}_{10}$ is generated by 2 and 5.

3 Describe the subgroup of $\mathbb{Z}_{12}$ generated by 6 and 9.

4 Describe the subgroup of $\mathbb{Z}$ generated by 10 and 15.

5 Show that $\mathbb{Z}$ is generated by 5 and 7.

6 Show that $\mathbb{Z}_2 \times \mathbb{Z}_3$ is a cyclic group. Show that $\mathbb{Z}_3 \times \mathbb{Z}_4$ is a cyclic group.

7 Show that $\mathbb{Z}_2 \times \mathbb{Z}_4$ is *not* a cyclic group, but is generated by $(1, 1)$ and $(1, 2)$.

8 Suppose a group G is generated by two elements a and b. If $ab = ba$, prove that G is abelian.

F. Groups Determined by Generators and Defining Equations

1 Let G be the group $\{e, a, b, b^2, ab, ab^2\}$ whose generators satisfy: $a^2 = e$, $b^3 = e$, $ba = ab^2$. Write the table of G.

2 Let G be the group $\{e, a, b, b^2, b^3, ab, ab^2, ab^3\}$ whose generators satisfy: $a^2 = e$, $b^4 = e$, $ba = ab^3$. Write the table of G. (G is called the *dihedral group* D_4.)

3 Let G be the group $\{e, a, b, b^2, b^3, ab, ab^2, ab^3\}$ whose generators satisfy: $a^4 = e$, $a^2 = b^2$, $ba = ab^3$. Write the table of G. (G is called the *quaternion group*.)

4 Let G be the *commutative* group $\{e, a, b, c, ab, bc, ac, abc\}$ whose generators satisfy: $a^2 = b^2 = c^2 = e$. Write the table of G.

G. Cayley Diagrams

Every finite group may be represented by a diagram known as a Cayley diagram. A Cayley diagram consists of points joined by arrows.

There is one point for every element of the group.
The arrows represent the result of multiplying by a generator.

For example, if G has only one generator a (that is, G is the cyclic group $\langle a \rangle$), then the arrow $\rightarrow$ represents the operation "multiply by a":

$$e \rightarrow a \rightarrow a^2 \rightarrow a^3 \rightarrow \cdots$$

If the group has two generators, say a and b, we need *two kinds of arrows*, say $\dashrightarrow$ and $\rightarrow$, where $\dashrightarrow$ means "multiply by a," and $\rightarrow$ means "multiply by b."

For example, the group $G = \{e, a, b, b^2, ab, ab^2\}$ where $a^2 = e$, $b^3 = e$, and $ba = ab^2$ (see page 48) has the following Cayley diagram:

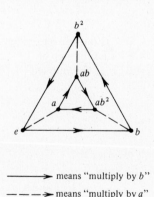

———————▶ means "multiply by b"

— — — ▶ means "multiply by a"

Moving in the *forward* direction of the arrow $\rightarrow$ means multiplying by b,

$$x \longrightarrow xb$$

whereas moving in the *backward* direction of the arrow means multiplying by b^{-1}:

$$x \longleftarrow xb^{-1}$$

(Note that "multiplying x by b" is understood to mean multiplying *on the right* by b: it means xb, not bx.) It is also a convention that if $a^2 = e$ (hence $a = a^{-1}$), then no arrowhead is used:

$$x \text{-----} xa$$

for if $a = a^{-1}$, then multiplying by a is the same as multiplying by a^{-1}.

The Cayley diagram of a group contains the same information as the group's table. For instance, to find the product $(ab)(ab^2)$ in the figure on page 52, we start at ab and follow the path corresponding to ab^2 (multiplying by a, then by b, then again by b), which is

This path leads to b, hence $(ab)(ab^2) = b$.

As another example, the inverse of ab^2 is the path which leads from ab^2 back to e. We note instantly that this is ba.

A point-and-arrow diagram is the Cayley diagram of a group iff it has the following two properties: (*a*) For each point x and generator a, there is exactly one a-arrow starting at x, and exactly one a-arrow ending at x; furthermore, at most one arrow goes from x to another point y. (*b*) If two different paths starting at x lead to the same destination, then these two paths, starting at any point y, lead to the same destination.

Cayley diagrams are a useful way of finding new groups.

Write the table of the groups having the following Cayley diagrams: (REMARK: You may take any point to represent e, because there is perfect symmetry in a Cayley diagram. Choose e, then label the diagram and proceed.)

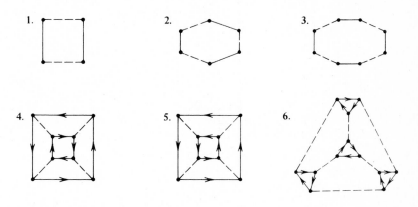

SIX

FUNCTIONS

The concept of a *function* is one of the most basic mathematical ideas and enters into almost every mathematical discussion. A function is generally defined as follows: If A and B are sets, then a function from A to B is a rule which to every element x in A assigns a unique element y in B. To indicate this connection between x and y we usually write $y = f(x)$, and we call y the *image* of x under the function f.

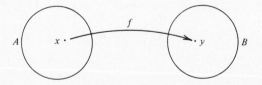

There is nothing inherently mathematical about this notion of function. For example, imagine A to be a set of married men and B to be the set of their wives. Let f be the rule which to each man assigns his wife. Then f is a perfectly good function from A to B; under this function, each wife is the image of her husband. (No pun is intended.)

Take care, however, to note that if there were a bachelor in A then f would not qualify as a function from A to B; for a function from A to B must assign a value in B to *every* element of A, without exception. Now, suppose the members of A and B are Ashanti, among whom polygamy is common; in the land of the Ashanti, f does not necessarily qualify as a function, for it may assign to a given member of A several wives. If f is a function from A to B, it must assign exactly *one* image to each element of A.

If f is a function from A to B it is customary to describe it by writing

$$f : A \rightarrow B$$

The set A is called the *domain* of f. The *range* of f is the subset of B which consists of *all the images of elements of A*. In the case of the function illustrated here, $\{a, b, c\}$ is the domain of f, and $\{x, y\}$ is the range of f

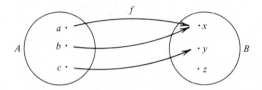

(z is not in the range of f). Incidentally, this function f may be represented in the simplified notation

$$f = \begin{pmatrix} a & b & c \\ x & x & y \end{pmatrix}$$

This notation is useful whenever A is a finite set: the elements of A are listed in the top row, and beneath each element of A is its image.

It may perfectly well happen, if f is a function from A to B, that two or more elements of A *have the same image*. For example, if we look at the function immediately above, we observe that a and b both have the same image x. If f is a function for which this kind of situation *does not occur*, then f is called an *injective* function. Thus,

Definition 1 *A function* $f : A \rightarrow B$ *is called* **injective** *if each element of B is the image of no more than one element of A.*

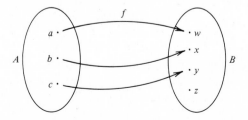

The intended meaning, of course, is that each element y in B is the image of no two *distinct* elements of A. So if

$$x_1 \searrow$$
$$\qquad\qquad y = f(x_1) = f(x_2)$$
$$x_2 \nearrow$$

that is, x_1 and x_2 have the same image y, we must require that x_1 be *equal* to x_2. Thus, a convenient definition of "injective" is this: a function $f : A \to B$ is *injective* if and only if

$$f(x_1) = f(x_2) \qquad \text{implies} \qquad x_1 = x_2$$

If f is a function from A to B, there may be elements in B which are not images of elements of A. If this does *not happen*, that is, if every element of B is the image of some element of A, we say that f is *surjective*.

Definition 2 *A function $f : A \to B$ is called **surjective** if each element of B is the image of at least one element of A.*

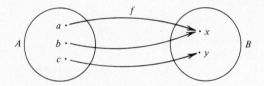

This is the same as saying that B is the range of f.

Now, suppose that f is both injective and surjective. By Definitions 1 and 2, each element of B is the image of *at least one* element of A, and *no more than one* element of A. So each element of B is the image of *exactly one* element of A. In this case, f is called a *bijective* function, or a *one-to-one correspondence*.

Definition 3 *A function $f : A \to B$ is called **bijective** if it is both injective and surjective.*

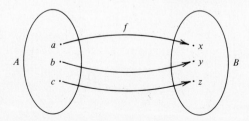

It is obvious that under a bijective function, each element of A has exactly one "partner" in B and each element of B has exactly one partner in A.

The most natural way of *combining* two functions is to form their "composite." The idea is this: suppose f is a function from A to B, and g is a function from B to C. We apply f to an element x in A and get an element y in B; then we apply g to y and get an element z in C. Thus, z is obtained from x by applying f and g in succession. The function which

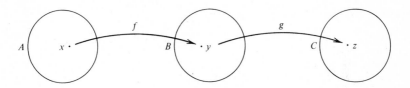

consists of *applying f and g in succession* is a function from A to C, and is called the *composite* of f and g. More precisely,

*Let $f : A \rightarrow B$ and $g : B \rightarrow C$ be functions. The **composite function** denoted by $g \circ f$ is a function from A to C defined as follows:*

$$[g \circ f](x) = g(f(x)) \quad \text{for every } x \in A$$

For example, consider once again a set A of married men and the set B of their wives. Let C be the set of all the mothers of members of B. Let $f : A \rightarrow B$ be the rule which to each member of A assigns his wife, and $g : B \rightarrow C$ the rule which to each woman in B assigns her mother. Then $g \circ f$ is the "mother-in-law function," which assigns to each member of A his wife's mother:

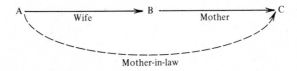

For another, more conventional, example, let f and g be the following functions from $\mathbb{R}$ to $\mathbb{R}$: $f(x) = 2x$; $g(x) = x + 1$. (In other words, f is the rule "multiply by 2" and g is the rule "add 1.") Their composites are the functions $g \circ f$ and $f \circ g$ given by

$$[f \circ g](x) = f(g(x)) = 2(x + 1)$$

and
$$[g \circ f](x) = g(f(x)) = 2x + 1$$

$f \circ g$ and $g \circ f$ are different: $f \circ g$ is the rule "add 1, then multiply by 2," whereas $g \circ f$ is the rule "multiply by 2 and then add 1."

It is an important fact that the composite of two injective functions is injective, the composite of two surjective functions is surjective, and the composite of two bijective functions is bijective. In other words, if $f : A \rightarrow B$ and $g : B \rightarrow C$ are functions, then the following are true:

> *If f and g are injective, then g ∘ f is injective.*
> *If f and g are surjective, then g ∘ f is surjective.*
> *If f and g are bijective, then g ∘ f is bijective.*

Let us tackle each of these claims in turn. We will suppose that f and g are injective, and prove that $g \circ f$ is injective. (That is, we will prove that if $[g \circ f](x) = [g \circ f](y)$, then $x = y$.)

Suppose $[g \circ f](x) = [g \circ f](y)$, that is,

$$g(f(x)) = g(f(y))$$

Because g is injective, we get

$$f(x) = f(y)$$

and because f is injective,

$$x = y$$

Next, let us suppose that f and g are surjective, and prove that $g \circ f$ is surjective. What we need to show here is that *every* element of C is $g \circ f$ of some element of A. Well, if $z \in C$, then (because g is surjective) $z = g(y)$ for some $y \in B$; but f is surjective, so $y = f(x)$ for some $x \in A$. Thus,

$$z = g(y) = g(f(x)) = [g \circ f](x)$$

Finally, if f and g are bijective, they are both injective and surjective. By what we have already proved, $g \circ f$ is injective and surjective, hence bijective.

A function f from A to B may have an inverse, but it doesn't have to. The inverse of f, if it exists, is a function f^{-1} ("f inverse") from B to A such that

$$x = f^{-1}(y) \qquad \text{if and only if} \qquad y = f(x)$$

Roughly speaking, if f carries x to y then f^{-1} carries y to x. For instance, returning (for the last time) to the example of a set A of husbands and the set B of their wives, if $f : A \to B$ is the rule which to each husband assigns his wife, then $f^{-1} : B \to A$ is the rule which to each wife assigns her husband:

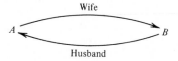

If we think of functions as rules, then f^{-1} is the rule which *undoes* whatever f does. For instance, if f is the real-valued function $f(x) = 2x$, then f^{-1} is the function $f^{-1}(x) = x/2$ [or, if preferred, $f^{-1}(y) = y/2$]. Indeed, the rule "divide by 2" undoes what the rule "multiply by 2" does.

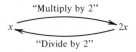

Which functions have inverses, and which others do not? If f, a function from A to B, is *not* injective, it *cannot* have an inverse; for "not injective" means there are at least two distinct elements x_1 and x_2 with the same image y:

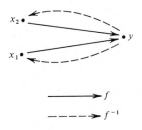

Clearly, $x_1 = f^{-1}(y)$ and $x_2 = f^{-1}(y)$ so $f^{-1}(y)$ is *ambiguous* (it has two different values), and this is not allowed for a function.

If f, a function from A to B, is *not* surjective, there is an element y in B which is not an image of any element of A; thus $f^{-1}(y)$ does not exist. So f^{-1} cannot be a function from B (that is, with domain B) to A.

It is therefore obvious that if f^{-1} exists, f *must be* injective and surjective, that is, bijective. On the other hand, if f is a bijective function from A to B, its inverse clearly exists and is determined by the rule that if $y = f(x)$ then $f^{-1}(y) = x$.

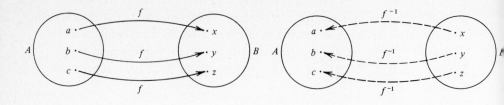

Furthermore, it is easy to see that the inverse of f is also a bijective function. To sum up:

> *A function $f : A \to B$ has an inverse if and only if it is bijective.*
>
> *In that case, the inverse f^{-1} is a bijective function from B to A.*

EXERCISES

A. Examples of Injective and Surjective Functions

Each of the following is a function $f : \mathbb{R} \to \mathbb{R}$. Determine
(*a*) whether or not f is injective, and
(*b*) whether or not f is surjective.
Prove your answer in either case.

Example 1 $f(x) = 2x$

f is injective.

PROOF Suppose $f(a) = f(b)$, that is

$$2a = 2b$$

Then $$a = b$$

Therefore f is injective.

f is surjective.

PROOF Take any element $y \in \mathbb{R}$. Then $y = 2(y/2) = f(y/2)$.
Thus, every $y \in \mathbb{R}$ is equal to $f(x)$ for $x = y/2$.
Therefore f is surjective.

Example 2 $f(x) = x^2$

f is not injective.

PROOF By exhibiting a counterexample: $f(2) = 4 = f(-2)$, although $2 \neq -2$.

f is not surjective.

PROOF By exhibiting a counterexample: -1 is not equal to $f(x)$ for any $x \in \mathbb{R}$.

1 $f(x) = 3x + 4$

2 $f(x) = x^3 + 1$

3 $f(x) = |x|$

4 $f(x) = x^3 - 3x$

5 $f(x) = \begin{cases} x & \text{if } x \text{ is rational} \\ 2x & \text{if } x \text{ is irrational} \end{cases}$

6 $f(x) = \begin{cases} 2x & \text{if } x \text{ is an integer} \\ x & \text{otherwise} \end{cases}$

7 Determine the range of each of the functions in parts 1 to 6.

B. Functions on $\mathbb{R}$ and $\mathbb{Z}$

Determine whether each of the following functions is or is not (a) injective, and (b) surjective. Proceed as in Exercise A.

1 $f: \mathbb{R} \to (0, \infty)$, defined by $f(x) = e^x$.

2 $f: (0, 1) \to \mathbb{R}$, defined by $f(x) = \tan x$.

3 $f: \mathbb{R} \to \mathbb{Z}$, defined by $f(x) =$ the least integer greater than or equal to x.

4 $f: \mathbb{Z} \to \mathbb{Z}$, defined by $f(n) = \begin{cases} n + 1 & \text{if } n \text{ is even} \\ n - 1 & \text{if } n \text{ is odd} \end{cases}$

5 Find a bijective function f from the set $\mathbb{Z}$ of the integers to the set E of the *even* integers.

C. Functions on Arbitrary Sets and Groups

Determine whether each of the following functions is or is not (a) injective, (b) surjective. Proceed as in Exercise A.

In parts 1 to 3 A and B are sets, and $A \times B$ denotes the set of all the ordered pairs (x, y) as x ranges over A and y over B.

1 $f: A \times B \to A$, defined by $f(x, y) = x$.

2 $f: A \times B \to B \times A$, defined by $f(x, y) = (y, x)$.

3 $f: B \to A \times B$, defined by $f(x) = (x, b)$, where b is a fixed element of B.

4 G is a group, $a \in G$, and $f: G \to G$ is defined by $f(x) = ax$.

5 G is a group and $f: G \to G$ is defined by $f(x) = x^{-1}$.

6 G is a group and $f: G \to G$ is defined by $f(x) = x^2$.

D. Composite Functions

Find the composite function, as indicated.

1 $f: \mathbb{R} \to \mathbb{R}$ is defined by $f(x) = \sin x$.
  $g: \mathbb{R} \to \mathbb{R}$ is defined by $g(x) = e^x$.
Find $f \circ g$ and $g \circ f$.

2 A and B are sets; $f: A \times B \to B \times A$ is given by $f(x, y) = (y, x)$.
  $g: B \times A \to B$ is given by $g(y, x) = y$.
Find $g \circ f$.

3 $f: (0, 1) \to \mathbb{R}$ is defined by $f(x) = 1/x$.
  $g: \mathbb{R} \to \mathbb{R}$ is defined by $g(x) = \ln x$.
Find $g \circ f$.

4 In school, Jack and Sam exchanged notes in a code f which consisted of spelling every word backwards and interchanging every letter s with t. Alternatively, they used a code g which interchanged the letters a with o, i with u, e with y, and s with t. Describe the codes $f \circ g$ and $g \circ f$. Are they the same?

5 $A = \{a, b, c, d\}$; f and g are functions from A to A; in the tabular form described on page 55, they are given by

$$f = \begin{pmatrix} a & b & c & d \\ a & c & a & c \end{pmatrix} \qquad g = \begin{pmatrix} a & b & c & d \\ b & a & b & a \end{pmatrix}$$

Give $f \circ g$ and $g \circ f$ in the same tabular form.

6 G is a group, and a and b are elements of G.
  $f: G \to G$ is defined by $f(x) = ax$.
  $g: G \to G$ is defined by $g(x) = bx$.
Find $f \circ g$ and $g \circ f$.

7 Indicate the domain and range of each of the composite functions you found in parts 1 to 6.

E. Inverses of Functions

Each of the following functions f is bijective. *Describe its inverse.*

1 $f: (0, \infty) \to (0, \infty)$, defined by $f(x) = 1/x$.

2 $f: \mathbb{R} \to (0, \infty)$, defined by $f(x) = e^x$.

3 $f: \mathbb{R} \to \mathbb{R}$, defined by $f(x) = x^3 + 1$.

4 $f: \mathbb{R} \to \mathbb{R}$, defined by $f(x) = \begin{cases} 2x & \text{if } x \text{ is rational} \\ 3x & \text{if } x \text{ is irrational} \end{cases}$

5 $A = \{a, b, c, d\}$, $B = \{1, 2, 3, 4\}$, and $f: A \to B$ is given by

$$f = \begin{pmatrix} a & b & c & d \\ 3 & 1 & 2 & 4 \end{pmatrix}$$

6 G is a group, $a \in G$, and $f: G \to G$ is defined by $f(x) = ax$.

F. Functions on Finite Sets

1 The members of the UN Peace Committee must choose, from among themselves, a presiding officer of their committee. For each member x, let $f(x)$ designate that member's choice for officer. If no two members vote alike, what is the range of f?

2 Let A be a finite set. Explain why any injective function $f: A \to A$ is necessarily surjective. (Look at part 1.)

3 If A is a finite set, explain why any surjective function $f: A \to A$ is necessarily injective.

4 Are the statements in parts 3 and 4 true when A is an infinite set? If not, give a counterexample.

5 If A has n elements, how many functions are there from A to A? How many bijective functions are there from A to A?

G. Some General Properties of Functions

In parts 1 to 3, let A, B, and C be sets, and let $f: A \to B$ and $g: B \to C$ be functions.

1 Prove that if $g \circ f$ is injective, then f is injective.

2 Prove that if $g \circ f$ is surjective, then g is surjective.

3 Parts 1 and 2, together, tell us that if $g \circ f$ is bijective, then f is injective and g is surjective. Is the converse of this statement true: If f is injective and g surjective, is $g \circ f$ bijective? (If "yes," prove it; if "no," give a counterexample.)

4 Let $f: A \to B$ and $g: B \to A$ be functions. Suppose that $y = f(x)$ iff $x = g(y)$. Prove that f is bijective, and $g = f^{-1}$.

GROUPS OF PERMUTATIONS

In this chapter we continue our discussion of functions, but we confine our discussions to functions *from a set to itself.* In other words, we consider only functions $f: A \to A$ whose domain is a set A and whose range is in the same set A.

To begin with, we note that any two functions f and g (from A to A) are *equal* if and only if $f(x) = g(x)$ for every element x in A.

If f and g are functions from A to A, their composite $f \circ g$ is also a function from A to A. We recall that it is the function defined by

$$[f \circ g](x) = f(g(x)) \qquad \textit{for every } x \textit{ in } A \qquad (1)$$

It is a very important fact that the *composition of functions is associative.* Thus, if f, g, and h are three functions from A to A, then

$$f \circ (g \circ h) = (f \circ g) \circ h$$

To prove that the functions $f \circ (g \circ h)$ and $(f \circ g) \circ h$ are equal, one must show that for every element x in A,

$$\{f \circ [g \circ h]\}(x) = \{[f \circ g] \circ h\}(x)$$

We get this by repeated use of Equation 1:

$$\{f \circ [g \circ h]\}(x) = f([g \circ h](x)) = f(g(h(x)))$$

$$= [f \circ g](h(x)) = \{[f \circ g] \circ h\}(x)$$

By a *permutation* of a set A we mean a *bijective function from A to A*, that is, a one-to-one correspondence between A and itself. In elementary algebra we learned to think of a permutation as a *rearrangement* of the elements of a set. Thus, for the set $\{1, 2, 3, 4, 5\}$, we may consider the rearrangement which changes $(1, 2, 3, 4, 5)$ to $(3, 2, 1, 5, 4)$; this rearrangement may be identified with the function

$$
\begin{array}{ccc}
1 & \longrightarrow & 3 \\
2 & \longrightarrow & 2 \\
3 & \longrightarrow & 1 \\
4 & \longrightarrow & 5 \\
5 & \longrightarrow & 4
\end{array}
$$

which is obviously a one-to-one correspondence between the set $\{1, 2, 3, 4, 5\}$ and itself. It is clear, therefore, that there is no real difference between the new definition of permutation and the old. The new definition, however, is more general in a very useful way since it allows us to speak of permutations of sets A even when A has infinitely many elements.

In Chapter 6 we saw that the composite of any two bijective functions is a bijective function. Thus, *the composite of any two permutations of A is a permutation of A*. It follows that we may regard the operation $\circ$ of composition as *an operation on the set of all the permutations of A*. We have just seen that composition is an associative operation. Is there a neutral element for composition?

For any set A, the *identity function on A*, symbolized by ε_A or simply ε, is the function $x \rightarrow x$ which carries every element of A to itself. That is, it is defined by

$$\varepsilon(x) = x \qquad \textit{for every element} \qquad x \in A$$

It is easy to see that ε is a permutation of A (it is a one-to-one correspondence between A and itself); and if f is any other permutation of A, then

$$f \circ \varepsilon = f \qquad \text{and} \qquad \varepsilon \circ f = f$$

The first of these equations asserts that $[f \circ \varepsilon](x) = f(x)$ for every element x in A, which is quite obvious, since $[f \circ \varepsilon](x) = f(\varepsilon(x)) = f(x)$. The second equation is proved analogously.

We saw in Chapter 6 that the inverse of any bijective function exists and is a bijective function. Thus, *the inverse of any permutation of A is a permutation of A*. Furthermore, if f is any permutation of A and f^{-1} is its inverse, then

$$f^{-1} \circ f = \varepsilon \qquad \text{and} \qquad f \circ f^{-1} = \varepsilon$$

The first of these equations asserts that for any element x in A,

$$[f^{-1} \circ f](x) = \varepsilon(x)$$

that is, $f^{-1}(f(x)) = x$:

This is obviously true, by the definition of the inverse of a function. The second equation is proved analogously.

Let us recapitulate: The operation $\circ$ of composition of functions qualifies as an operation on the set of all the permutations of A. This operation is associative. There is a permutation ε such that $\varepsilon \circ f = f$ and $f \circ \varepsilon = f$ for any permutation f of A. Finally, for every permutation f of A there is another permutation f^{-1} of A such that $f \circ f^{-1} = \varepsilon$ and $f^{-1} \circ f = \varepsilon$. Thus, *the set of all the permutations of A, with the operation $\circ$ of composition, is a group.*

For any set A, the group of all the permutations of A is called the *symmetric group on A*, and it is represented by the symbol S_A. For any positive integer n, the symmetric group on the set $\{1, 2, 3, \ldots, n\}$ is called the *symmetric group on n elements*, and is denoted by S_n.

Let us take a look at S_3. First, we list all the permutations of the set $\{1, 2, 3\}$:

$$\varepsilon = \begin{pmatrix} 1 & 2 & 3 \\ 1 & 2 & 3 \end{pmatrix} \qquad \alpha = \begin{pmatrix} 1 & 2 & 3 \\ 1 & 3 & 2 \end{pmatrix} \qquad \beta = \begin{pmatrix} 1 & 2 & 3 \\ 3 & 1 & 2 \end{pmatrix}$$

$$\gamma = \begin{pmatrix} 1 & 2 & 3 \\ 2 & 1 & 3 \end{pmatrix} \qquad \delta = \begin{pmatrix} 1 & 2 & 3 \\ 2 & 3 & 1 \end{pmatrix} \qquad \kappa = \begin{pmatrix} 1 & 2 & 3 \\ 3 & 2 & 1 \end{pmatrix}$$

This notation for functions was explained on page 55; for example,

$$\beta = \begin{pmatrix} 1 & 2 & 3 \\ 3 & 1 & 2 \end{pmatrix}$$

is the function such that $\beta(1) = 3$, $\beta(2) = 1$, and $\beta(3) = 2$. A more graphic way of representing the same function would be

$$\beta = \begin{pmatrix} 1 & 2 & 3 \\ \downarrow & \downarrow & \downarrow \\ 3 & 1 & 2 \end{pmatrix}$$

The operation on elements of S_3 is composition. To find $\alpha \circ \beta$, we note that

$$[\alpha \circ \beta](1) = \alpha(\beta(1)) = \alpha(3) = 2$$

$$[\alpha \circ \beta](2) = \alpha(\beta(2)) = \alpha(1) = 1$$

$$[\alpha \circ \beta](3) = \alpha(\beta(3)) = \alpha(2) = 3$$

Thus,

$$\alpha \circ \beta = \begin{pmatrix} 1 & 2 & 3 \\ 2 & 1 & 3 \end{pmatrix} = \gamma$$

Note that in $\alpha \circ \beta$, β is applied first and α next. A graphic way of representing this is:

$$\beta = \begin{pmatrix} 1 & 2 & 3 \\ \downarrow & \downarrow & \downarrow \\ 3 & 1 & 2 \end{pmatrix}$$

$$\alpha = \begin{pmatrix} 1 & 2 & 3 \\ \downarrow & \downarrow & \downarrow \\ 1 & 3 & 2 \end{pmatrix}$$

The other combinations of elements of S_3 may be computed in the same fashion. The student should check the following table, which is the table of the group S_3 :

$\circ$	ε	α	β	γ	δ	κ
ε	ε	α	β	γ	δ	κ
α	α	ε	γ	β	κ	δ
β	β	κ	δ	α	ε	γ
γ	γ	δ	κ	ε	α	β
δ	δ	γ	ε	κ	β	α
κ	κ	β	α	δ	γ	ε

By a *group of permutations* we mean any group S_A or S_n, or any subgroup of one of these groups. Among the most interesting groups of permutations are the groups of symmetries of geometric figures. We will see how such groups arise by considering the *group of symmetries of the square.*

We may think of a symmetry of the square as any way of moving a square to make it coincide with its former position. Every time we do this, vertices will coincide with vertices, so a symmetry is completely described by its effect on the vertices.

Let us number the vertices as in the following diagram:

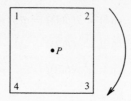

The most obvious symmetries are obtained by rotating the square clockwise about its center P, through angles of 90°, 180°, and 270°, respectively. We indicate each symmetry as a permutation of the vertices; thus a clockwise rotation of 90° yields the symmetry

$$R_1 = \begin{pmatrix} 1 & 2 & 3 & 4 \\ 2 & 3 & 4 & 1 \end{pmatrix}$$

for this rotation carries vertex 1 to 2, 2 to 3, 3 to 4, and 4 to 1. Rotations of 180° and 270° yield the following symmetries, respectively:

$$R_2 = \begin{pmatrix} 1 & 2 & 3 & 4 \\ 3 & 4 & 1 & 2 \end{pmatrix} \quad \text{and} \quad R_3 = \begin{pmatrix} 1 & 2 & 3 & 4 \\ 4 & 1 & 2 & 3 \end{pmatrix}$$

The remaining symmetries are flips of the square about its axes A, B, C, and D:

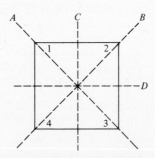

For example, when we flip the square about the axis A, vertices 1 and 3 stay put, but 2 and 4 change places; so we get the symmetry

$$R_4 = \begin{pmatrix} 1 & 2 & 3 & 4 \\ 1 & 4 & 3 & 2 \end{pmatrix}$$

In the same way, the other flips are

$$R_5 = \begin{pmatrix} 1 & 2 & 3 & 4 \\ 3 & 2 & 1 & 4 \end{pmatrix} \quad R_6 = \begin{pmatrix} 1 & 2 & 3 & 4 \\ 2 & 1 & 4 & 3 \end{pmatrix}$$

and

$$R_7 = \begin{pmatrix} 1 & 2 & 3 & 4 \\ 4 & 3 & 2 & 1 \end{pmatrix}$$

One last symmetry is the *identity*

$$R_0 = \begin{pmatrix} 1 & 2 & 3 & 4 \\ 1 & 2 & 3 & 4 \end{pmatrix}$$

which leaves the square as it was.

The operation on symmetries is composition: $R_i \circ R_j$ is the result of first performing R_j, and then R_i. For example, $R_1 \circ R_4$ is the result of first flipping the square about its axis A, then rotating it clockwise 90°:

$$R_1 \circ R_4 = \begin{pmatrix} 1 & 2 & 3 & 4 \\ 2 & 3 & 4 & 1 \end{pmatrix} \circ \begin{pmatrix} 1 & 2 & 3 & 4 \\ 1 & 4 & 3 & 2 \end{pmatrix}$$

$$= \begin{pmatrix} 1 & 2 & 3 & 4 \\ 2 & 1 & 4 & 3 \end{pmatrix} = R_6$$

Thus, the net effect is the same as if the square had been flipped about its axis C.

The eight symmetries of the square form a group under the operation $\circ$ of composition, called the *group of symmetries of the square*.

For every positive integer $n \geqslant 3$, the regular polygon with n sides has a group of symmetries, symbolized by D_n, which may be found as we did here. These groups are called the *dihedral groups*. For example, the group of the square is D_4, the group of the pentagon is D_5, and so on.

Every plane figure which exhibits regularities has a group of symmetries. For example, the following figure, has a group of symmetries con-

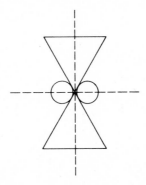

sisting of two rotations (180° and 360°) and two flips about the indicated axes. Artificial as well as natural objects often have a surprising number of symmetries.

Far more complicated than the plane symmetries are the symmetries of objects in space. Modern-day crystallography and crystal physics, for example, rely very heavily on knowledge about groups of symmetries of three-dimensional shapes.

Groups of symmetry are widely employed also in the theory of electron structure and of molecular vibrations. In elementary particle physics, such groups have been used to predict the existence of certain elementary particles before they were found experimentally!

Symmetries and their groups arise everywhere in nature: in quantum physics, flower petals, cell division, the work habits of bees in the hive, snowflakes, music, and Romanesque cathedrals.

EXERCISES

A. Computing Elements of S_6

1 Consider the following permutations f, g, and h in S_6 :

$$f = \begin{pmatrix} 1 & 2 & 3 & 4 & 5 & 6 \\ 6 & 1 & 3 & 5 & 4 & 2 \end{pmatrix} \qquad g = \begin{pmatrix} 1 & 2 & 3 & 4 & 5 & 6 \\ 2 & 3 & 1 & 6 & 5 & 4 \end{pmatrix}$$

$$h = \begin{pmatrix} 1 & 2 & 3 & 4 & 5 & 6 \\ 5 & 1 & 6 & 4 & 5 & 2 \\ 3 & & & & & \end{pmatrix}$$

Compute the following:

$$f^{-1} = \begin{pmatrix} 1 & 2 & 3 & 4 & 5 & 6 \\ & & & & & \end{pmatrix} \qquad g^{-1} = \begin{pmatrix} 1 & 2 & 3 & 4 & 5 & 6 \\ & & & & & \end{pmatrix}$$

$$h^{-1} = \begin{pmatrix} 1 & 2 & 3 & 4 & 5 & 6 \\ & & & & & \end{pmatrix}$$

$$f \circ g = \begin{pmatrix} 1 & 2 & 3 & 4 & 5 & 6 \\ & & & & & \end{pmatrix} \qquad g \circ f = \begin{pmatrix} 1 & 2 & 3 & 4 & 5 & 6 \\ & & & & & \end{pmatrix}$$

2 $f \circ (g \circ h) =$
3 $g \circ h^{-1} =$
4 $h \circ g^{-1} \circ f^{-1} =$
5 $g \circ g \circ g =$

B. Examples of Groups of Permutations

1 Let G be the subset of S_4 consisting of the permutations

$$\varepsilon = \begin{pmatrix} 1 & 2 & 3 & 4 \\ 1 & 2 & 3 & 4 \end{pmatrix} \qquad f = \begin{pmatrix} 1 & 2 & 3 & 4 \\ 2 & 1 & 4 & 3 \end{pmatrix}$$

$$g = \begin{pmatrix} 1 & 2 & 3 & 4 \\ 3 & 4 & 1 & 2 \end{pmatrix} \qquad h = \begin{pmatrix} 1 & 2 & 3 & 4 \\ 4 & 3 & 2 & 1 \end{pmatrix}$$

Show that G is a group of permutations, and write its table.

$\circ$	ε	f	g	h
ε				
f				
g				
h				

2 List the elements of the cyclic subgroup of S_6 generated by

$$f = \begin{pmatrix} 1 & 2 & 3 & 4 & 5 & 6 \\ 2 & 3 & 4 & 1 & 6 & 5 \end{pmatrix}$$

3 Find a four-element abelian subgroup of S_4. Write its table.

4 The subgroup of S_5 generated by

$$f = \begin{pmatrix} 1 & 2 & 3 & 4 & 5 \\ 2 & 1 & 3 & 4 & 5 \end{pmatrix} \qquad g = \begin{pmatrix} 1 & 2 & 3 & 4 & 5 \\ 1 & 2 & 4 & 5 & 3 \end{pmatrix}$$

has six elements. List them, then write the table of this group:

$$\varepsilon = \begin{pmatrix} 1 & 2 & 3 & 4 & 5 \\ 1 & 2 & 3 & 4 & 5 \end{pmatrix}$$

$$f = \begin{pmatrix} 1 & 2 & 3 & 4 & 5 \\ 2 & 1 & 3 & 4 & 5 \end{pmatrix}$$

$$g = \begin{pmatrix} 1 & 2 & 3 & 4 & 5 \\ 1 & 2 & 4 & 5 & 3 \end{pmatrix}$$

$$h = \begin{pmatrix} 1 & 2 & 3 & 4 & 5 \\ & & & & \end{pmatrix}$$

$$k = \begin{pmatrix} 1 & 2 & 3 & 4 & 5 \\ & & & & \end{pmatrix}$$

$$l = \begin{pmatrix} 1 & 2 & 3 & 4 & 5 \\ & & & & \end{pmatrix}$$

$\circ$	ε	f	g	h	k	l
ε						
f						
g						
h						
k						
l						

C. Groups of Permutations of $\mathbb{R}$

In each of the following, A is a subset of $\mathbb{R}$ and G is a set of permutations of A. *Show that G is a subgroup of S_A, and write the table of G.*

1 A is the set of all $x \in \mathbb{R}$ such that $x \neq 0,\ 1$. $G = \{\varepsilon, f, g\}$, where $f(x) = 1/(1-x)$ and $g(x) = (x-1)/x$.

2 A is the set of all the nonzero real numbers. $G = \{\varepsilon, f, g, h\}$, where $f(x) = 1/x$, $g(x) = -x$, and $h(x) = -1/x$.

3 A is the set of all the real numbers $x \neq 0,\ 1$. $G = \{\varepsilon, f, g, h, j, k\}$, where $f(x) = 1 - x$, $g(x) = 1/x$, $h(x) = 1/(1-x)$, $j(x) = (x-1)/x$, and $k(x) = x/(x-1)$.

4 A is the set of all the real numbers $x \neq 0,\ 1,\ 2$. G is the subgroup of S_A generated by $f(x) = 2 - x$ and $g(x) = 2/x$. (G has eight elements. List them, and write the table of G.)

† D. A Cyclic Group of Permutations

For each integer n, define f_n by: $f_n(x) = x + n$. *Prove the following:*

1 For each integer n, f_n is a permutation of $\mathbb{R}$, that is, $f_n \in S_{\mathbb{R}}$.

2 $f_n \circ f_m = f_{n+m}$ and $f_n^{-1} = f_{-n}$.

3 Let $G = \{f_n : n \in \mathbb{Z}\}$. Prove that G is a subgroup of $S_{\mathbb{R}}$.

4 Prove that G is cyclic. (Indicate a generator of G.)

† E. A Subgroup of $S_{\mathbb{R}}$

For any pair of real numbers $a \neq 0$ and b, define a function $f_{a,b}$ as follows:

$$f_{a,b}(x) = ax + b$$

1 Prove that $f_{a,b}$ is a permutation of $\mathbb{R}$, that is, $f_{a,b} \in S_{\mathbb{R}}$.

2 Prove that $f_{a,b} \circ f_{c,d} = f_{ac,\, ad+b}$.

3 Prove that $f_{a,b}^{-1} = f_{1/a,\, -b/a}$.

4 Let $G = \{f_{a,b} : a \in \mathbb{R},\ b \in \mathbb{R},\ a \neq 0\}$. Show that G is a subgroup of $S_{\mathbb{R}}$.

F. Symmetries of Geometric Figures

1 Let G be the group of symmetries of the regular hexagon. List the elements of G (there are 12 of them), then write the table of G.

$$R_0 = \begin{pmatrix} 1 & 2 & 3 & 4 & 5 & 6 \\ 1 & 2 & 3 & 4 & 5 & 6 \end{pmatrix} \qquad R_1 = \begin{pmatrix} 1 & 2 & 3 & 4 & 5 & 6 \\ 2 & 3 & 4 & 5 & 6 & 1 \end{pmatrix}$$

$$R_2 = \begin{pmatrix} 1 & 2 & 3 & 4 & 5 & 6 \\ 3 & 4 & 5 & 6 & 1 & 2 \end{pmatrix} \qquad R_3 = \begin{pmatrix} 1 & 2 & 3 & 4 & 5 & 6 \\ & & & & & \end{pmatrix}$$

etc. ...

2 Let G be the group of symmetries of the rectangle. List the elements of G (there are four of them), and write the table of G.

3 List the symmetries of the letter **Z** and give the table of this group of symmetries. Do the same for the letters **V** and **H**.

4 List the symmetries of the following shape, and give the table of their group.

(Assume that the three arms are of equal length, and the three central angles are equal.)

G. Symmetries of Polynomials

Consider the polynomial $p = (x_1 - x_2)^2 + (x_3 - x_4)^2$. It is unaltered when the subscripts undergo any of the following permutations:

$$\begin{pmatrix} 1 & 2 & 3 & 4 \\ 2 & 1 & 3 & 4 \end{pmatrix} \quad \begin{pmatrix} 1 & 2 & 3 & 4 \\ 1 & 2 & 4 & 3 \end{pmatrix} \quad \begin{pmatrix} 1 & 2 & 3 & 4 \\ 2 & 1 & 4 & 3 \end{pmatrix} \quad \begin{pmatrix} 1 & 2 & 3 & 4 \\ 3 & 4 & 1 & 2 \end{pmatrix}$$

$$\begin{pmatrix} 1 & 2 & 3 & 4 \\ 4 & 3 & 1 & 2 \end{pmatrix} \quad \begin{pmatrix} 1 & 2 & 3 & 4 \\ 3 & 4 & 2 & 1 \end{pmatrix} \quad \begin{pmatrix} 1 & 2 & 3 & 4 \\ 4 & 3 & 2 & 1 \end{pmatrix} \quad \begin{pmatrix} 1 & 2 & 3 & 4 \\ 1 & 2 & 3 & 4 \end{pmatrix}$$

For example, the first of these permutations replaces p by

$$(x_2 - x_1)^2 + (x_3 - x_4)^2$$

the second permutation replaces p by $(x_1 - x_2)^2 + (x_4 - x_3)^2$; and so on. The *symmetries of a polynomial p* are all the permutations of the subscripts which leave p unchanged. They form a group of permutations.

List the symmetries of each of the following polynomials, and write their group table.

1 $p = x_1 x_2 + x_2 x_3$
2 $p = (x_1 - x_2)(x_2 - x_3)(x_1 - x_3)$
3 $p = x_1 x_2 + x_2 x_3 + x_1 x_3$
4 $p = (x_1 - x_2)(x_3 - x_4)$

H. Properties of Permutations of a Set A

1 Let A be a set and $a \in A$. Let G be the subset of S_A consisting of all the permutations f of A such that $f(a) = a$. Prove that G is a subgroup of S_A.

2 If f is a permutation of A and $a \in A$, we say that f *moves a* if $f(a) \neq a$. Let A be an infinite set, and let G be the subset of S_A which consists of all the permutations f of A which move *only a finite number of elements* of A. Prove that G is a subgroup of S_A.

3 Let A be a *finite* set, and B a subset of A. Let G be the subset of S_A consisting of all the permutations f of A such that $f(x) \in B$ for every $x \in B$. Prove that G is a subgroup of S_A.

4 Give an example to show that the conclusion of part 3 is not necessarily true if A is an infinite set.

EIGHT

PERMUTATIONS OF A FINITE SET

Permutations of finite sets are used in every branch of mathematics—for example, in geometry, in statistics, in elementary algebra—and they have a myriad of applications in science and technology. Because of their practical importance, this chapter will be devoted to the study of a few special properties of permutations of finite sets.

If n is a positive integer, consider a set of n elements. It makes no difference which specific set we consider, just as long as it has n elements; so let us take the set $\{1, 2, \ldots, n\}$. We have already seen that the group of all the permutations of this set is called the *symmetric group on n elements* and is denoted by S_n. In the remainder of this chapter, when we say *permutation* we will invariably mean a permutation of the set $\{1, 2, \ldots, n\}$ for an arbitrary positive integer n.

One of the most characteristic activities of science (*any* kind of science) is to try to separate complex things into their simplest component parts. This intellectual "divide and conquer" helps us to understand complicated processes and solve difficult problems. The savvy mathematician never misses the chance of doing this whenever the opportunity presents itself. We will see now that every permutation can be decomposed into simple parts called "cycles," and these cycles are, in a sense, the most basic kind of permutations.

We begin with an example: take, for instance, the permutation

$$f = \begin{pmatrix} 1 & 2 & 3 & 4 & 5 & 6 & 7 & 8 & 9 \\ 3 & 1 & 6 & 9 & 8 & 2 & 4 & 5 & 7 \end{pmatrix}$$

and look at how f moves the elements in its domain:

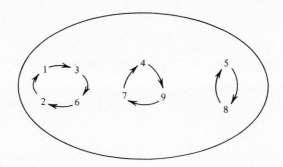

Notice how f decomposes its domain into three separate subsets, so that, in each subset, the elements are permuted cyclically so as to form a closed chain. These closed chains may be considered to be the component parts of the permutation; they are called "cycles." (This word will be carefully defined in a moment.) Every permutation breaks down, just as this one did, into separate cycles.

Let $a_1, a_2, \ldots, a_s$ be distinct elements of the set $\{1, 2, \ldots, n\}$. By the cycle $(a_1 a_2 \cdots a_s)$ we mean the permutation

$$\underbrace{a_1 \rightarrow a_2 \rightarrow a_3 \rightarrow \cdots \rightarrow a_{s-1} \rightarrow a_s}$$

of $\{1, 2, \ldots, n\}$ which carries a_1 to a_2, a_2 to a_3, ..., a_{s-1} to a_s, and a_s to a_1, while leaving all the remaining elements of $\{1, 2, \ldots, n\}$ fixed.

For instance, in S_6, the cycle (1426) is the permutation

$$\begin{pmatrix} 1 & 2 & 3 & 4 & 5 & 6 \\ 4 & 6 & 3 & 2 & 5 & 1 \end{pmatrix}$$

In S_5, the cycle (254) is the permutation

$$\begin{pmatrix} 1 & 2 & 3 & 4 & 5 \\ 1 & 5 & 3 & 2 & 4 \end{pmatrix}$$

Because cycles are permutations, we may form the *composite* of two cycles in the usual manner. The composite of cycles is generally called their *product* and it is customary to omit the symbol $\circ$. For example, in S_5,

$$(245)(124) = \begin{pmatrix} 1 & 2 & 3 & 4 & 5 \\ 1 & 4 & 3 & 5 & 2 \end{pmatrix} \circ \begin{pmatrix} 1 & 2 & 3 & 4 & 5 \\ 2 & 4 & 3 & 1 & 5 \end{pmatrix}$$

$$= \begin{pmatrix} 1 & 2 & 3 & 4 & 5 \\ 4 & 5 & 3 & 1 & 2 \end{pmatrix}$$

Actually, it is very easy to compute the product of two cycles by reasoning in the following manner: Let us continue with the same example,

$$\underbrace{(2 \quad 4 \quad 5)}_{\alpha}\underbrace{(1 \quad 2 \quad 4)}_{\beta}$$

Remember that the permutation on the right is applied first, and the permutation on the left is applied next. Now,

β carries 1 to 2, and α carries 2 to 4, hence $\alpha\beta$ carries 1 to 4;
β carries 2 to 4, and α carries 4 to 5, hence $\alpha\beta$ carries 2 to 5;
β leaves 3 fixed and so does α, hence $\alpha\beta$ leaves 3 fixed;
β carries 4 to 1 and α leaves 1 fixed, so $\alpha\beta$ carries 4 to 1;
β leaves 5 fixed and α carries 5 to 2, hence $\alpha\beta$ carries 5 to 2.

If $(a_1 a_2 \cdots a_s)$ is a cycle, the integer s is called its *length*; thus, $(a_1 a_2 \cdots a_s)$ is a *cycle of length s*. For example, (1532) is a cycle of length 4.

If two cycles have no elements in common they are said to be *disjoint*. For example, (132) and (465) are disjoint cycles, but (132) and (453) are not disjoint. *Disjoint cycles commute*: that is, if $(a_1 \cdots a_r)$ and $(b_1 \cdots b_s)$ are disjoint, then

$$\underbrace{(a_1 \cdots a_r)}_{\alpha}\underbrace{(b_1 \cdots b_s)}_{\beta} = \underbrace{(b_1 \cdots b_s)}_{\beta}\underbrace{(a_1 \cdots a_r)}_{\alpha}$$

It is easy to see why this is true: α moves the a's but not the b's, while β moves the b's but not the a's. Thus, if β carries b_i to b_j, then $\alpha\beta$ does the same, and so does $\beta\alpha$. Similarly, if α carries a_h to a_k then $\beta\alpha$ does the same, and so does $\alpha\beta$.

We are now ready to prove what was asserted at the beginning of this chapter: Every permutation can be decomposed into cycles—in fact, into *disjoint* cycles. More precisely.

Theorem 1 *Every permutation is either the identity, a single cycle, or a product of disjoint cycles.*

We begin with an example, because the proof uses the same technique as the example. Consider the permutation

$$f = \begin{pmatrix} 1 & 2 & 3 & 4 & 5 & 6 \\ 3 & 4 & 5 & 2 & 1 & 6 \end{pmatrix}$$

and let us write f as a product of disjoint cycles. We begin with 1 and note that

$$1 \xrightarrow{\ f\ } 3 \xrightarrow{\ f\ } 5 \xrightarrow{\ f\ } 1$$

We have come a complete circle and found our first cycle, which is (135). Next, we take the first number which hasn't yet been used, namely 2. We see that

$$2 \xrightarrow{\ f\ } 4 \xrightarrow{\ f\ } 2$$

again we have come a complete circle and found another cycle, which is (24). The only remaining number is 6, which f leaves fixed. We are done:

$$f = (135)(24)$$

The proof for *any* permutation f follows the same pattern as the example. Let a_1 be the first number in $\{1, \ldots, n\}$ such that $f(a_1) \neq a_1$. Let $a_2 = f(a_1)$, $a_3 = f(a_2)$, and so on in succession until we come to our first repetition, that is, until $f(a_k)$ is equal to one of the numbers $a_1, a_2, \ldots, a_{k-1}$. Say $f(a_k) = a_i$. If a_i is not a_1, we have

$$a_1 \rightarrow a_2 \rightarrow \cdots \rightarrow a_{i-1} \rightarrow \underbrace{a_i \cdots \rightarrow a_k}$$

so a_i is the image of two elements, a_k and a_{i-1}, which is impossible because f is bijective. Thus, $a_i = a_1$, and therefore $f(a_k) = a_1$. We have come a complete circle and found our first cycle, namely $(a_1 a_2 \cdots a_k)$.

Next, let b_1 be the first number which has not yet been examined and such that $f(b_1) \neq b_1$. We let $b_2 = f(b_1)$, $b_3 = f(b_2)$, and proceed as before to obtain the next cycle, say $(b_1 \cdots b_t)$. Obviously $(b_1 \cdots b_t)$ is disjoint from $(a_1 \cdots a_k)$. We continue this process until all the numbers in $\{1, \ldots, n\}$ have been exhausted.

Incidentally, it is easy to see that this product of cycles is unique, except for the order of the factors.

Now our curiosity may prod us to ask: once a permutation has been written as a product of disjoint cycles, *has it been simplified as much as possible*? Or is there some way of simplifying it further?

A cycle of length 2 is called a *transposition*. In other words, a transposition is a cycle $(a_i a_j)$ which interchanges the two numbers a_i and a_j. It is a fact both remarkable and trivial that every cycle can be expressed as a product of one or more transpositions. In fact,

$$(a_1 a_2 \cdots a_r) = (a_r a_{r-1})(a_r a_{r-2}) \cdots (a_r a_3)(a_r a_2)(a_r a_1)$$

which may be verified by direct computation. For example,

$$(12345) = (54)(53)(52)(51)$$

However, there is more than one way to write a given permutation as a product of transpositions. For example, (12345) may also be expressed as a product of transpositions in the following ways,

$$(12345) = (15)(14)(13)(12)$$

$$(12345) = (54)(52)(51)(14)(32)(41)$$

as well as in many other ways.

Thus, every permutation, after it has been decomposed into disjoint cycles, may be broken down further and expressed as a product of transpositions. However, the expression as a product of transpositions is not unique, and even the *number of transpositions* involved is not unique.

Nevertheless, when a permutation π is written as a product of transpositions, *one* property of this expression is unique: the number of transpositions involved is either *always even*, or *always odd*. (This fact will be proved in a moment.) For example, we have just seen that (12345) can be written as a product of four transpositions and also as a product of six transpositions; it can be written in many other ways, but always as a product of an *even* number of transpositions. Likewise, (1234) can be decomposed in many ways into transpositions, but always an *odd* number of transpositions.

A permutation is called *even* if it is a product of an even number of transpositions, and *odd* if it is a product of an odd number of transpositions. What we are asserting, therefore, is that every permutation is unambiguously either odd or even.

This may seem like a pretty useless fact—but actually the very opposite is true. A number of great theorems of mathematics depend for their proof (at that crucial step when the razor of logic makes its decisive cut) on none other but the distinction between even and odd permutations.

We begin by showing that the identity permutation, ε, is an even permutation.

Theorem 2 *No matter how ε is written as a product of transpositions, the number of transpositions is even.*

Let $t_1, t_2, \ldots, t_m$ be m transpositions, and suppose that

$$\varepsilon = t_1 t_2 \cdots t_m \tag{*}$$

We aim to prove that ε *can be rewritten as a product of $m - 2$ transpositions.* We will then be done: for if ε were equal to a product of an odd number of transpositions, and we were able to rewrite this product repeatedly, each time with two fewer transpositions, then eventually we would get ε equal to a single transposition (ab), and this is impossible.

Let x be any numeral appearing in one of the transpositions $t_2, \ldots, t_m$. Let $t_k = (xa)$, and suppose t_k is the last transposition in (*) (reading from left to right), in which x appears:

$$\varepsilon = t_1 t_2 \cdots t_{k-1} \underset{\underset{= (xa)}{\smile}}{t_k} \underbrace{t_{k+1} \cdots t_m}_{\substack{x \text{ does not} \\ \text{appear here}}}$$

Now, t_{k-1} is a transposition which is either equal to (xa), or else one or both of its components are different from x and a. This gives four possibilities, which we now treat as four separate cases.

Case I $t_{k-1} = (xa)$.

Then $t_{k-1}t_k = (xa)(xa)$, which is equal to the identity permutation. Thus, $t_{k-1}t_k$ may be removed without changing (*). As a result, ε is a product of $m - 2$ transpositions, as required.

Case II $t_{k-1} = (xb)$, *where $b \neq x, a$.*

Then $$t_{k-1}t_k = (xb)(xa)$$

But $$(xb)(xa) = (xa)(ab)$$

We replace $t_{k-1}t_k$ by $(xa)(ab)$ in (*). As a result, the last occurrence of x is one position further left than it was at the start.

Case III $t_{k-1} = (ca)$, *where $c \neq x, a$.*

Then $$t_{k-1}t_k = (ca)(xa)$$

But $$(ca)(xa) = (xc)(ca)$$

We replace $t_{k-1}t_k$ by $(xc)(ca)$ in (∗), as in Case II.

Case IV $t_{k-1} = (bc)$, *where* $b \neq x, a$ *and* $c \neq x, a$.

Then $$t_{k-1}t_k = (bc)(xa)$$

But $$(bc)(xa) = (xa)(bc)$$

We replace $t_{k-1}t_k$ by $(xa)(bc)$ in (∗), as in Cases II and III.

In Case I, we are done. In Cases II, III, or IV, we repeat the argument one or more times. Each time, the last appearance of x is one position further left than the time before. This must eventually lead to Case I. For otherwise, we end up with the last (hence the only) appearance of x being in t_1. This cannot be: for if $t_1 = (xa)$ and x does not appear in $t_2, \ldots, t_m$, then $\varepsilon(x) = a$, which is impossible!

Our conclusion is contained in the next theorem.

Theorem 3 *If* $\pi \in S_n$, *then* π *cannot be both an odd permutation and an even permutation.*

Suppose π can be written as the product of an even number of transpositions, and differently as the product of an odd number of transpositions. Then the same would be true for π^{-1}. But $\varepsilon = \pi \circ \pi^{-1}$: thus, writing π^{-1} as a product of an even number of transpositions and π as a product of an odd number of transpositions, we get an expression for ε as a product of an odd number of transpositions. This is impossible by Theorem 2.

The set of all the even permutations in S_n is a subgroup of S_n. It is denoted by A_n, and is called the *alternating group* on the set $\{1, 2, \ldots, n\}$.

EXERCISES

A. Practice in Multiplying and Factoring Permutations

1 Compute each of the following products in S_9. (Write your answer as a single permutation.)

(a) (145)(37)(682)   (b) (17)(628)(9354)
(c) (71825)(36)(49)   (d) (12)(347)
(e) (147)(1678)(74132)
(f) (6148)(2345)(12493)

2 Write each of the following permutations in S_9 as a product of disjoint cycles:

(a) $\begin{pmatrix} 1 & 2 & 3 & 4 & 5 & 6 & 7 & 8 & 9 \\ 4 & 9 & 2 & 5 & 1 & 7 & 6 & 8 & 3 \end{pmatrix}$ (b) $\begin{pmatrix} 1 & 2 & 3 & 4 & 5 & 6 & 7 & 8 & 9 \\ 7 & 4 & 9 & 2 & 3 & 8 & 1 & 6 & 5 \end{pmatrix}$

(c) $\begin{pmatrix} 1 & 2 & 3 & 4 & 5 & 6 & 7 & 8 & 9 \\ 7 & 9 & 5 & 3 & 1 & 2 & 4 & 8 & 6 \end{pmatrix}$ (d) $\begin{pmatrix} 1 & 2 & 3 & 4 & 5 & 6 & 7 & 8 & 9 \\ 9 & 8 & 7 & 4 & 3 & 6 & 5 & 1 & 2 \end{pmatrix}$

3 Express each of the following as a product of transpositions in S_8.

(a) (137428)   (b) (416)(8235)

(c) (123)(456)(1574)   (d) $\pi = \begin{pmatrix} 1 & 2 & 3 & 4 & 5 & 6 & 7 & 8 \\ 3 & 1 & 4 & 2 & 8 & 7 & 6 & 5 \end{pmatrix}$

4 If $\alpha = (3714)$, $\beta = (123)$, and $\gamma = (24135)$ in S_7, express each of the following as a product of disjoint cycles:
(a) $\alpha^{-1}\beta$ (b) $\gamma^{-1}\alpha$ (c) $\alpha^2\beta$ (d) $\beta^2\alpha\gamma$ (e) γ^4 (f) $\gamma^3\alpha^{-1}$
(g) $\beta^{-1}\gamma$ (h) $\alpha^{-1}\gamma^2\alpha$

5 In S_5, write (12345) in five different ways as a cycle, and in five different ways as a product of transpositions.

6 In S_5, express each of the following as the square of a cycle (that is, express as α^2 where α is a cycle):
(a) (132)   (b) (12345)   (c) (13)(24)

B. Powers of Permutations

If π is any permutation, we write $\pi \circ \pi = \pi^2$, $\pi \circ \pi \circ \pi = \pi^3$, etc. The convenience of this notation is evident.

1 Compute α^{-1}, α^2, α^3, α^4, α^5 where

(i) $\alpha = (123)$ (ii) $\alpha = (1234)$ (iii) $\alpha = (123456)$.

In the following problems, let α be a cycle of length s, say $\alpha = (a_1 a_2 \cdots a_s)$.

2 Describe all the *distinct* powers of α. How many are there?

3 Find the inverse of α, and show that $\alpha^{-1} = \alpha^{s-1}$.

Prove each of the following:

4 α^2 is a cycle iff s is odd.

5 If s is odd, α is the square of some cycle of length s. (Find it.)

6 If s is even, say $s = 2t$, then α^2 is the product of two cycles of length t. (Find them.)

7 If s is a multiple of k, say $s = kt$, then α^k is the product of k cycles of length t.

8 If s is a prime number, every power of α is a cycle.

X C. Even and Odd Permutations

1 Determine which of the following permutations is even, and which is odd.

(a) $\pi = \begin{pmatrix} 1 & 2 & 3 & 4 & 5 & 6 & 7 & 8 \\ 7 & 4 & 1 & 5 & 6 & 2 & 3 & 8 \end{pmatrix}$ (b) (71864)

(c) (12)(76)(345)   (d) (1276)(3241)(7812)

(e) (123)(2345)(1357)

Prove each of the following:

2 (a) The product of two even permutations is even.
    (b) The product of two odd permutations is even.
    (c) The product of an even permutation and an odd permutation is odd.

3 (a) A cycle of length l is even if l is odd.
    (b) A cycle of length l is odd if l is even.
    (c) A cycle of length l is odd or even depending on whether $l - 1$ is odd or even.

4 (a) If α and β are cycles of length l and m, respectively, then $\alpha\beta$ is even or odd depending on whether $l + m - 2$ is even or odd.
    (b) If $\pi = \beta_1 \cdots \beta_r$ where each β_i is a cycle of length l_i, then π is even or odd depending on whether $l_1 + l_2 + \cdots + l_r - r$ is even or odd.

D. Disjoint Cycles

In each of the following, let α and β be disjoint cycles, say

$$\alpha = (a_1 a_2 \cdots a_s) \quad \text{and} \quad \beta = (b_1 b_2 \cdots b_r)$$

Prove each of the following:

1 For every positive integer n, $(\alpha\beta)^n = \alpha^n \beta^n$.

2 If $\alpha\beta = \varepsilon$, then $\alpha = \varepsilon$ and $\beta = \varepsilon$.

3 If $(\alpha\beta)^t = \varepsilon$, then $\alpha^t = \varepsilon$ and $\beta^t = \varepsilon$ (where t is any positive integer). [Use (2).]

4 Find a transposition γ such that $\alpha\beta\gamma$ is a cycle.

5 Let γ be the same transposition as in the preceding exercise. Show that $\alpha\gamma\beta$ and $\gamma\alpha\beta$ are cycles.

6 Let α and β be cycles (not necessarily disjoint). If $\alpha^2 = \beta^2$, then $\alpha = \beta$.

† E. Conjugate Cycles

Prove each of the following in S_n:

1 Let $\alpha = (a_1 \cdots a_s)$ and $\beta = (b_1 \cdots b_s)$ be cycles of the same length, and let π be any permutation. If $\pi(a_i) = b_i$ for $i = 1, \ldots, s$, then $\pi\alpha\pi^{-1} = \beta$.

If α is any cycle and π any permutation, $\pi\alpha\pi^{-1}$ is called a *conjugate* of α.

2 If α and β are any two cycles of the same length s, there is a permutation $\pi \in S_n$ such that $\beta = \pi\alpha\pi^{-1}$.

3 Conclude: Any two cycles of the same length are conjugates of each other.

4 If α and β are disjoint cycles, then $\pi\alpha\pi^{-1}$ and $\pi\beta\pi^{-1}$ are disjoint cycles.

5 Let σ be a product $\alpha_1 \cdots \alpha_t$ of t disjoint cycles of lengths $l_1, \ldots, l_t$, respectively. Then $\pi\sigma\pi^{-1}$ is also a product of t disjoint cycles of lengths $l_1, \ldots, l_t$.

6 Let α_1 and α_2 be cycles of the same length. Let β_1 and β_2 be cycles of the same length. Let α_1 and β_1 be disjoint, and let α_2 and β_2 be disjoint. There is a permutation $\pi \in S_n$ such that $\alpha_1\beta_1 = \pi\alpha_2\beta_2\pi^{-1}$.

† F. Order of Cycles

Prove each of the following in S_n.

1 If $\alpha = (a_1 \cdots a_s)$ is a cycle of length s, then $\alpha^s = \varepsilon$, $\alpha^{2s} = \varepsilon$, and $\alpha^{3s} = \varepsilon$. Is $\alpha^k = \varepsilon$ for any positive integer $k < s$? (Explain.)

If α is any permutation, the least positive integer n such that $\alpha^n = \varepsilon$ is called the *order* of α.

2 If $\alpha = (a_1 \cdots a_s)$ is any cycle of length s, the order of α is s.

3 Find the order of each of the following permutations:
    (a) (12)(345)   (b) (12)(3456)
    (c) (1234)(567890)

4 What is the order of $\alpha\beta$, if α and β are disjoint cycles of lengths 4 and 6, respectively? (Explain why. Use the fact that disjoint cycles commute.)

5 What is the order of $\alpha\beta$, if α and β are disjoint cycles of lengths r and s, respectively. (Venture a guess, explain, but do not attempt a rigorous proof.)

† G. Even/Odd Permutations in Subgroups of S_n

Prove each of the following in S_n:

1 Let $\alpha_1, \ldots, \alpha_r$ be even permutations, and β an odd permutation. Then $\alpha_1\beta, \ldots, \alpha_r\beta$ are r *distinct* odd permutations. (See Exercise C2.)

2 If $\beta_1, \ldots, \beta_r$ are odd permutations, then $\beta_1\beta_1, \beta_1\beta_2, \ldots, \beta_1\beta_r$ are r *distinct* even permutations.

3 In S_n, there are the same number of odd permutations as even permutations. (HINT: Use part 1 to prove that the number of even permutations is $\le$ the number of odd permutations. Use part 2 to prove the reverse of that inequality.)

4 The set of all the even permutations is a subgroup of S_n. (It is denoted by A_n and is called the *alternating group* on n symbols.)

5 Let H be any subgroup of S_n. H either contains only even permutations, or H contains the same number of odd as even permutations. (Use parts 1 and 2.)

† **H. Generators of A_n and S_n**

Remember that in any group G, a set S of elements of G is said to *generate* G if every element of G can be expressed as a product of elements in S and inverses of elements in S. (See page 48.)

1 Prove that the set T of all the transpositions in S_n generates S_n.

2 Prove that the set $T_1 = \{(12), (13), \ldots, (1n)\}$ generates S_n.

3 Prove that every even permutation is a product of one or more cycles of length 3. [HINT: $(13)(12) = (123)$; $(12)(34) = (321)(134)$.] Conclude that the set U of all cycles of length 3 generates A_n.

4 Prove that the set $U_1 = \{(123), (124), \ldots, (12n)\}$ generates A_n. [HINT: $(abc) = (1ca)(1ab)$, $(1ab) = (1b2)(12a)(12b)$, and $(1b2) = (12b)^2$.]

5 The pair of cycles (12) and $(12 \cdots n)$ generates S_n. [HINT: $(1 \cdots n)(12)(1 \cdots n)^{-1} = (23)$; $(12)(23)(12) = (13)$.]

6 If α is any cycle of length n, and β is any transposition, then $\{\alpha, \beta\}$ generates S_n.

NINE

ISOMORPHISM

Human perception, as well as the "perception" of so-called intelligent machines, is based on the ability to recognize the same structure in different guises. It is the faculty for discerning, in *different* objects, the same relationships between their parts.

The dictionary tells us that two things are "isomorphic" if they *have the same structure*. The notion of isomorphism—of having the same structure—is central to every branch of mathematics and permeates all of abstract reasoning. It is an expression of the simple fact that objects may be different in substance but identical in form.

In geometry there are several kinds of isomorphism, the simplest being congruence and similarity. Two geometric figures are congruent if there exists a plane motion which makes one figure coincide with the other; they are similar if there exists a transformation of the plane, magnifying or shrinking lengths in a fixed ratio, which (again) makes one figure coincide with the other.

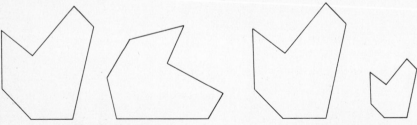

These two figures are congruent   These two figures are similar

We do not even need to venture into mathematics to meet some simple examples of isomorphism. For instance, the two palindromes

M A D A M   R O T O R

A   and   O

M A D   R O T

are different, but obviously isomorphic; indeed, the first one coincides with the second if we replace M by R, A by O, and D by T.

Here is an example from applied mathematics: A flow network is a set of points, with arrows joining some of the points. Such networks are used to represent flows of cash or goods, channels of communication, electrical circuits, and so on. The flow networks (A) and (B), below, are different, but

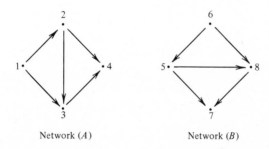

Network (A) Network (B)

can be shown to be isomorphic. Indeed, (A) can be made to coincide with (B) if we superimpose point 1 on point 6, point 2 on point 5, point 3 on point 8, and point 4 on point 7. (A) and (B) then coincide in the sense of having the same points joined by arrows in the same direction. Thus, *network (A) is transformed into network (B)* if we replace points 1 by 6, 2 by 5, 3 by 8, and 4 by 7. The one-to-one correspondence which carries out this transformation, namely,

$$\begin{pmatrix} 1 & 2 & 3 & 4 \\ \downarrow & \downarrow & \downarrow & \downarrow \\ 6 & 5 & 8 & 7 \end{pmatrix}$$

is called an *isomorphism* from network (A) to network (B), for it transforms (A) into (B).

Incidentally, the one-to-one correspondence

$$\begin{pmatrix} M & A & D \\ \downarrow & \downarrow & \downarrow \\ R & O & T \end{pmatrix}$$

is an *isomorphism* between the two palindromes of the preceding example, for it transforms the first palindrome into the second.

Our next and final example is from algebra. Consider the two groups G_1 and G_2 described below:

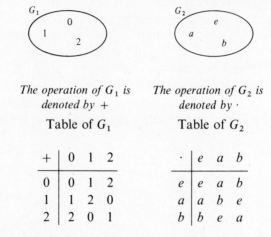

The operation of G_1 is denoted by $+$

The operation of G_2 is denoted by $\cdot$

Table of G_1

Table of G_2

$+$	0	1	2
0	0	1	2
1	1	2	0
2	2	0	1

$\cdot$	e	a	b
e	e	a	b
a	a	b	e
b	b	e	a

G_1 and G_2 are different, but isomorphic. Indeed, if in G_1 we replace 0 by e, 1 by a, and 2 by b, then G_1 coincides with G_2, the table of G_1 being transformed into the table of G_2. In other words, the one-to-one correspondence

$$\begin{pmatrix} 0 & 1 & 2 \\ \downarrow & \downarrow & \downarrow \\ e & a & b \end{pmatrix}$$

transforms G_1 to G_2. It is called an *isomorphism* from G_1 to G_2. Finally, because there exists an isomorphism from G_1 to G_2, G_1 and G_2 are isomorphic to each other.

In general, by an isomorphism between two groups we mean a one-to-one correspondence between them which transforms one of the groups into the other. If there *exists* an isomorphism from one of the groups to the other, we say they are isomorphic. Let us be more specific:

If G_1 and G_2 are any groups, an *isomorphism* from G_1 to G_2 is a one-to-one correspondence f from G_1 to G_2 with the following property: For every pair of elements a and b in G_1,

$$\text{If } f(a) = a' \text{ and } f(b) = b', \text{ then } f(ab) = a'b' \qquad (*)$$

In other words, if f matches a with a' and b with b', it *must* match ab with $a'b'$.

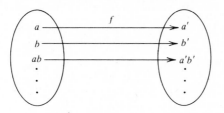

It is easy to see that if f has this property it transforms the table of G_1 into the table of G_2 :

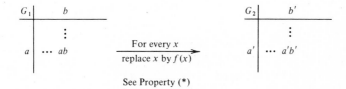

There is another, equivalent way of looking at this situation: If two groups G_1 and G_2 are isomorphic, we can say the two groups are actually the same, except that the elements of G_1 have *different names* from the elements of G_2. G_1 becomes exactly G_2 if we *rename* its elements. The function which does the renaming is an isomorphism from G_1 to G_2. Thus, in our last example, if 0 is renamed e, 1 is renamed a, and 2 is renamed b, G_1 becomes exactly G_2, with the same table. (Note that we have also renamed the operation: it was called $+$ in G_1 and $\cdot$ in G_2.)

By the way, Property ($*$) may be written more concisely as follows:

$$f(ab) = f(a)f(b) \qquad\qquad (**)$$

So we may sum up our definition of isomorphism in the following way:

Definition *Let G_1 and G_2 be groups. A bijective function $f : G_1 \to G_2$ with the property that for any two elements a and b in G_1,*

$$f(ab) = f(a)f(b) \qquad\qquad (**)$$

*is called an **isomorphism** from G_1 to G_2.*

*If there exists an isomorphism from G_1 to G_2, we say that G_1 is **isomorphic** to G_2.*

If there exists an isomorphism f from G_1 to G_2, in other words, if G_1 is isomorphic to G_2, we symbolize this fact by writing

$$G_1 \cong G_2$$

to be read, "G_1 is isomorphic to G_2."

How does one recognize if two groups are isomorphic? This is an important question, and not quite so easy to answer as it may appear. There is no way of spontaneously recognizing whether two groups G_1 and G_2 are isomorphic. Rather, the groups must be carefully tested according to the above definition.

G_1 and G_2 are isomorphic if *there exists* an isomorphism from G_1 to G_2. Therefore, the burden of proof is upon us to *find* an isomorphism from G_1 to G_2, and show that it *is* an isomorphism. In other words, we must go through the following steps:

1. Make an educated guess, and come up with a function $f: G_1 \rightarrow G_2$ which looks as though it might be an isomorphism.
2. Check that f is *injective* and *surjective* (hence bijective).
3. Check that f satisfies the identity

$$f(ab) = f(a)f(b)$$

Here's an example: $\mathbb{R}$ is the group of the real numbers with the operation of *addition*. $\mathbb{R}^+$ is the group of the *positive* real numbers with the operation of *multiplication*. It is an interesting fact that $\mathbb{R}$ and $\mathbb{R}^+$ are isomorphic. To see this, let us go through the steps outlined above:

1. The educated guess: The exponential function $f(x) = e^x$ is a function from $\mathbb{R}$ to $\mathbb{R}^+$ which, if we recall its properties, might do the trick.
2. f is injective: Indeed, if $f(a) = f(b)$, that is, $e^a = e^b$, then, taking the natural log on both sides, we get $a = b$.
   f is surjective: Indeed, if $y \in \mathbb{R}^+$, that is, if y is any positive real number, then $y = e^{\ln y} = f(\ln y)$; thus, $y = f(x)$ for $x = \ln y$.
3. It is well known that $e^{a+b} = e^a \cdot e^b$, that is,

$$f(a + b) = f(a) \cdot f(b)$$

Incidentally, note carefully that the operation of $\mathbb{R}$ is $+$, whereas the operation of $\mathbb{R}^+$ is $\cdot$. That is the reason we have to use $+$ on the left side of the preceding equation, and $\cdot$ on the right side of the equation.

How does one recognize when two groups are not isomorphic? In practice it is usually easier to show that two groups are *not* isomorphic than to show they *are*. Remember that if two groups are isomorphic they are replicas of each other; their elements (and their operation) may be named differently, but in all other respects they are the same and share the same properties. Thus, if a group G_1 has a property which group G_2 does not have (or vice versa), they are not isomorphic! Here are some examples of properties to look out for:

1. Perhaps G_1 is commutative, and G_2 is not.
2. Perhaps G_1 has an element which is its own inverse, and G_2 does not.
3. Perhaps G_1 is generated by two elements whereas G_2 is not generated by any choice of two of its elements.
4. Perhaps every element of G_1 is the square of an element of G_1, whereas G_2 does not have this property.

This list is by no means exhaustive; it merely illustrates the kind of things to be on the lookout for. Incidentally, the kind of properties to watch for are properties which do not depend merely on the *names* assigned to individual elements; for instance, in our last example, $0 \in G_1$ and $0 \notin G_2$, but nevertheless G_1 and G_2 are isomorphic.

Finally, let us state the obvious: if G_1 and G_2 cannot be put in one-to-one correspondence (say, G_1 has more elements that G_2), clearly they cannot be isomorphic.

In the early days of modern algebra the word "group" had a different meaning from the meaning it has today. In those days a group always meant a *group of permutations*. The only groups mathematicians used were groups whose elements were permutations of some fixed set and whose operation was composition.

There is something comforting about working with tangible, concrete things, such as groups of permutations of a set. At all times we have a clear picture of what it is we are working with. Later, as the axiomatic method reshaped algebra, a group came to mean *any* set with *any* associative operation having a neutral element and allowing each element an inverse. The new notion of group pleases mathematicians because it is simpler and more lean and sparing than the old notion of groups of permutations; it is also more general because it allows many new things to be groups which are not groups of permutations. However, it is harder to visualize, precisely because so many different things can be groups.

It was therefore a great revelation when, about 100 years ago, Arthur Cayley discovered that *every group is isomorphic to a group of permutations*. Roughly, this means that the groups of permutations are actually all the groups there are! Every group is (or is a carbon copy of) a group of permutations. This great result is a classic theorem of modern algebra. As a bonanza, its proof is not very difficult.

Cayley's Theorem *Every group is isomorphic to a group of permutations.*

Let G be a group; we wish to show that G is isomorphic to a group of permutations. The first question to ask is, "*What* group of permutations? Permutations of *what* set?" (After all, every permutation must be a permutation of some fixed set.) Well, the one set we have at hand is the set G, so we had better fix our attention on *permutations of G*. The way we match up elements of G with permutations of G is quite interesting:

With each element a in G we associate a function $\pi_a : G \to G$ defined by

$$\pi_a(x) = ax$$

In other words, π_a is the function whose rule may be described by the words "multiply on the left by a." We will now show that π_a is a permutation of G:

1. π_a *is injective*: Indeed, if $\pi_a(x_1) = \pi_a(x_2)$, then $ax_1 = ax_2$, so by the cancelation law, $x_1 = x_2$.
2. π_a *is surjective*: For if $y \in G$, then $y = a(a^{-1}y) = \pi_a(a^{-1}y)$. Thus, each y in G is equal to $\pi_a(x)$ for $x = a^{-1}y$.
3. Since π_a is an injective and surjective function from G to G, π_a *is a permutation of G.*

Let us remember that we have a permutation π_a for *each* element a in G; for example, if b and c are other elements in G, π_b is the permutation "multiply on the left by b," π_c is the permutation "multiply on the left by c," and so on. In general, let G^* denote the set of *all* the permutations π_a as a ranges over all the elements of G:

$$G^* = \{\pi_a : a \in G\}$$

Observe now that G^* is a set consisting of permutations of G—but not necessarily *all* the permutations of G. In Chapter 7 we used the symbol S_G to designate the group of *all* the permutations of G. We must show now that G^* is a *subgroup* of S_G, for that will prove that G^* is a *group of permutations*.

To prove that G^* is a subgroup of S_G, we must show that G^* is closed with respect to composition, and closed with respect to inverses. That is, we must show that if π_a and π_b are any elements of G^*, their composite $\pi_a \circ \pi_b$ is also in G^*; and if π_a is any element of G^*, its inverse is in G^*.

First, we claim that if a and b are any elements of G, then

$$\pi_a \circ \pi_b = \pi_{ab} \qquad\qquad (\text{***})$$

To show that $\pi_a \circ \pi_b$ and π_{ab} are the same, we must show that they have the same effect on every element x; that is, we must prove the identity $[\pi_a \circ \pi_b](x) = \pi_{ab}(x)$.   Well, $[\pi_a \circ \pi_b](x) = \pi_a(\pi_b(x)) = \pi_a(bx) = a(bx) = (ab)x = \pi_{ab}(x)$. Thus, $\pi_a \circ \pi_b = \pi_{ab}$; this proves that the composite of any two members π_a and π_b of G^* is another member π_{ab} of G^*. Thus, G^* is *closed with respect to composition.*

It is easy to see that π_e is the identity function: indeed,

$$\pi_e(x) = ex = x$$

In other words, π_e is the identity element of S_G.

Finally, by (***),

$$\pi_a \circ \pi_{a-1} = \pi_{aa-1} = \pi_e$$

So by Theorem 2 of Chapter 4, the inverse of π_a is π_{a-1}. This proves that the inverse of any member π_a of G^* is another member π_{a-1} of G^*. Thus, G^* is *closed with respect to inverses.*

Since G^* is closed with respect to composition and inverses, G^* *is a subgroup of* S_G.

We are now in the final lap of our proof. We have a group of permutations G^*, and it remains only to show that G is isomorphic to G^*. To do this, we must find an isomorphism $f : G \to G^*$. Let f be the function

$$f(a) = \pi_a$$

In other words, f matches each element a in G with the permutation π_a in G^*. We can quickly show that f is an isomorphism:

1. f *is injective*: Indeed, if $f(a) = f(b)$ then $\pi_a = \pi_b$. Thus, $\pi_a(e) = \pi_b(e)$, that is, $ae = be$, so, finally, $a = b$.
2. f *is surjective*: Indeed, every element of G^* is some π_a, and $\pi_a = f(a)$.
3. Lastly, $f(ab) = \pi_{ab} = \pi_a \circ \pi_b = f(a) \circ f(b)$.

Thus, f is an isomorphism, and so $G \cong G^*$.

EXERCISES

A. Isomorphism Is an Equivalence Relation among Groups

The following three facts about isomorphism are true for all groups:

1. Every group is isomorphic to itself.
2. If $G_1 \cong G_2$, then $G_2 \cong G_1$.
3. If $G_1 \cong G_2$ and $G_2 \cong G_3$, then $G_1 \cong G_3$.

Fact 1 asserts that for any group G, there exists an isomorphism from G to G.

Fact 2 asserts that, if there is an isomorphism f from G_1 to G_2, there must be some isomorphism from G_2 to G_1. Well, the inverse of f is such an isomorphism.

Fact 3 asserts that, if there are isomorphisms $f: G_1 \rightarrow G_2$ and $g: G_2 \rightarrow G_3$, there must be an isomorphism from G_1 to G_3. One can easily guess that $g \circ f$ is such an isomorphism. The details of facts 1, 2, and 3 are left as exercises.

1 Let G be any group. If $\varepsilon: G \rightarrow G$ is the identity function, $\varepsilon(x) = x$, show that ε is an isomorphism.

2 Let G_1 and G_2 be groups, and $f: G_1 \rightarrow G_2$ an isomorphism. Show that $f^{-1}: G_2 \rightarrow G_1$ is an isomorphism. [HINT: Review the discussion of inverse functions at the end of Chapter 6. Then, for arbitrary elements $a, b \in G_1$, let $c = f(a)$ and $d = f(b)$. Note that $a = f^{-1}(c)$ and $b = f^{-1}(d)$. Show that $f^{-1}(cd) = f^{-1}(c)f^{-1}(d)$.]

3 Let G_1, G_2, and G_3 be groups, and let $f: G_1 \rightarrow G_2$ and $g: G_2 \rightarrow G_3$ be isomorphisms. Prove that $g \circ f: G_1 \rightarrow G_3$ is an isomorphism.

B. Elements Which Correspond under an Isomorphism

Recall that an isomorphism f from G_1 to G_2 is a one-to-one correspondence between G_1 and G_2 satisfying $f(ab) = f(a)f(b)$. f matches every element of G_1 with a corresponding element of G_2. It is important to note that:

(i) f matches the neutral element of G_1 with the neutral element of G_2.

(ii) If f matches an element x in G_1 with y in G_2, then, necessarily, f matches x^{-1} with y^{-1}. That is, if $x \leftrightarrow y$, then $x^{-1} \leftrightarrow y^{-1}$.

(iii) f matches a generator of G_1 with a generator of G_2.

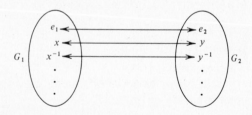

The details of these statements are now left as an exercise. Let G_1 and G_2 be groups, and let $f: G_1 \rightarrow G_2$ be an isomorphism.

1 If e_1 denotes the neutral element of G_1 and e_2 denotes the neutral element of G_2, prove that $f(e_1) = e_2$. [HINT: In any group, there is exactly one neutral element; show that $f(e_1)$ is the neutral element of G_2.]

2 Prove that for each element a in G_1, $f(a^{-1}) = [f(a)]^{-1}$. (HINT: You may use Theorem 2 of Chapter 4.)

3 If G_1 is a cyclic group with generator a, prove that G_2 is also a cyclic group, with generator $f(a)$.

C. Isomorphism of Some Finite Groups

In each of the following, G and H are finite groups. Determine whether or not $G \cong H$. Prove your answer in either case.

To find an isomorphism from G to H will require a little ingenuity. For example, if G and H are cyclic groups, it is clear that we must match a generator a of G with a generator b of H; that is, $f(a) = b$. Then $f(aa) = bb$, $f(aaa) = bbb$, and so on. If G and H are not cyclic, we have other ways: for example, if G has an element which is its own inverse, it must be matched with an element of H having the same property. Often, the specifics of a problem will suggest an isomorphism, if we keep our eyes open.

To prove that a specific one-to-one correspondence $f: G \rightarrow H$ is an isomorphism, we may check that it transforms the table of G into the table of H.

1 G is the checkerboard game group of Chapter 3, Exercise D. H is the group of the complex numbers $\{i, -i, 1, -1\}$ under multiplication.

2 G is the same as in part 1. $H = \mathbb{Z}_4$.

3 G is the group P_2 of subsets of a two-element set. (See Chapter 3, Exercise C.) H is as in part 1.

4 G is S_3. H is the group of matrices described on page 30 of the text.

5 G is the coin game group of Chapter 3, Exercise E. H is D_4, the group of symmetries of the square.

6 G is the group of symmetries of the rectangle. H is as in part 1.

D. Separating Groups into Isomorphism Classes

Each of the following is a set of four groups. In each set, determine which groups are isomorphic to which others. Prove your answers, and use Exercise A3 where convenient.

1 $\mathbb{Z}_4$ $\mathbb{Z}_2 \times \mathbb{Z}_2$ P_2 V

[P_2 denotes the group of subsets of a two-element set. (See Chapter 3, Exercise C.) V denotes the group of the four complex numbers $\{i, -i, 1, -1\}$ with respect to multiplication.]

2 S_3 $\mathbb{Z}_6$ $\mathbb{Z}_3 \times \mathbb{Z}_2$ $\mathbb{Z}_7^*$

[$\mathbb{Z}_7^*$ denotes the group $\{1, 2, 3, 4, 5, 6\}$ with multiplication modulo 7. The product modulo 7 of a and b is the remainder of ab after division by 7.]

3 $\mathbb{Z}_8$ P_3 $\mathbb{Z}_2 \times \mathbb{Z}_2 \times \mathbb{Z}_2$ D_4

(D_4 is the group of symmetries of the square.)

4 The groups having the following Cayley diagrams:

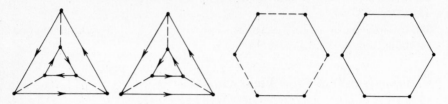

E. Isomorphism of Infinite Groups

1 Let E designate the group of all the even integers, with respect to addition. Prove that $\mathbb{Z} \cong E$.

2 Let G be the group $\{10^n : n \; \varepsilon \; \mathbb{Z}\}$ with respect to multiplication. Prove that $G \cong \mathbb{Z}$. (Remember that the operation of $\mathbb{Z}$ is addition.)

3 Prove that $\mathbb{C} \cong \mathbb{R} \times \mathbb{R}$.

4 We have seen in the text that $\mathbb{R}$ is isomorphic to $\mathbb{R}^+$. Prove that $\mathbb{R}$ is *not* isomorphic to $\mathbb{R}^*$ (the multiplicative group of the nonzero real numbers). (HINT: Consider the properties of the number -1 in $\mathbb{R}^*$. Does $\mathbb{R}$ have *any* element with those properties?)

5 Prove that $\mathbb{Z}$ is not isomorphic to $\mathbb{Q}$.

6 We have seen that $\mathbb{R} \cong \mathbb{R}^+$. However, prove that $\mathbb{Q}$ is *not* isomorphic to $\mathbb{Q}^+$. ($\mathbb{Q}^+$ is the multiplicative group of positive rational numbers.)

F. Isomorphism of Groups Given by Generators and Defining Equations

If a group G is generated, say, by a, b, and c, then a set of equations involving a, b, and c is called a set of *defining equations* for G if these equations completely determine the table of G. (See end of Chapter 5.) If G' is another group, generated by elements a', b', and c' satisfying the same defining equations as a, b, and c, then G' has the same table as G (because the tables of G and G' are completely determined by the defining equations, which are the same for G as for G').

Consequently, if we know generators and defining equations for two groups G and G', and if we are able to match the generators of G with those of G' so that the defining equations are the same, we may conclude that $G \cong G'$.

Prove that the following pairs of groups G, G' are isomorphic.

1 G = the subgroup of S_4 generated by (24) and (1234); $G' = \{e, a, b, b^2, b^3, ab, ab^2, ab^3\}$ where $a^2 = e$, $b^4 = e$, and $ba = ab^3$.

2 $G = S_3$; $G' = \{e, a, b, ab, aba, abab\}$ where $a^2 = e$, $b^2 = e$, and $bab = aba$.

3 $G = D_4$; $G' = \{e, a, b, ab, aba, (ab)^2, ba, bab\}$ where $a^2 = b^2 = e$ and $(ab)^4 = e$.

4 $G = \mathbb{Z}_2 \times \mathbb{Z}_2 \times \mathbb{Z}_2$; $G' = \{e, a, b, c, ab, ac, bc, abc\}$ where $a^2 = b^2 = c^2 = e$ and $(ab)^2 = (bc)^2 = (ac)^2 = e$.

G. Isomorphic Groups on the Set $\mathbb{R}$

Prove that the following groups are isomorphic.

1 G is the set $\{x \in \mathbb{R} : x \neq -1\}$ with the operation $x * y = x + y + xy$. Show that $f(x) = x - 1$ is an isomorphism from $\mathbb{R}^*$ to G. Thus, $\mathbb{R}^* \cong G$.

2 G is the set of the real numbers with the operation $x * y = x + y + 1$. Find an isomorphism $f : \mathbb{R} \to G$.

3 G is the set of the nonzero real numbers with the operation $x * y = xy/2$. Show that $f(x) = 2x$ is an isomorphism from $\mathbb{R}^*$ to G.

4 Show that $f(x, y) = (-1)^y x$ is an isomorphism from $\mathbb{R}^+ \times \mathbb{Z}_2$ to $\mathbb{R}^*$. (REMARK: To combine elements of $\mathbb{R}^+ \times \mathbb{Z}_2$, one multiplies first components, adds second components.) Conclude that $\mathbb{R}^* \cong \mathbb{R}^+ \times \mathbb{Z}_2$.

H. Some General Properties of Isomorphism

1 Let G and H be groups. Prove that $G \times H \cong H \times G$.

2 If $G_1 \cong G_2$ and $H_1 \cong H_2$, then $G_1 \times H_1 \cong G_2 \times H_2$.

3 Let G be any group. Prove that G is abelian iff the function $f(x) = x^{-1}$ is an isomorphism from G to G.

4 Let G be any group, with its operation denoted multiplicatively. Let H be a group with the same set as G and let its operation be defined by $x * y = y \cdot x$ (where $\cdot$ is the operation of G). Prove that $G \cong H$.

5 Let c be a fixed element of G. Let H be a group with the same set as G, and with the operation $x * y = xcy$. Prove that the function $f(x) = c^{-1}x$ is an isomorphism from G to H.

I. Group Automorphisms

If G is a group, an *automorphism* of G is an isomorphism from G to G. We have seen (Exercise A1) that the identity function $\varepsilon(x) = x$, is an automorphism of G. However, many groups have *other* automorphisms besides this obvious one.

1 Verify that

$$f = \begin{pmatrix} 0 & 1 & 2 & 3 & 4 & 5 \\ 0 & 5 & 4 & 3 & 2 & 1 \end{pmatrix}$$

is an automorphism of $\mathbb{Z}_6$.

2 Verify that

$$f_1 = \begin{pmatrix} 0 & 1 & 2 & 3 & 4 \\ 0 & 2 & 4 & 1 & 3 \end{pmatrix} \qquad f_2 = \begin{pmatrix} 0 & 1 & 2 & 3 & 4 \\ 0 & 3 & 1 & 4 & 2 \end{pmatrix}$$

and

$$f_3 = \begin{pmatrix} 0 & 1 & 2 & 3 & 4 \\ 0 & 4 & 3 & 2 & 1 \end{pmatrix}$$

are all automorphisms of $\mathbb{Z}_5$.

3 If G is any group, and a is any element of G, prove that $f(x) = axa^{-1}$ is an automorphism of G.

4 Since each automorphism of G is a bijective function from G to G, it is a *permutation* of G. Prove the set

$$\text{Aut}(G)$$

of all the automorphisms of G is a subgroup of S_G. (Remember that the operation is composition.)

J. Regular Representation of Groups

By Cayley's theorem, every group G is isomorphic to a group G^* of permutations of G. Recall that we match each element a in G with the permutation π_a defined by $\pi_a(x) = ax$, that is, the rule "multiply on the left by a." We let $G^* = \{\pi_a : a \in G\}$; with the operation $\circ$ of composition it is a group of permutations, called the *left regular representation* of G. (It is called a "representation" of G because it is isomorphic to G.)

Instead of using the permutations π_a, we could just as well have used the permutations ρ_a defined by $\rho_a(x) = xa$, that is, "multiply on the right by a." The group $G^\rho = \{\rho_a : a \in G\}$ is called the *right regular representation* of G.

If G is commutative, there is no difference between right and left multiplication, so G^* and G^ρ are the same, and are simply called the *regular representation* of G. Also, if the operation of G is denoted by $+$, the permutation corresponding to a is "add a" instead of "multiply by a."

Example The regular representation of $\mathbb{Z}_3$ consists of the following permutations:

$$\pi_0 = \begin{pmatrix} 0 & 1 & 2 \\ 0 & 1 & 2 \end{pmatrix} \qquad \text{that is, the identity permutation}$$

$$\pi_1 = \begin{pmatrix} 0 & 1 & 2 \\ 1 & 2 & 0 \end{pmatrix} \qquad \text{that is, the rule "add 1"}$$

$$\pi_2 = \begin{pmatrix} 0 & 1 & 2 \\ 2 & 0 & 1 \end{pmatrix} \qquad \text{that is, the rule "add 2"}$$

The regular representation of $\mathbb{Z}_3$ has the following table:

$\circ$	π_0	π_1	π_2
π_0	π_0	π_1	π_2
π_1	π_1	π_2	π_0
π_2	π_2	π_0	π_1

The function

$$f = \begin{pmatrix} 0 & 1 & 2 \\ \pi_0 & \pi_1 & \pi_2 \end{pmatrix}$$

is easily seen to be an isomorphism from $\mathbb{Z}_3$ to its regular representation.

Find the right and left regular representation of each of the following groups, and compute their tables. (If the group is abelian, find its regular representation.)

1 P_2, the group of subsets of a two-element set. (See Chapter 3, Exercise C.)

2 $\mathbb{Z}_4$

3 The group G of matrices described on page 30 of the text.

TEN

ORDER OF GROUP ELEMENTS

Let G be an arbitrary group, with its operation denoted multiplicatively. *Exponential notation* is a convenient shorthand: for any positive integer n, we will agree to let

$$a^n = \underbrace{a \cdot a \cdots a}_{n \text{ times}}$$

$$a^{-n} = \underbrace{a^{-1}a^{-1} \cdots a^{-1}}_{n \text{ times}}$$

and $\qquad\qquad a^0 = e$

Take care to observe that we are considering only integer exponents, *not* rational or real exponents. Raising a to a positive power means multiplying a by itself the given number of times. Raising a to a negative power means multiplying a^{-1} by itself the given number of times. Raising a to the power 0 yields the group's identity element.

These are the same conventions used in elementary algebra, and they lead to the same familiar "laws of exponents."

Theorem 1: Laws of Exponents *If G is a group and $a \in G$, the following identities hold for all integers m and n:*

(i) $a^m a^n = a^{m+n}$

(ii) $(a^m)^n = a^{mn}$

(iii) $a^{-n} = (a^{-1})^n = (a^n)^{-1}$

These laws are very easy to prove when m and n are positive integers. Indeed,

(i) $a^m a^n = \underbrace{aa \cdots a}_{m \text{ times}} \cdot \underbrace{aa \cdots a}_{n \text{ times}} = a^{m+n}$

(ii) $(a^m)^n = \underbrace{a^m a^m \cdots a^m}_{n \text{ times}} = \underbrace{aa \cdots a}_{mn \text{ times}} = a^{mn}$

Next, by definition $a^{-n} = a^{-1} \cdots a^{-1} = (a^{-1})^n$. Finally, since the inverse of a product is the product of the inverses in reverse order,

$$(a^n)^{-1} = (aa \cdots a)^{-1} = a^{-1} a^{-1} \cdots a^{-1} = a^{-n}$$

To prove Theorem 1 completely, we need to check the other cases, where each of the integers m and n is allowed to be zero or negative. This routine case-by-case verification is left to the student as Exercise A at the end of this chapter.

In order to delve more deeply into the behavior of exponents we must use an elementary but very important property of the integers: From elementary arithmetic we know that we can divide any integer by any positive integer to get an integer quotient and an integer remainder. The remainder is nonnegative and less than the dividend. For example, 25 may be divided by 8, giving a quotient of 3, and leaving a remainder of 1:

$$25 = 8 \times \underset{q}{3} + \underset{r}{1}$$

Similarly, -25 may be divided by 8, with a quotient of -4 and a remainder of 7:

$$-25 = 8 \times \underset{q}{(-4)} + \underset{r}{7}$$

This important principle is called the division algorithm. In its precise form, it may be stated as follows:

Theorem 2: Division Algorithm *If m and n are integers and n is positive,* *there exist unique integers q and r such that*

$$m = nq + r \qquad \text{and} \qquad 0 \leqslant r < n$$

We call q the *quotient*, and r the *remainder*, in the division of m by n.

At this stage we will take the division algorithm to be a *postulate* of the system of the integers. Later, in Chapter 19, we will turn things around, starting with a simpler premise and proving the division algorithm from it.

Let G be a group, and a an element of G. Let us observe that

if there exists a nonzero integer m such that $a^m = e$, then there exists a positive integer n such that $a^n = e$.

Indeed, if $a^m = e$ where m is negative, then $a^{-m} = (a^m)^{-1} = e^{-1} = e$. Thus, $a^{-m} = e$ where $-m$ is positive. This simple observation is crucial in our next definition. Let G be an arbitrary group, and a an element of G:

Definition *If there exists a nonzero integer m such that $a^m = e$, then the **order** of the element a is defined to be the **least positive integer** n such that $a^n = e$.*

*If there does not exist any nonzero integer m such that $a^m = e$, we say that a has order **infinity**.*

Thus, in any group G, every element .has an order which is either a positive integer or infinity. If the order of a is a positive integer, we say that a has *finite order*; otherwise, a has *infinite order*. For example, let us glance at S_3, whose table appears on page 67. (It is customary to use exponential notation for composition: for instance, $\beta \circ \beta = \beta^2$, $\beta \circ \beta \circ \beta = \beta^3$, and so on.) The order of α is 2, because $\alpha^2 = \varepsilon$ and 2 is the smallest positive integer which satisfies that equation. The order of β is 3, because $\beta^3 = \varepsilon$, and 3 is the lowest positive power of β equal to ε. It is clear, similarly, that γ has order 2, δ has order 3, and κ has order 2. What is the order of ε?

It is important to note that one speaks of *powers* of a only when the group's operation is called multiplication. When we use additive notation, we speak of *multiples* of a instead of powers of a. The *positive multiples* of a are a, $a + a$, $a + a + a$, and so on, while the *negative multiples* of a are $-a$, $(-a) + (-a)$, $(-a) + (-a) + (-a)$, and so on. In $\mathbb{Z}_6$, the number 2 has order 3, because $2 + 2 + 2 = 0$, and no smaller multiple of 2 is equal to 0. Similarly, the order of 3 is 2, the order of 4 is 3, and the order of 5 is 6.

In $\mathbb{Z}$, the number 2 has infinite order, because no nonzero multiple of 2 is equal to 0. As a matter of fact, in $\mathbb{Z}$, every nonzero number has infinite order.

The main fact about the order of elements is given in the next two theorems. In each of the following theorems, G is an arbitrary group and a is any element of G.

Theorem 3: Powers of a, if a has finite order *If the order of a is n, there are exactly n different powers of a, namely*

$$a^0, a, a^2, a^3, \ldots, a^{n-1}$$

What this theorem asserts is that every positive or negative power of a is equal to one of the above, and the above are all different from one another.

Before going on, remember that the order of a is n, hence

$$a^n = e$$

and n is the *smallest* positive integer which satisfies this equation.

Let us begin by proving that every power of a is equal to one of the powers $a^0, a, a^2, \ldots, a^{n-1}$. Let a^m be any power of a. Use the division algorithm to divide m by n:

$$m = nq + r \qquad 0 \leqslant r < n$$

Then $\qquad\qquad a^m = a^{nq+r} = a^{nq}a^r = (a^n)^q a^r = e^q a^r = a^r$

Thus, $a^m = a^r$, and r is one of the integers $0, 1, 2, \ldots, n-1$.

Next, we will prove that $a^0, a^1, a^2, \ldots, a^{n-1}$ are all different. Suppose not; suppose $a^r = a^s$, where r and s are distinct integers between 0 and $n-1$. Either $r < s$ or $s < r$, say $s < r$. Then $0 \leqslant s < r < n$, and consequently,

$$0 < r - s < n \qquad\qquad (*)$$

But $a^r = a^s$, hence $\qquad\qquad a^r(a^s)^{-1} = a^s(a^s)^{-1}$

Therefore, $\qquad\qquad a^r a^{-s} = e$

so $\qquad\qquad a^{r-s} = e$

However, this is impossible, because by $(*)$, $r - s$ is a positive integer less than n, whereas n (the order of a) is the *smallest* positive integer such that $a^n = e$.

This proves that we cannot have $a^r = a^s$ where $r \neq s$. Thus, $a^0, a^1, a^2, \ldots, a^{n-1}$ are all different!

Theorem 4: Powers of *a*, if *a* has infinite order *If a has order infinity, then all the powers of a are different. That is, if r and s are distinct integers, then* $a^r \neq a^s$.

Let r and s be integers, and suppose $a^r = a^s$.

Then $$a^r(a^s)^{-1} = a^s(a^s)^{-1}$$

hence $$a^{r-s} = e$$

But a has order infinity, and this means that a^m is not equal to e for any integer m except 0. Thus, $r - s = 0$, so $r = s$.

This chapter concludes with a technical property of exponents, which is of tremendous importance in applications.

If a is an element of a group, the order of a is the *least positive integer n* such that $a^n = e$. But there are *other* integers, say t, such that $a^t = e$. How are they related to n? The answer is very simple:

Theorem 5 *Suppose an element a in a group has order n. Then* $a^t = e$ *iff t is a multiple of n* ("*t is a multiple of n*" *means that t = nq for some integer q*).

If $t = nq$, then $a^t = a^{nq} = (a^n)^q = e^q = e$. Conversely, suppose $a^t = e$. Divide t by n using the division algorithm:

$$t = nq + r \qquad 0 \leqslant r < n$$

Then

$$e = a^t = a^{nq+r} = (a^n)^q a^r = e^q a^r = a^r$$

Thus, $a^r = e$, where $0 \leqslant r < n$. If $r \neq 0$, then r is a positive integer less than n, whereas n is the *smallest* positive integer such that $a^n = e$. Thus $r = 0$, and therefore $t = nq$.

If a is any element of a group, we will denote the order of a by

$$\mathrm{ord}(a)$$

EXERCISES

A. Laws of Exponents

Let G be a group and $a \in G$.

1 Prove that $a^m a^n = a^{m+n}$ in the following cases:
   (i) $m = 0$ (ii) $m < 0$ and $n > 0$ (iii) $m < 0$ and $n < 0$
2 Prove that $(a^m)^n = a^{mn}$ in the following cases:
   (i) $m = 0$ (ii) $n = 0$ (iii) $m < 0$ and $n > 0$ (iv) $m > 0$ and $n < 0$
(v) $m < 0$ and $n < 0$
3 Prove that $(a^n)^{-1} = a^{-n}$ in the following cases:
   (i) $n = 0$ (ii) $n < 0$

B. Examples of Orders of Elements

1 What is the order of 10 in $\mathbb{Z}_{25}$?
2 What is the order of 6 in $\mathbb{Z}_{16}$?
3 What is the order of

$$f = \begin{pmatrix} 1 & 2 & 3 & 4 & 5 & 6 \\ 6 & 1 & 3 & 2 & 5 & 4 \end{pmatrix}$$

in S_6 ?
4 What is the order of 1 in $\mathbb{R}^*$? What is the order of 1 in $\mathbb{R}$?
5 If A is the set of all the real numbers $x \neq 0, 1, 2$, what is the order of

$$f(x) = \frac{2}{2 - x}$$

in S_A ?
6 Can an element of an *infinite* group have *finite* order? Explain
7 In $\mathbb{Z}_{24}$, list all the elements (a) of order 2; (b) of order 3; (c) of order 4; (d) of order 6.

C. Elementary Properties of Order

Let a, b, and c be elements of a group G. *Prove the following*:

1 Ord$(a) = 1$ iff $a = e$
2 If ord$(a) = n$, then $a^{n-r} = (a^r)^{-1}$
3 If $a^k = e$ where k is odd, then the order of a is odd.
4 Ord$(a) = $ ord(bab^{-1}).
5 The order of a^{-1} is the same as the order of a.
6 The order of ab is the same as the order of ba. [HINT: If

$$(ba)^n = \underbrace{baba \cdots ba}_{x} = e$$

then a is the inverse of x. Thus, $ax = e$.]
7 Ord$(abc) = $ ord$(cab) = $ ord(bca)
8 Let $x + a_1 a_2 \cdots a_n$, and let y be a product of the same factors, permuted cyclically. (That is, $y = a_k a_{k+1} \cdots a_n a_1 \cdots a_{k-1}$.) Then ord$(x) = $ ord(y).

D. Further Properties of Order

Let a be any element of a group G. *Prove the following*:

1 If $a^p = e$ where p is a prime number, then a has order p.

2 The order of a^k is a divisor (factor) of the order of a.

3 If $\text{ord}(a) = km$, then $\text{ord}(a^k) = m$.

4 If $\text{ord}(a) = n$ where n is odd, then $\text{ord}(a^2) = n$.

5 If a has order n, and $a^r = a^s$, then n is a factor of $r - s$.

6 If a is the *only* element of order k in G, then a is in the center of G. (HINT: Use Exercise C4. Also, see Chapter 4, Exercise C6.)

7 If the order of a is not a multiple of m, then the order of a^k is not a multiple of m. (HINT: Use part 2.)

8 If $\text{ord}(a) = mk$ and $a^{rk} = e$, then r is a multiple of m.

† E. Relationship between ord(ab), ord(a), and ord(b)

Let a and b be elements of a group G. Let $\text{ord}(a) = m$ and $\text{ord}(b) = n$; $\text{lcm}(m, n)$ denotes the least common multiple of m and n. *Prove the following*:

1 If a and b commute, then $\text{ord}(ab)$ is a divisor of $\text{lcm}(m, n)$.

2 If m and n are relatively prime, then no power of a can be equal to any power of b (except for e). (REMARK: Two integers are said to be relatively prime if they have no common factors except ± 1.) (HINT: Use Exercise D2.)

3 If m and n are relatively prime, then the products $a^i b^j$ ($0 \leqslant i \leqslant m$, $0 \leqslant j \leqslant n$) are all distinct.

4 Let a and b commute. If m and n are relatively prime, then $\text{ord}(ab) = mn$. (HINT: Use part 2.)

5 Let a and b commute. There is an element c in G whose order is $\text{lcm}(m, n)$. (HINT: Use part 4, above, together with Exercise D3. Let c be a certain power of a.)

6 Give an example to show that part 1 is not true if a and b do not commute.

Thus, there is no simple relationship between $\text{ord}(ab)$, $\text{ord}(a)$, and $\text{ord}(b)$ if a and b fail to commute.

† F. Orders of Powers of Elements

Let a be an element of order 12 in a group G.

1 What is the smallest positive integer k such that $a^{8k} = e$? (HINT: Use Theorem 5 and explain carefully!)

2 What is the order of a^8 ?

3 What are the orders of a^9, a^{10}, a^5 ?

4 Which powers of a have the same order as a? [That is, for what values of k is $\text{ord}(a^k) = 12$?]

5 Let a be an element of order m in any group G. What is the order of a^k ? (Look at the preceding examples, and generalize. Do not prove.)

6 Let a be an element of order m in any group G. For what values of k is $\text{ord}(a^k) = m$? (Look at the preceding examples. Do not prove.)

† G. Relationship between ord(a) and ord(a^k)

From elementary arithmetic we know that every integer may be written uniquely as a product of prime numbers. Two integers m and n are said to be *relatively prime* if they have no prime factors in common. (For example, 15 and 8 are relatively prime.) Here is a useful fact: If m and n are relatively prime, and m is a factor of nk, then m is a factor of k. (Indeed, all the prime factors of m are factors of nk but not of n, hence are factors of k.)

Let a be an element of order n in a group G. *Prove the following:*

1 If m and n are relatively prime, then a^m has order n. (HINT: If $a^{mk} = e$, use Theorem 5 and explain why n must be a factor of k.)

2 If a^m has order n, then m and n are relatively prime. [HINT: Assume m and n have a common factor $q > 1$, hence we can write $m = m'q$ and $n = n'q$. Explain why $(a^m)^{n'} = e$, and proceed from there.]

3 Conclude from parts 1 and 2 that: $\text{ord}(a^m) = n$ iff n and m are relatively prime.

4 Let l be the least common multiple of m and n. Let $l/m = k$. Explain why $(a^m)^k = e$.

5 Prove: If $(a^m)^t = e$, then n is a factor of mt. (Thus, mt is a common multiple of m and n.) Conclude that

$$\underset{= mk}{\underbrace{\frac{l}{\,}}} \leqslant mt$$

6 Use parts 4 and 5 to prove that the order of a^m is $[\text{lcm}(m, n)]/m$

† H. Relationship between the Order of a and the Order of any kth Root of a

Let a denote an element of a group G.

1 Let a have order 12. Prove that if a has a cube root, say $a = b^3$ for some $b \in G$, then b has order 36. {HINT: Show that $b^{36} = e$; then show that for each factor k of 36, $b^k = e$ is impossible. [Example: If $b^{12} = e$, then $b^{12} = (b^3)^4 = a^4 = e$.] Derive your conclusion from these facts.}

2 Let a have order 6. If a has a fourth root in G, say $a = b^4$, what is the order of b?

3 Let a have order 10. If a has a sixth root in G, say $a = b^6$, what is the order of b?

4 Let a have order n, and suppose a has a kth root in G, say $a = b^k$. Explain why the order of b is a factor of nk. Let

$$\text{ord}(b) = \frac{nk}{l}$$

5 Prove that n and l are relatively prime. [HINT: Suppose n and l have a common factor $q > 1$. Then $n = qn'$ and $l = ql'$, so

$$\text{ord}(b) = \frac{qn'k}{ql'} = \frac{n'k}{l'}$$

Thus $b^{n'k} = e$ (why?) Draw your conclusion from these facts.]

Thus, if a has order n and a has a kth root b, then b has order nk/l, where n and l are relatively prime.

6 Let a have order n. Let k be an integer such that every prime factor of k is a factor of n. Prove: If a has a kth root b, then $\text{ord}(b) = nk$.

ELEVEN

CYCLIC GROUPS

If G is a group and $a \in G$, it may happen that *every* element of G is a power of a. In other words, G may consist of all the powers of a, and nothing else:

$$G = \{a^n : n \in \mathbb{Z}\}$$

In that case, G is called a *cyclic group*, and a is called its *generator*. We write

$$G = \langle a \rangle$$

and say that G is the cyclic group generated by a.

If $G = \langle a \rangle$ is the cyclic group generated by a, and a has order n, we say that G is a *cyclic group of order n*. We will see in a moment that, in that case, G has exactly n elements. If the generator of G has order infinity, we say that G is a *cyclic group of order infinity*. In that case, we will see that G has infinitely many elements.

The simplest example of a cyclic group is $\mathbb{Z}$, which consists of all the multiples of 1. (Remember that in additive notation we speak of "multiples" instead of "powers.") $\mathbb{Z}$ is a cyclic group of order infinity; its generator is 1. Another example of a cyclic group is $\mathbb{Z}_6$, which consists of all the multiples of 1, added modulo 6. $\mathbb{Z}_6$ is a cyclic group of order 6; 1 is a generator of $\mathbb{Z}_6$, but $\mathbb{Z}_6$ has another generator too. What is it?

Suppose $\langle a \rangle$ is a cyclic group whose generator a has order n. Since $\langle a \rangle$

is the set of all the powers of a, it follows from Theorem 3 of Chapter 10 that

$$\langle a \rangle = \{e, a, a^2, \ldots, a^{n-1}\}$$

If we compare this group with $\mathbb{Z}_n$, we notice a remarkable resemblance! For one thing, they are obviously in one-to-one correspondence:

$$\langle a \rangle = \{a^0, a^1, a^2, \ldots, a^{n-1}\}$$

$$\mathbb{Z}_n = \{0, \quad 1, \quad 2, \quad \ldots, n-1\}$$

In other words, the function

$$f(i) = a^i$$

is a one-to-one correspondence from $\mathbb{Z}_n$ to $\langle a \rangle$. But this function has an additional property, namely

$$f(i + j) = a^{i+j} = a^i a^j = f(i)f(j)$$

Thus, f is an *isomorphism* from $\mathbb{Z}_n$ to $\langle a \rangle$. In conclusion,

$$\mathbb{Z}_n \cong \langle a \rangle$$

Let us review this situation in the case where a has *order infinity*. In this case, by Theorem 4 of Chapter 10,

$$\langle a \rangle = \{\ldots, a^{-2}, a^{-1}, e, a, a^2, \ldots\}$$

There is obviously a one-to-one correspondence between this group and $\mathbb{Z}$:

$$\langle a \rangle = \{\ldots, \quad a^{-2}, \quad a^{-1}, a^0, a^1, a^2, \ldots\}$$

$$\mathbb{Z} = \{\ldots, -2, \quad -1, \quad 0, 1, 2, \ldots\}$$

In other words, the function

$$f(i) = a^i$$

is a one-to-one correspondence from $\mathbb{Z}$ to $\langle a \rangle$. As before, f is an isomorphism, and therefore

$$\mathbb{Z} \cong \langle a \rangle$$

What we have just proved is a very important fact about cyclic groups; let us state it as a theorem.

Theorem 1: Isomorphism of Cyclic Groups

(i) *For every positive integer n, every cyclic group of order n is isomorphic to $\mathbb{Z}_n$. Thus, any two cyclic groups of order n are isomorphic to each other.*
(ii) *Every cyclic group of order infinity is isomorphic to $\mathbb{Z}$, and therefore any two cyclic groups of order infinity are isomorphic to each other.*

If G is any group and $a \in G$, it is easy to see that

(i) the product of any two powers of a is a power of a; for $a^m a^n = a^{m+n}$. Furthermore,
(ii) the inverse of any power of a is a power of a, because $(a^n)^{-1} = a^{-n}$. It therefore follows that
(iii) the set of all the powers of a is a subgroup of G.

This subgroup is called the *cyclic subgroup of G generated by a*. It is obviously a cyclic group, and therefore we denote it by $\langle a \rangle$. If the element a has order n, then, as we have seen, $\langle a \rangle$ contains the n elements $\{e, a, a^2, \ldots, a^{n-1}\}$. If a has order infinity, then $\langle a \rangle = \{\ldots, a^{-2}, a^{-1}, e, a, a^2, \ldots\}$ and has infinitely many elements.

For example, in $\mathbb{Z}$, $\langle 2 \rangle$ is the cyclic subgroup of $\mathbb{Z}$ which consists of all the multiples of 2. In $\mathbb{Z}_{15}$, $\langle 3 \rangle$ is the cyclic subgroup $\{0, 3, 6, 9, 12\}$ which contains all the multiples of 3. In S_3, $\langle \beta \rangle = \{\varepsilon, \beta, \delta\}$, and contains all the powers of β.

Can a cyclic group, such as $\mathbb{Z}$, have a *subgroup which is not cyclic*? This question is of great importance in our understanding of cyclic groups. Its answer is not obvious, and certainly not self-evident:

Theorem 2 *Every subgroup of a cyclic group is cyclic.*

Let $G = \langle a \rangle$ be a cyclic group, and let H be any subgroup of G. We wish to prove that H is cyclic.

Now, G has a generator a; and when we say that H is cyclic, what we mean is that H too has a generator (call it b), and H consists of all the powers of b. The gist of this proof, therefore, is to *find* a generator of H, and then check that every element of H is a power of this generator.

Here is the idea: G is the cyclic group generated by a, and H is a subgroup of G. Every element of H is therefore in G, which means that every element of H is some power of a. The generator of H which we are searching for is therefore one of the powers of a—one of the powers of a

which happens to be in H; but which one? Obviously the lowest one! More accurately, the lowest *positive* power of a in H.

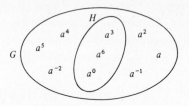

And now, carefully, here is the proof:

Let m be the smallest positive integer such that $a^m \in H$. We will show that every element of H is a power of a^m, hence a^m is a generator of H.

Let a^t be *any* element of H. Divide t by m using the division algorithm:

$$t = mq + r \qquad 0 \leqslant r < m$$

Then $$a^t = a^{mq+r} = a^{mq}a^r$$

Solving for a^r, $$a^r = (a^{mq})^{-1}a^t = (a^m)^{-q}a^t$$

But $a^m \in H$ and $a^t \in H$, thus $(a^m)^{-q}a^t \in H$.

It follows that $a^r \in H$. But $r < m$ and m is the smallest positive integer such that $a^m \in H$. So $r = 0$, and therefore $t = mq$.

We conclude that every element $a^t \in H$ is of the form $a^t = (a^m)^q$, that is, a power of a^m. Thus, H is the cyclic group generated by a^m.

This chapter ends with a final comment regarding the different uses of the word "order" in algebra.

Let G be a group; as we have seen, the *order of an element a in G* is the least positive integer n such that

$$\underbrace{aaa \cdots a}_{n \text{ factors}} = e$$

(The order of a is *infinity* if there is no such n.)

Earlier, we defined the *order of the group G* to be the number of elements in G. Remember that the order of G is denoted by $|G|$.

These are two separate and distinct definitions, not to be confused with one another. Nevertheless, there is a connection between them: Let a be an element of order n in a group. By Chapter 10, Theorem 3, there are exactly n different powers of a, hence $\langle a \rangle$ has n elements. Thus,

$$\text{If} \quad \text{ord}(a) = n \quad \text{then} \quad |\langle a \rangle| = n$$

That is, the order of a cyclic group is the same as the order of its generator.

EXERCISES

A. Examples of Cyclic Groups

1 List the elements of $\langle 6 \rangle$ in $\mathbb{Z}_{16}$.

2 List the elements of $\langle f \rangle$ in S_6, where

$$f = \begin{pmatrix} 1 & 2 & 3 & 4 & 5 & 6 \\ 6 & 1 & 3 & 2 & 5 & 4 \end{pmatrix}$$

3 Describe the cyclic subgroup $\langle \frac{1}{2} \rangle$ in $\mathbb{R}^*$.

4 If $f(x) = x + 1$, describe the cyclic subgroup $\langle f \rangle$ of $S_{\mathbb{R}}$.

5 If $f(x) = x + 1$, describe the cyclic subgroup $\langle f \rangle$ of $\mathscr{F}(\mathbb{R})$.

6 Show that -1, as well as 1, is a generator of $\mathbb{Z}$. Are there any other generators of $\mathbb{Z}$? Explain! What are the generators of an arbitrary infinite cyclic group $\langle a \rangle$?

7 Is $\mathbb{R}^*$ cyclic? Try to prove your answer.

(HINT: If $k < 1$, then $k > k^2 > k^3 > \dots$;

if $k > 1$, then $k < k^2 < k^3 < \dots$.)

B. Elementary Properties of Cyclic Groups

Prove each of the following:

1 If G is a group of order n, G is cyclic iff G has an element of order n.

2 Every cyclic group is abelian. (HINT: Show that any two powers of a commute.)

3 If $G = \langle a \rangle$ and $b \in G$, the order of b is a factor of the order of a.

4 In any cyclic group of order n, there are elements of order k for every integer k which divides n.

5 Let G be an abelian group of order mn, where m and n are relatively prime. If G has an element of order m and an element of order n, G is cyclic. (See Chapter 10, Exercise E4.)

6 Let $\langle a \rangle$ be a cyclic group of order n. If n and m are relatively prime, then the function $f(x) = x^m$ is an automorphism of $\langle a \rangle$. (HINT: Use Exercise B3, and Chapter 10, Theorem 5.)

C. Generators of Cyclic Groups

For any positive integer n, let $\phi(n)$ denote the number of positive integers less than n and relatively prime to n. For example, 1, 2, 4, 5, 7, and 8 are relatively prime to 9, so $\phi(9) = 6$. Let a have order n. and *prove the following*:

1 a^r is a generator of $\langle a \rangle$ iff r and n are relatively prime. (HINT: See Chapter 10, Exercise G3.)

2 $\langle a \rangle$ has $\phi(n)$ different generators. [Use (1).]

3 For any factor m of n, let $C_m = \{x \in \langle a \rangle : x^m = e\}$. C_m is a subgroup of $\langle a \rangle$.

4 C_m has exactly m elements. (HINT: Use Exercise B4.)

5 An element x in $\langle a \rangle$ has order m iff x is a generator of C_m.

6 There are $\phi(m)$ elements of order m in $\langle a \rangle$. [Use (1) and (5).]

7 Let $n = mk$. a^r has order m iff $r = kl$ where l and m are relatively prime. (HINT: See Chapter 10, Exercise G3.)

8 If c is any generator of $\langle a \rangle$, then $\{c^r : r$ is relatively prime to $n\}$ is the set of all the generators of $\langle a \rangle$.

D. Elementary Properties of Cyclic Subgroups of Groups

Let G be a group and let $a, b \in G$. *Prove the following*:

1 If a is a power of b, say $a = b^k$, then $\langle a \rangle \subseteq \langle b \rangle$.

2 Suppose a is a power of b, say $a = b^k$. Then b is equal to a power of a iff $\langle a \rangle = \langle b \rangle$.

3 Suppose $a \in \langle b \rangle$. Then $\langle a \rangle = \langle b \rangle$ iff a and b have the same order.

4 Let ord$(a) = n$, and $b = a^k$. Then $\langle a \rangle = \langle b \rangle$ iff n and k are relatively prime.

5 Let ord$(a) = n$, and suppose a has a kth root, say $a = b^k$. Then $\langle a \rangle = \langle b \rangle$ iff k and n are relatively prime.

6 Any cyclic group of order mn has a unique subgroup of order n.

E. Direct Products of Cyclic Groups

Let G and H be groups, with $a \in G$ and $b \in H$. *Prove the following*:

1 If (a, b) is a generator of $G \times H$, then a is a generator of G and b is a generator of H.

2 If $G \times H$ is a cyclic group, then G and H are both cyclic.

3 The converse of part 2 is false. (Give an example to demonstrate this.)

4 Let ord$(a) = m$ and ord$(b) = n$. The order of (a, b) in $G \times H$ is the least common multiple of m and n. (HINT: Use Chapter 10, Theorem 5. Nothing else is needed!)

5 Conclude from part 4 that if m and n are relatively prime, then (a, b) has order mn. (HINT: If two numbers are relatively prime, their least common multiple is equal to their product.)

6 Suppose $(c, d) \in G \times H$, where c has order m and d has order n. If m and n are *not* relatively prime (hence have a common factor $q > 1$), then the order of (c, d) is less than mn.

7 Conclude from parts 5 and 6 that: $\langle a \rangle \times \langle b \rangle$ is cyclic iff ord(a) and ord(b) are relatively prime.

8 Let G be an abelian group of order mn, where m and n are relatively prime. If G has an element a of order m and an element b of order n, then $G \cong \langle a \rangle \times \langle b \rangle$. (HINT: See Chapter 10, Exercise E3.)

9 Let $\langle a \rangle$ be a cyclic group of order mn, where m and n are relatively prime. Then $\langle a \rangle \cong \langle a^m \rangle \times \langle a^n \rangle$.

† F. *k*th Roots of Elements in a Cyclic Group

Let $\langle a \rangle$ be a cyclic group of order n. For any integer k, we may ask: which elements in $\langle a \rangle$ have a kth root? The exercises which follow will answer this question.

1 Let a have order 10. For what integers k ($0 \leqslant k \leqslant 12$), does a have a kth root? For what integers k ($0 \leqslant k \leqslant 12$), does a^6 have a kth root?

Let k and n be any integers, and let gcd(k, n) denote the greatest common divisor of k and n. A linear combination of k and n is any expression $ck + dn$ where c and d are integers. It is a simple fact of number theory (the proof is given on page 218), that an integer m is equal to a linear combination of k and n iff m is a multiple of gcd(k, n). Use this fact to prove the following, where a is an element of order n in a group G.

2 If m is a multiple of gcd(k, n), then a^m has a kth root in $\langle a \rangle$. [HINT: Compute a^m, and show that $a^m = (a^c)^k$ for some $a^c \in \langle a \rangle$.]

3 If a^m has a kth root in $\langle a \rangle$, then m is a multiple of gcd(k, n). Thus, a^m has a kth root in $\langle a \rangle$ iff gcd(k, n) is a factor of m.

4 a has a kth root in $\langle a \rangle$ iff k and n are relatively prime.

5 Let p be a prime number.

    (i) If n is not a multiple of p, then every element in $\langle a \rangle$ has a pth root.

    (ii) If n is a multiple of p, and a^m has a pth root, then m is a multiple of p. (Thus, the only elements in $\langle a \rangle$ which have pth roots are e, a^p, a^{2p}, etc.)

6 The set of all the elements in $\langle a \rangle$ having a kth root is a subgroup of $\langle a \rangle$. (Prove this.) Explain why this subgroup is cyclic, say $\langle a^m \rangle$. What is this value of m? (Use part 3.)

TWELVE

PARTITIONS AND EQUIVALENCE RELATIONS

Imagine emptying a jar of coins onto a table and sorting them into separate piles, one with the pennies, one with the nickels, one with the dimes, one with the quarters and one with the half-dollars. This is a simple example of *partitioning* a set. Each separate pile is called a class of the partition; the jarful of coins has been partitioned into five classes.

Here are some other examples of partitions: The distributor of farm-fresh eggs usually sorts the daily supply according to size, and separates the eggs into three classes called "large," "medium," and "small."

The delegates to the Democratic national convention may be classified according to their home state, thus falling into 50 separate classes, one for each state.

A student files class notes according to subject matter; the notebook pages are separated into four distinct categories, marked (let us say) "algebra," "psychology," "English," and "American history."

Every time we file, sort, or classify, we are performing the simple act of partitioning a set. To partition a set A is to separate the elements of A into nonempty subsets, say A_1, A_2, A_3, etc., which are called the *classes* of the partition.

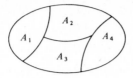

Any two distinct classes, say A_i and A_j, are *disjoint*, which means they have no elements in common. And the union of the classes is all of A.

Instead of dealing with the *process* of partitioning a set (which is awkward mathematically), it is more convenient to deal with the *result* of partitioning a set. Thus, $\{A_1, A_2, A_3, A_4\}$, in the illustration above, is called a *partition* of A. We therefore have the following definition:

> A **partition** of a set A is a family $\{A_i : i \in I\}$ of nonempty subsets of A *which are mutually disjoint and whose union is all of A.*

The notation $\{A_i : i \in I\}$ is the customary way of representing a family of sets $\{A_i, A_j, A_k, \dots\}$ consisting of one set A_i for each index i in I. (The elements of I are called *indices*; the notation $\{A_i : i \in I\}$ may be read: the family of sets A_i, as i ranges over I.)

Let $\{A_i : i \in I\}$ be a partition of the set A. We may think of the indices $i, j, k, \dots$ as labels for *naming* the classes $A_i, A_j, A_k, \dots$. Now, in practical problems, it is very inconvenient to insist that each class be named once and only once. It is simpler to allow repetition of indexing whenever convenience dictates. For example, the partition illustrated previously might also be represented like this, where A_1 is the same class as A_5, A_2 is the same as A_6, and A_3 is the same as A_7.

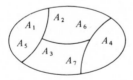

As we have seen, any two *distinct* classes of a partition are disjoint; this is the same as saying that *if two classes are not disjoint, they must be equal.* The other condition for the classes of a partition of A is that their union must be equal to A; this is the same as saying that *every element of A lies in one of the classes.* Thus, we have the following, more explicit definition of partition:

> By a **partition** of a set A we mean a family $\{A_i : i \in I\}$ of nonempty subsets of A such that

(i) *If any two classes, say A_i and A_j, have a common element x (that is, are not disjoint), then $A_i = A_j$, and*

(ii) *Every element x of A lies in one of the classes.*

We now turn to another elementary concept of mathematics. A *relation* on a set A is any statement which is either true or false for each ordered pair (x, y) of elements of A. Examples of relations, on appropriate sets, are "$x = y$," "$x < y$," "x is parallel to y," "x is the offspring of y," and so on. An especially important kind of relation on sets is an "equivalence relation." Such a relation will usually be represented by the symbol $\sim$, so that $x \sim y$ may be read "x is equivalent to y." Here is how equivalence relations are defined:

By an equivalence relation on a set A we mean a relation $\sim$ which is

Reflexive: that is, $x \sim x$ for every x in A;
Symmetric: that is, if $x \sim y$, then $y \sim x$; and
Transitive: that is, if $x \sim y$ and $y \sim z$, then $x \sim z$.

The most obvious example of an equivalence relation is equality, but there are many other examples, as we shall be seeing soon. Some examples from our everyday experience are: "x weighs the same as y," "x is the same color as y," "x is synonymous with y," and so on.

Equivalence relations also arise in a natural way out of partitions. Indeed, if $\{A_i : i \in I\}$ is a partition of A, we may define an equivalence relation $\sim$ on A be letting $x \sim y$ iff x and y are in the same class of the partition.

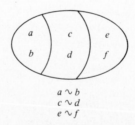

$$a \sim b$$
$$c \sim d$$
$$e \sim f$$

In other words, we call two elements "equivalent" if they are members of the same class. It is easy to see that this relation $\sim$ is an equivalence relation on A. Indeed, $x \sim x$ because x is in the same class as x; next, if x and y are in the same class, then y and x are in the same class; finally, if x and y are in the same class, and y and z are in the same class, then x and z are in the same class. Thus, $\sim$ is an equivalence relation on A; it is called the *equivalence relation determined by the partition* $\{A_i : i \in I\}$.

Let $\sim$ be an equivalence relation on A, and x an element of A. The set of all the elements equivalent to x is called the *equivalence class* of x, and is denoted by $[x]$. In symbols,

$$[x] = \{y \in A : y \sim x\}$$

A useful property of equivalence classes is this:

Lemma *If* $x \sim y$, *then* $[x] = [y]$.

In other words, if two elements are equivalent, they have the same equivalence class. The proof of this lemma is fairly obvious: for if $x \sim y$, then the elements equivalent to x are the same as the elements equivalent to y.

For example, let us return to the jarful of coins we discussed earlier. If A is the set of coins in the jar, call any two coins "equivalent" if they have the same value: thus, pennies are equivalent to pennies, nickels are equivalent to nickels, and so on. If x is a particular nickel in A, then $[x]$, the equivalence class of x, is the class of all the nickels in A. If y is a particular dime, then $[y]$ is the pile of all the dimes; and so forth. There are exactly five distinct equivalence classes. If we apply the lemma to this example, it states simply that if two coins are equivalent (that is, have the same value), they are in the same pile. By the way, the five equivalence classes obviously form a partition of A; this observation is expressed in the next theorem.

Theorem *If* $\sim$ *is an equivalence relation on* A, *the family of all the equivalence classes, that is,* $\{[x] : x \in A\}$, *is a partition of* A.

This theorem states that if $\sim$ is an equivalence relation on A and we sort the elements of A into distinct classes by placing each element with the ones equivalent to it, we get a partition of A.

To prove the theorem, we observe first that each equivalence class is a nonempty subset of A. (It is nonempty because $x \sim x$, so $x \in [x]$). Next, we need to show that any two distinct classes are disjoint—or, equivalently, that if two classes $[x]$ and $[y]$ have a common element, they are equal. Well, if $[x]$ and $[y]$ have a common element u, then $u \sim x$ and $u \sim y$. By the symmetric and transitive laws, $x \sim y$. Thus, $[x] = [y]$ by the lemma.

Finally, we must show that every element of A lies in some equivalence class. This is true because $x \in [x]$. Thus, the family of all the equivalence classes is a partition of A.

When $\sim$ is an equivalence relation on A and A is partitioned into its equivalence classes, we call this partition the *partition determined by the equivalence relation* $\sim$.

The student may have noticed by now that the two concepts of *partition* and *equivalence relation*, while superficially different, are actually twin aspects of the same structure on sets. Starting with an equivalence relation on A, we may partition A into equivalence classes, thus getting a partition of A. But from this partition we may retrieve the equivalence relation, for any two elements x and y are equivalent iff they lie in the same class of the partition.

Going the other way, we may begin with a partition of A and *define* an equivalence relation by letting any two elements be equivalent iff they lie in the same class. We may then retrieve the partition by partitioning A into equivalence classes.

As a final example, let A be a set of poker chips of various colors, say red, blue, green, and white. Call any two chips "equivalent" if they are the same color. This equivalence relation has four equivalence classes: the set of all the red chips, the set of blue chips, the set of green chips, and the set of white chips. These four equivalence classes are a partition of A.

Conversely, if we begin by partitioning the set A of poker chips into four classes according to their color, this partition determines an equivalence relation whereby chips are equivalent iff they belong to the same class. This is precisely the equivalence relation we had previously.

A final comment is in order. In general, there are many ways of partitioning a given set A; each partition determines (and is determined by) *exactly one specific equivalence relation* on A. Thus, if A is a set of three elements, say a, b, and c, there are five ways of partitioning A, as indicated by the accompanying illustration. Under each partition is written the equivalence relation determined by that partition.

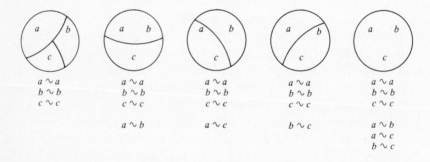

Once again, each partition of A determines, and is determined by, exactly one equivalence relation on A.

EXERCISES

A. Examples of Partitions

Prove that each of the following is a partition of the indicated set. Then describe the equivalence relation associated with that partition.

1 For each integer $r \in \{0, 1, 2, 3, 4\}$, let A_r be the set of all the integers which leave a remainder of r when divided by 5. (That is, $x \in A_r$ iff $x = 5q + r$ for some integer q.) Prove: $\{A_0, A_1, A_2, A_3, A_4\}$ is a partition of $\mathbb{Z}$.

2 For each integer n, let $A_n = \{x \in \mathbb{Q} : n \leqslant x < n + 1\}$. Prove $\{A_n : n \in \mathbb{Z}\}$ is a partition of $\mathbb{Q}$.

3 For each rational number r, let $A_r = \{(m, n) \in \mathbb{Z} \times \mathbb{Z} : m/n = r\}$. Prove that $\{A_r : r \in \mathbb{Q}\}$ is a partition of $\mathbb{Z} \times \mathbb{Z}$.

4 For $r \in \{0, 1, 2, \ldots, 9\}$, let A_r be the set of all the integers whose units digit (in decimal notation) is equal to r. Prove: $\{A_0, A_1, A_2, \ldots, A_9\}$ is a partition of $\mathbb{Z}$.

5 For any rational number x, we can write $x = q + n/m$ where q is an integer and $0 \leqslant n/m < 1$. Call n/m the *fractional part* of x. For each rational $r \in \{x : 0 \leqslant x < 1\}$, let $A_r = \{x \in \mathbb{Q} : \text{the fractional part of } x \text{ is equal to } r\}$. Prove: $\{A_r : 0 \leqslant r < 1\}$ is a partition of $\mathbb{Q}$.

6 For each $r \in \mathbb{R}$, let $A_r = \{(x, y) \in \mathbb{R} \times \mathbb{R} : x - y = r\}$. Prove: $\{A_r : r \in \mathbb{R}\}$ is a partition of $\mathbb{R} \times \mathbb{R}$.

B. Examples of Equivalence Relations

Prove that each of the following is an equivalence relation on the indicated set. Then describe the partition associated with that equivalence relation.

1 In $\mathbb{Z}$, $m \sim n$ iff $|m| = |n|$.

2 In $\mathbb{Q}$, $r \sim s$ iff $r - s \in \mathbb{Z}$.

3 Let $\lceil x \rceil$ denote the greatest integer $\leqslant x$. In $\mathbb{R}$, let $a \sim b$ iff $\lceil a \rceil = \lceil b \rceil$.

4 In $\mathbb{Z}$, let $m \sim n$ iff $m - n$ is a multiple of 10.

5 In $\mathbb{R}$, let $a \sim b$ iff $a - b \in \mathbb{Q}$.

6 In $\mathscr{F}(\mathbb{R})$, let $f \sim g$ iff $f(0) = g(0)$.

7 In $\mathscr{F}(\mathbb{R})$, let $f \sim g$ iff $f(x) = g(x)$ for all $x > c$, where c is some fixed real number.

8 If C is any set, P_C denotes the set of all the subsets of C. Let $D \subseteq C$. In P_C, let $A \sim B$ iff $A \cap D = B \cap D$.

9 In $\mathbb{R} \times \mathbb{R}$, let $(a, b) \sim (c, d)$ iff $a^2 + b^2 = c^2 + d^2$.

10 In $\mathbb{R}$, let $a \sim b$ iff $a/b \in \mathbb{Q}$.

C. Equivalence Relations and Partitions of $\mathbb{R} \times \mathbb{R}$

In parts 1 to 3, $\{A_r : r \in \mathbb{R}\}$ is a family of subsets of $\mathbb{R} \times \mathbb{R}$. Prove it is a partition, describe the partition geometrically, and give the corresponding equivalence relation.

1 For each $r \in \mathbb{R}$, $A_r = \{(x, y) : y = 2x + r\}$.

2 For each $r \in \mathbb{R}$, $A_r = \{(x, y) : x^2 + y^2 = r^2\}$.

3 For each $r \in \mathbb{R}$, $A_r = \{(x, y) : y = |x| + r\}$.

In parts 4 to 6, an equivalence relation on $\mathbb{R} \times \mathbb{R}$ is given. Prove it is an equivalence relation, describe it geometrically, and give the corresponding partition.

4 $(x, y) \sim (u, v)$ iff $ax^2 + by^2 = au^2 + bv^2$ (where $a, b > 0$)

5 $(x, y) \sim (u, v)$ iff $x + y = u + v$

6 $(x, y) \sim (u, v)$ iff $x^2 - y = u^2 - v$

D. Equivalence Relations on Groups

Let G be a group. In each of the following, a relation on G is defined. *Prove it is an equivalence relation. Then describe the equivalence class of e.*

1 If H is a subgroup of G, let $a \sim b$ iff $ab^{-1} \in H$.

2 If H is a subgroup of G, let $a \sim b$ iff $a^{-1}b \in H$. Is this the same equivalence relation as in part 1? Prove your answer.

3 Let $a \sim b$ iff there is an $x \in G$ such that $a = xbx^{-1}$.

4 Let $a \sim b$ iff there is an integer k such that $a^k = b^k$.

5 Let $a \sim b$ iff ab^{-1} commutes with every $x \in G$.

6 Let $a \sim b$ iff ab^{-1} is a power of c (where c is a fixed element of G).

E. General Properties of Equivalence Relations and Partitions

1 Let $\{A_i : i \in I\}$ be a partition of A. Let $\{B_j : j \in J\}$ be a partition of B. Prove that $\{A_i \times B_j : (i, j) \in I \times J\}$ is a partition of $A \times B$.

2 Let $\sim_I$ be the equivalence relation corresponding to the above partition of A, and let $\sim_J$ be the equivalence relation corresponding to the partition of B. Describe the equivalence relation corresponding to the above partition of $A \times B$.

3 Let $f : A \to B$ be a function. Define $\sim$ by: $a \sim b$ iff $f(a) = f(b)$. Prove that $\sim$ is an equivalence relation on A. Describe its equivalence classes.

4 Let $f : A \to B$ be a surjective function, and let $\{B_i : i \in I\}$ be a partition of B. Prove that $\{f^{-1}(B_i) : i \in I\}$ is a partition of A. If $\sim_I$ is the equivalence relation corresponding to the partition of B, describe the equivalence relation corresponding to the partition of A. [REMARK: For any $C \subseteq B$, $f^{-1}(C) = \{x \in A : f(x) \in C\}$.]

5 Let $\sim_1$ and $\sim_2$ be distinct equivalence relations on A. Define $\sim_3$ by: $a \sim_3 b$ iff $a \sim_1 b$ and $a \sim_2 b$. Prove that $\sim_3$ is an equivalence relation on A. If $[x]_i$ denotes the equivalence class of x for $\sim_i$ ($i = 1, 2, 3$), prove that $[x]_3 = [x]_1 \cap [x]_2$.

THIRTEEN

COUNTING COSETS

Just as there are great works in art and music, there are also great creations of mathematics. "Greatness," in mathematics as in art, is hard to define, but the basic ingredients are clear: a *great* theorem should contribute substantial new information, and it should be *unexpected*! That is, it should reveal something which common sense would not naturally lead us to expect. The most celebrated theorems of plane geometry, as may be recalled, come as a complete surprise; as the proof unfolds in simple, sure steps and we reach the conclusion—a conclusion we may have been skeptical about, but which is now established beyond a doubt—we feel a certain sense of awe not unlike our reaction to the ironic or tragic twist of a great story.

In this chapter we will consider a result of modern algebra which, by all standards, is a great theorem. It is something we would not likely have foreseen, and which brings new order and simplicity to the relationship between a group and its subgroups.

We begin by adding to our algebraic tool kit a new notion—a conceptual tool of great versatility which will serve us well in all the remaining chapters of this book. It is the concept of a *coset*.

Let G be a group, and H a subgroup of G. For any element a in G, the symbol

$$aH$$

*denotes the set of all products ah, as a remains fixed and h ranges over H.
aH is called a **left coset** of H in G.*

In similar fashion,

$$Ha$$

*denotes the set of all products ha, as a remains fixed and h ranges over H.
Ha is called a **right coset** of H in G.*

In practice, it will make no difference whether we use left cosets or right
cosets, just as long as we *remain consistent*. Thus, from here on, whenever
we use cosets we will use *right* cosets. To simplify our sentences, we will say
coset when we mean "right coset."

When we deal with cosets in a group G, we must keep in mind that
every coset in G *is a subset of G*. Thus, when we need to prove that two
cosets Ha and Hb are equal, we must show that they are *equal sets*. What
this means, of course, is that every element $x \in Ha$ is in Hb, and conversely,
every element $y \in Hb$ is in Ha. For example, let us prove the following
elementary fact:

$$\text{If } a \in Hb, \text{ then } Ha = Hb \qquad\qquad (*)$$

We are given that $a \in Hb$, which means that $a = h_1 b$ for some $h_1 \in H$. We
need to prove that $Ha = Hb$.

Let $x \in Ha$; this means that $x = h_2 a$ for some $h_2 \in H$. But $a = h_1 b$, so
$x = h_2 a = (h_2 h_1)b$, and the latter is clearly in Hb. This proves that every
$x \in Ha$ is in Hb; analogously, we may show that every $y \in Hb$ is in Ha, and
therefore $Ha = Hb$.

The first major fact about cosets now follows. Let G be a group and let
H be a fixed subgroup of G:

Theorem 1 *The family of all the cosets Ha, as a ranges over G, is a partition
of G.*

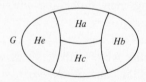

First, we must show that any two cosets, say Ha and Hb, are either
disjoint or equal. If they are disjoint, we are done. If not, let $x \in Ha \cap Hb$.
Because $x \in Ha$, $x = h_1 a$ for some $h_1 \in H$. Because $x \in Hb$, $x = h_2 b$ for

some $h_2 \in H$. Thus, $h_1 a = h_2 b$, and solving for a, we have

$$a = (h_1^{-1} h_2) b$$

Thus,

$$a \in Hb$$

It follows from (*) that $Ha = Hb$.

Next, we must show that *every* element $c \in G$ is in one of the cosets of H. But this is obvious, because $c = ec$ and $e \in H$; therefore,

$$c = ec \in Hc$$

Thus, the family of all the cosets of H is a partition of G!

Before going on, it is worth making a small comment: A given coset, say Hb, may be written in more than one way. By (*), *if a is any element in Hb, then Hb is the same as Ha.* Thus, for example, if a coset of H contains n different elements $a_1, a_2, \ldots, a_n$, it may be written in n different ways, namely $Ha_1, Ha_2, \ldots, Ha_n$.

The next important fact about cosets concerns finite groups. Let G be a finite group, and H a subgroup of G. We will show that *all the cosets of H have the same number of elements!* This fact is a consequence of the next theorem.

Theorem 2 *If Ha is any coset of H, there is a one-to-one correspondence from H to Ha.*

The most obvious function from H to Ha is the one which, for each $h \in H$, matches h with ha. Thus, let $f : H \to Ha$ be defined by

$$f(h) = ha$$

Remember that a remains fixed whereas h varies, and check that f is injective and surjective.

f *is injective*: Indeed, if $f(h_1) = f(h_2)$, then $h_1 a = h_2 a$, and therefore $h_1 = h_2$.
f *is surjective*, because every element of Ha is of the form ha for some $h \in H$, and $ha = f(h)$.

Thus, f is a one-to-one correspondence from H to Ha, as claimed.

By Theorem 2, any coset Ha has the same number of elements as H, and therefore all the cosets have the same number of elements!

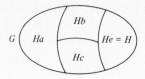

Let us take a careful look at what we have proved in Theorems 1 and 2. Let G be a finite group and H any subgroup of G. G has been partitioned into cosets of H, and all the cosets of H have the same number of elements (which is the same as the number of elements in H). Thus, *the number of elements in G is equal to the number of elements in H, multiplied by the number of distinct cosets of H.* This statement is known as Lagrange's theorem. (Remember that the number of elements in a group is called the group's *order*.)

Theorem 3: Lagrange's theorem *Let G be a finite group, and H any subgroup of G. The order of G is a multiple of the order of H.*

In other words, the order of any subgroup of a group G is a divisor of the order of G.

For example, if G has 15 elements, its proper subgroups may have either 3 or 5 elements. If G has 7 elements, it has *no* proper subgroups, for 7 has no factors other than 1 and 7. This last example may be generalized:

Let G be a group with a *prime* number p of elements. If $a \in G$ where $a \neq e$, then the order of a is some integer $m \neq 1$. But then the cyclic group $\langle a \rangle$ has m elements. By Lagrange's theorem, m must be a factor of p. But p is a prime number, and therefore $m = p$. It follows that $\langle a \rangle$ has p elements, and is therefore all of G! Conclusion:

Theorem 4 *If G is a group with a prime number p of elements, then G is a cyclic group. Furthermore, any element a $\neq$ e in G is a generator of G.*

Theorem 4, which is merely a consequence of Lagrange's theorem, is quite remarkable in itself. What it says is that *there is (up to isomorphism) only one group of any given prime order p.* For example, the only group (up to isomorphism) of order 7 is $\mathbb{Z}_7$, the only group of order 11 is $\mathbb{Z}_{11}$, and so on! So we now have complete information about all the groups whose order is a prime number.

By the way, if a is any element of a group G, the order of a is the same as the order of the cyclic subgroup $\langle a \rangle$, and by Lagrange's theorem this number is a divisor of the order of G. Thus,

Theorem 5 *The order of any element of a finite group divides the order of the group.*

Finally, if G is a group and H is a subgroup of G, the *index of H in G* is the number of cosets of H in G. We denote it by $(G : H)$. Since the number of elements in G is equal to the number of elements in H, multiplied by the number of cosets of H in G,

$$(G : H) = \frac{\text{order of } G}{\text{order of } H}$$

EXERCISES

A. Examples of Cosets in Finite Groups

In each of the following, H is a subgroup of G. *List the cosets of H. For each coset, list the elements of the coset.*

Example $G = \mathbb{Z}_4$, $H = \{0, 2\}$.

(REMARK: If the operation of G is denoted by $+$, it is customary to write $H + x$ for a coset, rather than Hx.) The cosets of H in this example are:

$$H = H + 0 = H + 2 = \{0, 2\} \qquad \text{and} \qquad H + 1 = H + 3 = \{1, 3\}$$

1 $G = S_3$, $H = \{\varepsilon, \beta, \delta\}$
2 $G = S_3$, $H = \{\varepsilon, \alpha\}$
3 $G = \mathbb{Z}_{15}$, $H = \langle 5 \rangle$
4 $G = D_4$, $H = \{R_0, R_4\}$ (For D_4, see page 69.)
5 $G = S_4$, $H = A_4$ (For A_4, see page 81.)
6 Indicate the order and index of each of the subgroups in parts 1 to 5.

B. Examples of Cosets in Infinite groups

Describe the cosets of the following subgroups.

1 The subgroup $\langle 3 \rangle$ of $\mathbb{Z}$.
2 The subgroup $\mathbb{Z}$ of $\mathbb{R}$.
3 The subgroup $H = \{2^n : n \in \mathbb{Z}\}$ of $\mathbb{R}^*$.
4 The subgroup $\langle \frac{1}{2} \rangle$ of $\mathbb{R}^*$.
5 The subgroup $H = \{(x, y) : x = y\}$ of $\mathbb{R} \times \mathbb{R}$.
6 For any positive integer m, what is the index of $\langle m \rangle$ in $\mathbb{Z}$?
7 Find a subgroup of $\mathbb{R}^*$ whose index is equal to 2.

C. Elementary Consequences of Lagrange's Theorem

Let G be a finite group. *Prove the following*:

1 If G has order n, then $x^n = e$ for every x in G.

2 Let G have order pq, where p and q are primes. Either G is cyclic, or every element $x \neq e$ in G has order p or q.

3 Let G have order 4. Either G is cyclic, or every element of G is its own inverse. Conclude that every group of order 4 is abelian.

4 If G has an element of order p and an element of order q, where p and q are distinct primes, then the order of G is a multiple of pq.

5 If G has an element of order k and an element of order m, then $|G|$ is a multiple of $\text{lcm}(k, m)$ [$\text{lcm}(k, m)$ is the least common multiple of k and m].

6 Let p be a prime number. In any finite group, the number of elements of order p is a multiple of $p - 1$.

D. Further Elementary Consequences of Lagrange's Theorem

Let G be a finite group, and let H and K be subgroups of G. *Prove the following*:

1 Suppose $H \subseteq K$ (therefore H is a subgroup of K). Then $(G : H) = (G : K)(K : H)$.

2 The order of $H \cap K$ is a common divisor of the order of H and the order of K.

3 Let H have order m and K have order n, where m and n are relatively prime. Then $H \cap K = \{e\}$.

4 Suppose H and K are not equal, and both have order the same prime number p. Then $H \cap K = \{e\}$.

5 Suppose H has index p and K has index q, where p and q are distinct primes. Then the index of $H \cap K$ is a multiple of pq.

6 If G is an abelian group of order n, and m is an integer such that m and n are relatively prime, then the function $f(x) = x^m$ is an automorphism of G.

E. Elementary Properties of Cosets

Let G be a group, and H a subgroup of G. Let a and b denote elements of G. *Prove the following*:

1 $Ha = Hb$ iff $ab^{-1} \in H$.

2 $Ha = H$ iff $a \in H$.

3 If $aH = Ha$ and $bH = Hb$, then $(ab)H = H(ab)$.

4 If $aH = Ha$, then $a^{-1}H = Ha^{-1}$.

5 If $(ab)H = (ac)H$, then $bH = cH$.

6 The number of right cosets of H is equal to the number of left cosets of H.

7 If J is a subgroup of G such that $J = H \cap K$, then for any $a \in G$, $Ja = Ha \cap Ka$. Conclude that if H and K are of finite index in G, then their intersection $H \cap K$ is also of finite index in G.

Theorem 5 of this chapter has a useful converse, which is the following:

Cauchy's theorem *If G is a finite group, and p is a prime divisor of $|G|$, then G has an element of order p.*

For example, a group of order 30 must have elements of orders 2, 3 and 5. Cauchy's theorem has an elementary proof, which may be found on page 340.

In the next few exercise sets, we will survey all possible groups whose order is $\leqslant 10$. By Theorem 4 of this chapter, if G is a group with a prime number p of elements, then $G \cong \mathbb{Z}_p$. This takes care of all groups of orders 2, 3, 5, and 7. In Exercise G6, page 152, it will be shown that if G is a group with p^2 elements (where p is a prime), then $G \cong \mathbb{Z}_{p^2}$ or $G \cong \mathbb{Z}_p \times \mathbb{Z}_p$. This will take care of all groups of orders 4 and 9. The remaining cases are examined in the next three exercise sets.

† F. Survey of All Six-Element Groups

Let G be any group of order 6. By Cauchy's theorem, G has an element a of order 2 and an element b of order 3. By Chapter 10, Exercise E3, the elements

$$e, a, b, b^2, ab, ab^2$$

are all distinct; and since G has only six elements, these are all the elements in G. Thus, ba is one of the elements $e, a, b, b^2, ab,$ or ab^2.

1 Prove that ba cannot be equal to either $e, a, b,$ or b^2. Thus, $ba = ab$ or $ba = ab^2$.

Either of these two equations completely determines the table of G. (See the discussion at the end of Chapter 5.)

2 If $ba = ab$, prove that $G \cong \mathbb{Z}_{10}$.
3 If $ba = ab^2$, prove that $G \cong S_3$.

It follows that $\mathbb{Z}_6$ and S_3 are (up to isomorphism), the only possible groups of order 6.

† G. Survey of All 10-Element Groups

Let G be any group of order 10.

1 Reason as in Exercise F to show that $G = \{e, a, b, b^2, b^3, b^4, ab, ab^2, ab^3, ab^4\}$, where a has order 2 and b has order 5.
2 Prove that ba cannot be equal to $e, a, b, b^2, b^3,$ or b^4.

3 Prove that if $ba = ab$, then $G \cong \mathbb{Z}_{10}$.

4 If $ba = ab^2$, prove that $ba^2 = a^2b^4$, and conclude that $b = b^4$. This is impossible because b has order 5, hence $ba \neq ab^2$. (HINT: The equation $ba = ab^2$ tells us that we may move a factor a from the right to the left of a factor b, but in so doing, we must square b. To prove an equation such as the preceding one, move all factors a to the left of all factors b.)

5 If $ba = ab^3$, prove that $ba^2 = a^2b^9 = a^2b^4$, and conclude that $b = b^4$. This is impossible (why?); hence $ba \neq ab^3$.

6 Prove that if $ba = ab^4$, then $G \cong D_5$ (where D_5 is the group of symmetries of the pentagon).

Thus, the only possible groups of order 10 (up to isomorphism), are $\mathbb{Z}_{10}$ and D_5.

† H. Survey of All Eight-Element Groups

Let G be any group of order 8. If G has an element of order 8, then $G \cong \mathbb{Z}_8$. Let us assume now that G has no element of order 8; hence all the elements $\neq e$ in G have order 2 or 4.

1 If every $x \neq e$ in G has order 2, let a, b, c be three such elements. Prove that $G = \{e, a, b, c, ab, bc, ac, abc\}$. Conclude that $G \cong \mathbb{Z}_2 \times \mathbb{Z}_2 \times \mathbb{Z}_2$.

In the remainder of this exercise set, assume G has an element a of order 4. Let $H = \langle a \rangle = \{e, a, a^2, a^3\}$. If $b \in G$ is not in H, then the coset $Hb = \{b, ab, a^2b, a^3b\}$. By Lagrange's theorem, G is the union of $He = H$ and Hb; hence

$$G = \{e, a, a^2, a^3, b, ab, a^2b, a^3b\}$$

2 Assume there is in Hb an element of order 2. (Let b be this element.) If $ba = a^2b$, prove that $b^2a = a^4b^2$, hence $a = a^4$, which is impossible. (Why?) Conclude that either $ba = ab$ or $ba = a^3b$.

3 Let b be as in part 2. Prove that if $ba = ab$, then $G \cong \mathbb{Z}_4 \times \mathbb{Z}_2$.

4 Let b be as in part 2. Prove that if $ba = a^3b$, then $G \cong D_4$.

5 Now assume the hypothesis in part 2 is false. Then b, ab, a^2b, and a^3b all have order 4. Prove that $b^2 = a^2$. (HINT: What is the order of b^2? What element in G has the same order?)

6 Prove: If $ba = ab$, then $(a^3b)^2 = e$, contrary to the assumption that ord$(a^3b) = 4$. If $ba = a^2b$, then $a = b^4a = e$, which is impossible. Thus, $ba = a^3b$.

7 The equations $a^4 = b^4 = e$, $a^2 = b^2$, and $ba = a^3b$ completely determine the table of G. Write this table. (G is known as the *quaternion group Q*.)

Thus, the only groups of order 8 (up to isomorphism) are $\mathbb{Z}_8$, $\mathbb{Z}_2 \times \mathbb{Z}_2 \times \mathbb{Z}_2$, $\mathbb{Z}_4 \times \mathbb{Z}_2$, D_4, and Q.

† I. Conjugate Elements

If $a \in G$, a *conjugate* of a is any element of the form xax^{-1}, with $x \in G$. (Roughly speaking, a conjugate of a is any product consisting of a sandwiched between any element and its inverse.) Prove each of the following:

1 The relation "a is equal to a conjugate of b" is an equivalence relation in G. (Write $a \sim b$ for "a is equal to a conjugate of b.")

This relation $\sim$ partitions any group G into classes called *conjugacy classes*. (The conjugacy class of a is $[a] = \{xax^{-1} : x \in G\}$.)

For any element $a \in G$, the *centralizer* of a, denoted by C_a, is the set of all the elements in G which commute with a. That is,

$$C_a = \{x \in G : xa = ax\} = \{x \in G : xax^{-1} = a\}$$

Prove:

2 For any $a \in G$, C_a is a subgroup of G.
3 $xax^{-1} = yay^{-1}$ iff xy^{-1} commutes with a iff $xy^{-1} \in C_a$.
4 $xax^{-1} = yay^{-1}$ iff $C_a x = C_a y$. (HINT: Use Exercise E1.)
5 There is a one-to-one correspondence between the set of all the conjugates of a and the set of all the cosets of C_a. (HINT: Use part 4.)
6 The number of distinct conjugates of a is equal to $(G : C_a)$, the index of C_a in G. Thus, *the size of every conjugacy class is a factor of* $|G|$.

† J. Group Acting on a Set

Let A be a set, and let G be any subgroup of S_A. G is a group of permutations of A; we say it is a *group acting on the set A*. Assume here that G is a finite group. If $u \in A$, the *orbit of u* (with respect to G) is the set

$$O(u) = \{g(u) : g \in G\}$$

1 Define a relation $\sim$ on A by: $u \sim v$ iff $g(u) = v$ for some $g \in G$. Prove that $\sim$ is an equivalence relation on A, and that the orbits are its equivalence classes.

If $u \in A$, the *stabilizer of u* is the set $G_u = \{g \in G : g(u) = u\}$, that is, the set of all the permutations in G which leave u fixed.
2 Prove that G_u is a subgroup of G.
3 Let $\alpha = (1\ 2)(3\ 4)(5\ 6)$ and $\beta = (2\ 3)$ in S_6. Let G be the following subgroup of S_6: $G = \{\varepsilon, \alpha, \beta, \alpha\beta, \beta\alpha, \alpha\beta\alpha, \beta\alpha\beta, (\alpha\beta)^2\}$. Find $O(1), O(2), O(5), G_1, G_2, G_4, G_5$.
4 Let $f, g \in G$. Prove that f and g are in the same coset of G_u iff $f(u) = g(u)$. (HINT: Use Exercise E1.)
5 Use part 4 to show that the number of elements in $O(u)$ is equal to the index of G_u in G. [HINT: If $f(u) = v$, match the coset of f with v.]
6 Conclude from part 5 that the size of every orbit (with respect to G) is a factor of the order of G. In particular, if $f \in S_A$, the length of each cycle of f is a factor of the order of f in S_A.

FOURTEEN

HOMOMORPHISMS

We have seen that if two groups are isomorphic, this means there is a one-to-one correspondence between them which transforms one of the groups into the other. Now if G and H are any groups, it may happen that there is a function which transforms G into H although this function is *not* a one-to-one correspondence. For example, $\mathbb{Z}_6$ is transformed into $\mathbb{Z}_3$ by

$$f = \begin{pmatrix} 0 & 1 & 2 & 3 & 4 & 5 \\ 0 & 1 & 2 & 0 & 1 & 2 \end{pmatrix}$$

as we may verify by comparing their tables:

+	0	1	2	3	4	5
0	0	1	2	3	4	5
1	1	2	3	4	5	0
2	2	3	4	5	0	1
3	3	4	5	0	1	2
4	4	5	0	1	2	3
5	5	0	1	2	3	4

Replace
x by $f(x)$

+	0	1	2	0	1	2
0	0	1	2	0	1	2
1	1	2	0	1	2	0
2	2	0	1	2	0	1
0	0	1	2	0	1	2
1	1	2	0	1	2	0
2	2	0	1	2	0	1

Eliminate duplicate
information

(For example, $2 + 2 = 1$
appears four separate
times in this table.)

+	0	1	2
0	0	1	2
1	1	2	0
2	2	0	1

If G and H are any groups, and there is a function f which transforms G into H, we say that H is a *homomorphic image* of G. The function f is called a *homomorphism* from G to H. This notion of homomorphism is one of the skeleton keys of algebra, and this chapter is devoted to explaining it and defining it precisely.

First, let us examine carefully what we mean by saying that "f transforms G into H." To begin with, f must be a function from G onto H; but that is not all, because f must also transform the table of G into the table of H. To accomplish this, f must have the following property: for any two elements a and b in G,

$$\text{if} \quad f(a) = a' \quad \text{and} \quad f(b) = b', \quad \text{then} \quad f(ab) = a'b' \qquad (*)$$

Graphically,

$$if \quad a \xrightarrow{\quad f \quad} a'$$

$$and \quad b \xrightarrow{\quad f \quad} b'$$

$$then \quad ab \xrightarrow{\quad f \quad} a'b'$$

Condition (*) may be written more succinctly as follows:

$$f(ab) = f(a)f(b) \qquad (**)$$

Thus,

Definition *If G and H are groups, a **homomorphism** from G to H is a function $f : G \to H$ such that for any two elements a and b in G,*

$$f(ab) = f(a)f(b)$$

*If there exists a homomorphism from G **onto** H, we say that H is a **homomorphic image** of G.*

Groups have a very important and privileged relationship with their homomorphic images, as the next few examples will show.

Let P denote the group consisting of two elements, e and o, with the table

+	e	o
e	e	o
o	o	e

We call this group the *parity group* of even and odd numbers. We should think of e as "even" and o as "odd." and the table as describing the rule for adding even and odd numbers. For example, even $+$ odd $=$ odd, odd $+$ odd $=$ even, and so on.

The function $f: \mathbb{Z} \to P$ which carries every even integer to e and every odd integer to o is clearly a homomorphism from $\mathbb{Z}$ to P. This is easy to check because there are only four different cases: for arbitrary integers r and s, r and s are either both even, both odd, or mixed. For example, if r and s are both odd, their sum is even, so $f(r) = o, f(s) = o$, and $f(r + s) = e$. Since $e = o + o$,

$$f(r + s) = f(r) + f(s)$$

This equation holds analogously in the remaining three cases; hence f is a homomorphism. (Note that the symbol $+$ is used on both sides of the above equation because the operation, in $\mathbb{Z}$ as well as in P, is denoted by $+$.)

It follows that P is a homomorphic image of $\mathbb{Z}$!

Now, what do P and $\mathbb{Z}$ have in common? P is a much smaller group than $\mathbb{Z}$, therefore it is not surprising that very few properties of the integers are to be found in P. Nevertheless, *one* aspect of the structure of $\mathbb{Z}$ is retained absolutely intact in P, namely the structure of the odd and even numbers. (The fact of being odd or even is called the *parity* of integers.) In other words, *as we pass from $\mathbb{Z}$ to P we deliberately lose every aspect of the integers except their parity*; their parity alone (with its arithmetic) is retained, and faithfully preserved.

Another example will make this point clearer. Remember that D_4 is the group of the symmetries of the square. Now, every symmetry of the square

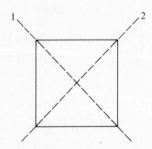

either interchanges the two diagonals here labeled 1 and 2, or leaves them as they were. In other words, every symmetry of the square brings about one of the permutations

$$\begin{pmatrix} 1 & 2 \\ 2 & 1 \end{pmatrix} \quad \text{or} \quad \begin{pmatrix} 1 & 2 \\ 1 & 2 \end{pmatrix}$$

of the diagonals.

For each $R_i \in D_4$, let $f(R_i)$ be the permutation of the diagonals produced by R_i. Then f is clearly a homomorphism from D_4 onto S_2. Indeed, it is clear on geometrical grounds that when we perform the motion R_i followed by the motion R_j on the square, we are, at the same time, carrying out the motions $f(R_i)$ followed by $f(R_j)$ on the diagonals. Thus,

$$f(R_j \circ R_i) = f(R_j) \circ f(R_i)$$

It follows that S_2 is a homomorphic image of D_4. Now, S_2 is a smaller group than D_4, and therefore very few of the features of D_4 are to be found in S_2. Nevertheless, *one* aspect of the structure of D_4 is retained absolutely intact in S_2, namely the diagonal motions. Thus, as we pass from D_4 to S_2, we deliberately lose every aspect of plane motions except the motions of the diagonals; these alone are retained and faithfully preserved.

A final example may be of some help; it relates to the group $\mathbb{B}^n$ described in Chapter 3, Exercise E. Here, briefly, is the context in which this group arises: The most basic way of transmitting information is to code it into strings of 0s and 1s, such as 0010111, 1010011, etc. Such strings are called *binary words*, and the number of 0s and 1s in any binary word is called its *length*. The symbol $\mathbb{B}^n$ designates the group consisting of all binary words of length n, with an operation of addition described in Chapter 3, Exercise E.

Consider the function $f : \mathbb{B}^7 \to \mathbb{B}^5$ which consists of *dropping the last two digits* of every seven-digit word. This kind of function arises in many practical situations: for example, it frequently happens that the first five digits of a word carry the message while the last two digits are an error check. Thus, f separates the message from the error check.

It is easy to verify that f is a homomorphism, hence $\mathbb{B}^5$ is a homomorphic image of $\mathbb{B}^7$. As we pass from $\mathbb{B}^7$ to $\mathbb{B}^5$, the message component of words in $\mathbb{B}^7$ is exactly preserved while the error check is deliberately lost.

These examples illustrate the basic idea inherent in the concept of a homomorphic image. The cases which arise in practice are not always so clear-cut as these, but the underlying idea is still the same: In a homomorphic image of G, some aspect of G is isolated and faithfully preserved while all else is deliberately lost.

The next theorem presents two elementary properties of homomorphisms.

Theorem 1 *Let G and H be groups, and* $f : G \rightarrow H$ *a homomorphism. Then*
(i) $f(e) = e$, *and*
(ii) $f(a^{-1}) = [f(a)]^{-1}$ *for every element* $a \in G$.

In the equation $f(e) = e$, the letter e on the left refers to the neutral element in G, whereas the letter e on the right refers to the neutral element in H.

To prove (i), we note that in any group,

$$\text{if} \quad yy = y \quad \text{then} \quad y = e$$

(Use the cancelation law on the equation $yy = ye$.) Now, $f(e)f(e) = f(ee) = f(e)$, hence $f(e) = e$.

To prove (ii), note that $f(a)f(a^{-1}) = f(aa^{-1}) = f(e)$. But $f(e) = e$, so $f(a)f(a^{-1}) = e$. It follows by Theorem 2 of Chapter 4 that $f(a^{-1})$ is the inverse of $f(a)$, that is, $f(a^{-1}) = [f(a)]^{-1}$.

Before going on with our study of homomorphisms, we must be introduced to an important new concept. If a is an element of a group G, a *conjugate* of a is any element of the form xax^{-1}, where $x \in G$. For example, the conjugates of α in S_3 are

$$\beta \circ \alpha \circ \beta^{-1} = \gamma$$

$$\gamma \circ \alpha \circ \gamma^{-1} = \kappa$$

$$\delta \circ \alpha \circ \delta^{-1} = \kappa$$

$$\kappa \circ \alpha \circ \kappa^{-1} = \gamma$$

as well as α itself, which may be written in two ways, as $\varepsilon \circ \alpha \circ \varepsilon^{-1}$ or as $\alpha \circ \alpha \circ \alpha^{-1}$. If H is any subset of a group G, we say that H is *closed with respect to conjugates* if every conjugate of every element of H is in H. Finally,

Definition *Let H be a subgroup of a group G. H is called a **normal** subgroup of G if it is closed with respect to conjugates; that is, if*

$$\text{for any} \quad a \in H \quad \text{and} \quad x \in G \quad xax^{-1} \in H$$

(Note that according to this definition, a normal subgroup of G is any nonempty subset of G which is closed with respect to products, with respect to inverses, and with respect to conjugates.)

We now return to our discussion of homomorphisms.

Definition Let $f : G \to H$ be a homomorphism. The **kernel** of f is the set K of all the elements of G which are carried by f onto the neutral element of H. That is,

$$K = \{x \in G : f(x) = e\}$$

Theorem 2 Let $f : G \to H$ be a homomorphism.
(i) The kernel of f is a normal subgroup of G, and
(ii) The range of f is a subgroup of H.

Let K denote the kernel of f. If $a, b \in K$, this means that $f(a) = e$ and $f(b) = e$. Thus, $f(ab) = f(a)f(b) = ee = e$, hence $ab \in K$.

If $a \in K$, then $f(a) = e$. Thus, $f(a^{-1}) = [f(a)]^{-1} = e^{-1} = e$, so $a^{-1} \in K$.

Finally, if $a \in K$ and $x \in G$, then $f(xax^{-1}) = f(x)f(a)f(x^{-1}) = f(x)f(a)[f(x)]^{-1} = e$, which shows that $xax^{-1} \in K$. Thus, K is a normal subgroup of G.

Now we must prove part (ii). If $f(a)$ and $f(b)$ are in the range of f, then their product $f(a)f(b) = f(ab)$ is also in the range of f.

If $f(a)$ is in the range of f, its inverse is $[f(a)]^{-1} = f(a^{-1})$, which is also in the range of f. Thus, the range of f is a subgroup of H.

If f is a homomorphism, we represent the kernel of f and the range of f with the symbols

$$\ker(f) \qquad \text{and} \qquad \operatorname{ran}(f)$$

EXERCISES

A. Examples of Homomorphisms of Finite Groups

1 Consider the function $f : \mathbb{Z}_8 \to \mathbb{Z}_4$ given by

$$f = \begin{pmatrix} 0 & 1 & 2 & 3 & 4 & 5 & 6 & 7 \\ 0 & 1 & 2 & 3 & 0 & 1 & 2 & 3 \end{pmatrix}$$

Verify that f is a homomorphism, find its kernel K, and list the cosets of K. [REMARK: To verify that f is a homomorphism, you must show that $f(a + b) = f(a) + f(b)$ for all choices of a and b in $\mathbb{Z}_8$; there are 64 choices. This may be accomplished by checking that f transforms the table of $\mathbb{Z}_8$ to the table of $\mathbb{Z}_4$, as on page 132.]

2. Consider the function $f : S_3 \to \mathbb{Z}_2$ given by

$$f = \begin{pmatrix} \varepsilon & \alpha & \beta & \gamma & \delta & \kappa \\ 0 & 1 & 0 & 1 & 0 & 1 \end{pmatrix}$$

Verify that f is a homomorphism, find its kernel K, and list the cosets of K.

3. Find a homomorphism $f : \mathbb{Z}_{15} \to \mathbb{Z}_5$, and indicate its kernel. (Do not actually verify that f is a homomorphism.)

4 Imagine a square as a piece of paper lying on a table. The side facing you is side A. The side hidden from view is side B. Every motion of the square either inter-

Side A

changes the two sides (that is, side B becomes visible and side A hidden) or leaves the sides as they were. In other words, every motion R_i of the square brings about one of the permutations

$$\begin{pmatrix} A & B \\ A & B \end{pmatrix} \quad \text{or} \quad \begin{pmatrix} A & B \\ B & A \end{pmatrix}$$

of the sides; call it $g(R_i)$. Verify that $g : D_4 \to S_2$ is a homomorphism, and give its kernel.

5 Every motion of the regular hexagon brings about a permutation of its diagonals, labeled 1, 2, and 3. For each $R_i \in D_6$, let $f(R_i)$ be the permutation of the diagonals

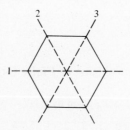

produced by R_i. Argue informally (appealing to geometric intuition) to explain why $f : D_6 \to S_3$ is a homomorphism. Then complete the following:

$$f\begin{pmatrix} 1 & 2 & 3 & 4 & 5 & 6 \\ 1 & 2 & 3 & 4 & 5 & 6 \end{pmatrix} = \varepsilon \quad f\begin{pmatrix} 1 & 2 & 3 & 4 & 5 & 6 \\ 2 & 3 & 4 & 5 & 6 & 1 \end{pmatrix} = \delta \quad \cdots$$

(That is, find the value of f on all 12 elements of D_6.)

6 Let $B \subset A$. Let $h : P_A \to P_B$ be defined by $h(C) = C \cap B$. For $A = \{1, 2, 3\}$ and $B = \{1, 2\}$, complete the following:

$$h = \begin{pmatrix} \emptyset & \{1\} & \{2\} & \{3\} & \{1, 2\} & \{1, 3\} & \{2, 3\} & A \end{pmatrix}$$

For any A and $B \subset A$, argue as in Chapter 3, Exercise C to show that h is a homomorphism.

B. Examples of Homomorphisms of Infinite Groups

Prove that each of the following is a homomorphism, and describe its kernel.

1 The function $\phi : \mathscr{F}(\mathbb{R}) \to \mathbb{R}$ given by $\phi(f) = f(0)$.
2 The function $\phi : \mathscr{D}(\mathbb{R}) \to \mathscr{F}(\mathbb{R})$ given by $\phi(f) = f'$. $\mathscr{D}(R)$ is the group of differentiable functions from $\mathbb{R}$ to $\mathbb{R}$; f' is the derivative of f.
3 The function $f : \mathbb{R} \times \mathbb{R} \to \mathbb{R}$ given by $f(x, y) = x + y$.
4 The function $f : \mathbb{R}^* \to \mathbb{R}^+$ defined by $f(x) = |x|$.
5 The function $f : \mathbb{C}^* \to \mathbb{R}^+$ defined by $f(a + bi) = \sqrt{a^2 + b^2}$.
6 Let G be the multiplicative group of all 2×2 matrices

$$\begin{pmatrix} a & b \\ c & d \end{pmatrix}$$

satisfying $ad - bc \neq 0$. Let $f : G \to \mathbb{R}^*$ be given by $f(A) =$ determinant of $A = ad - bc$.

C. Elementary Properties of Homomorphisms

Let G, H, and K be groups. *Prove the following:*

1 If $f : G \to H$ and $g : H \to K$ are homomorphisms, then their composite $g \circ f : G \to K$ is a homomorphism.
2 If $f : G \to H$ is a homomorphism with kernel K, then f is injective iff $K = \{e\}$.
3 If $f : G \to H$ is a homomorphism and K is any subgroup of G, then $f(K) = \{f(x) : x \in K\}$ is a subgroup of H.
4 If $f : G \to H$ is a homomorphism and J is any subgroup of H, then

$$f^{-1}(J) = \{x \in G : f(x) \in J\}$$

is a subgroup of G. Furthermore, $\ker f \subseteq f^{-1}(J)$.
5 If $f : G \to H$ is a homomorphism with kernel K, and J is a subgroup of G, let f_J designate the restriction of f to J. (In other words, f_J is the same function as f, except that its domain is restricted to J.) Prove that $\ker f_J = J \cap K$.
6 For any group G, the function $f : G \to G$ defined by $f(x) = e$ is a homomorphism.
7 For any group G, $\{e\}$ and G are homomorphic images of G.
8 The function $f : G \to G$ defined by $f(x) = x^2$ is a homomorphism iff G is abelian.
9 The functions $f_1(x, y) = x$ and $f_2(x, y) = y$, from $G \times H$ to G and H, respectively, are homomorphisms.

D. Basic Properties of Normal Subgroups

In the following, let G denote an arbitrary group.

1 Find all the normal subgroups   (a) of S_3 (b) of D_4.

Prove the following:

2 Every subgroup of an abelian group is normal.

3 The center of any group G is a normal subgroup of G.

4 Let H be a subgroup of G. H is normal iff it has the following property: For all a and b in G, $ab \in H$ iff $ba \in H$.

5 Let H be a subgroup of G. H is normal iff $aH = Ha$ for every $a \in G$.

6 Any intersection of normal subgroups of G is a normal subgroup of G.

E. Further Properties of Normal Subgroups

Let G denote a group, and H a subgroup of G. *Prove the following*:

1 If H has index 2 in G, then H is normal. (HINT: Use Exercise D5.)

2 Suppose an element $a \in G$ has order 2. Then $\langle a \rangle$ is a normal subgroup of G iff a is in the center of G.

3 If a is any element of G, $\langle a \rangle$ is a normal subgroup of G iff a has the following property: For any $x \in G$, there is a positive integer k such that $xa = a^k x$.

4 In a group G, a *commutator* is any product of the form $aba^{-1}b^{-1}$, where a and b are any elements of G. If a subgroup H of G contains *all* the commutators of G, then H is normal.

5 If H and K are subgroups of G, and K is normal, then HK is a subgroup of G. (HK denotes the set of all products hk as h ranges over H and k ranges over K.)

6 Let S be the union of all the cosets Ha such that $Ha = aH$. Then S is a normal subgroup of G.

F. Homomorphism and the Order of Elements

If $f: G \to H$ is a homomorphism, *prove each of the following*:

1 For each element $a \in G$, the order of $f(a)$ is a divisor of the order of a.

2 The order of any element $b \neq e$ in the range of f is a common divisor of $|G|$ and $|H|$. [Use (1).]

3 If the range of f has n elements, then $x^n \in \ker f$ for every $x \in G$.

4 Let m be an integer such that m and $|H|$ are relatively prime. For any $x \in G$, if $x^m \in \ker f$, then $x \in \ker f$.

5 Let the range of f have m elements. If $a \in G$ has order n, where m and n are relatively prime, then a is in the kernel of f. [Use (1).]

6 Let p be a prime. If H has an element of order p, then G has an element of order p.

G. Properties Preserved under Homomorphism

A property of groups is said to be "preserved under homomorphism" if, whenever a group G has that property, every homomorphic image of G does also. In this exercise set, we will survey a few typical properties preserved under homomorphism. If $f : G \rightarrow H$ is a homomorphism of G onto H, *prove each of the following*:

1 If G is abelian, then H is abelian.

2 If G is cyclic, then H is cyclic.

3 If every element of G has finite order, then every element of H has finite order.

4 If every element of G is its own inverse, every element of H is its own inverse.

5 If every element of G has a square root, then every element of H has a square root.

6 If G is finitely generated, then H is finitely generated. (A group is said to be "finitely generated" if it is generated by finitely many of its elements.)

† H. Inner Direct Products

If G is any group, let H and K be normal subgroups of G such that $H \cap K = \{e\}$. *Prove the following*:

1 Let h_1 and h_2 be any two elements of H, and k_1 and k_2 any two elements of K.

$$h_1 k_1 = h_2 k_2 \qquad \text{implies} \qquad h_1 = h_2 \qquad \text{and} \qquad k_1 = k_2$$

(HINT: If $h_1 k_1 = h_2 k_2$, then $h_2^{-1} h_1 \in H \cap K$ and $k_2 k_1^{-1} \in H \cap K$. Explain why.)

2 For any $h \in H$ and $k \in K$, $hk = kh$. (HINT: $hk = kh$ iff $hkh^{-1}k^{-1} = e$. Use the fact that H and K are normal.)

3 Now, make the additional assumption that $G = HK$, that is, every x in G can be written as $x = hk$ for some $h \in H$ and $k \in K$. Prove that the function $\phi(h, k) = hk$ is an isomorphism from $H \times K$ onto G.

We have thus proved the following: *If H and K are normal subgroups of G, such that $H \cap K = \{e\}$ and $G = HK$, then $G \cong H \times K$. G is sometimes called the inner direct product of H and K.*

† I. Conjugate Subgroups

Let H be a subgroup of G. For any $a \in G$, let $aHa^{-1} = \{axa^{-1} : x \in H\}$; aHa^{-1} is called a conjugate of H. *Prove the following*:

1 For each $a \in G$, aHa^{-1} is a subgroup of G.

2 For each $a \in G$, $H \cong aHa^{-1}$.

3 H is a normal subgroup of G iff $H = aHa^{-1}$ for every $a \in G$.

In the remaining exercises of this set, let G be a finite group. By the normalizer of H we mean the set $N(H) = \{a \in G : axa^{-1} \in H \text{ for every } x \in H\}$.

4 If $a \in N(H)$, then $aHa^{-1} = H$. (Remember that G is now a finite group.)

5 $N(H)$ is a subgroup of G.

6 $H \subseteq N(H)$. Furthermore, H is a normal subgroup of $N(H)$.

In (7) $-$ (10), let $N = N(H)$.

7 For any $a, b \in G$, $aHa^{-1} = bHb^{-1}$ iff $ab^{-1} \in N$ iff $Na = Nb$.

8 There is a one-to-one correspondence between the set of conjugates of H and the set of cosets of N. (Thus, there are as many conjugates of H as cosets of N.)

9 H has exactly $(G : N)$ conjugates. In particular, the number of distinct conjugates of H is a divisor of $|G|$.

10 Let K be any subgroup of G, let $K^* = \{Na : a \in K\}$, and let

$$X_K = \{aHa^{-1} : a \in K\}$$

Argue as in (8) to prove that X_K is in one-to-one correspondence with K^*. Conclude that the number of elements in X_K is a divisor of $|K|$.

FIFTEEN
QUOTIENT GROUPS

In Chapter 14 we learned to recognize when a group H is a homomorphic image of a group G. Now we will make a great leap forward by learning a method for actually *constructing all the homomorphic images of any group*. This is a remarkable procedure, of great importance in algebra. In many cases this construction will allow us to deliberately select *which* properties of a group G we wish to preserve in a homomorphic image, and which other properties we wish to discard.

The most important instrument to be used in this construction is the notion of a normal subgroup. Remember that a *normal* subgroup of G is any subgroup of G which is closed with respect to conjugates. We begin by giving an elementary property of normal subgroups.

Theorem 1 *If H is a normal subgroup of G, then $aH = Ha$ for every $a \in G$.*

(In other words, there is no distinction between left and right cosets for a normal subgroup.)

Indeed, if x is any element of aH, then $x = ah$ for some $h \in H$. But H is closed with respect to conjugates, hence $aha^{-1} \in H$. Thus, $x = ah = (aha^{-1})a$ is an element of Ha. This shows that every element of aH is in Ha; analogously, every element of Ha is in aH. Thus, $aH = Ha$.

Let G be a group and let H be a subgroup of G. There is a way of combining cosets, called *coset multiplication,* which works as follows: *the coset of a, multiplied by the coset of b, is defined to be the coset of ab.* In symbols,

$$Ha \cdot Hb = H(ab)$$

This definition is deceptively simple, for it conceals a fundamental difficulty. Indeed, it is not at all clear that the product of two cosets Ha and Hb, multiplied together in this fashion, is *uniquely defined.* Remember that Ha may be the same coset as Hc (this happens iff c is in Ha), and, similarly, Hb may be the same coset as Hd. Therefore, the product $Ha \cdot Hb$ is the same as the product $Hc \cdot Hd$. Yet it may easily happen that $H(ab)$ is *not* the same coset as $H(cd)$. Graphically,

$$Ha \cdot Hb = H(ab)$$
$$\| \quad \| \quad \nparallel$$
$$Hc \cdot Hd = H(cd)$$

For example, if $G = S_3$ and $H = \{\varepsilon, \alpha\}$, then

$$H\beta = \{\beta, \gamma\} = H\gamma$$
$$H\delta = \{\delta, \kappa\} = H\kappa$$

and yet

$$H(\beta \circ \delta) = H\varepsilon \neq H\beta = H(\gamma \circ \kappa)$$

Thus, coset multiplication *does not work* as an operation on the cosets of $H = \{\varepsilon, \alpha\}$ in S_3. The reason is that, although H is a subgroup of S_3, H is *not a normal subgroup of S_3.* If H were a normal subgroup, coset multiplication would work. The next theorem states exactly that!

Theorem 2 *Let H be a normal subgroup of G. If $Ha = Hc$ and $Hb = Hd$, then $H(ab) = H(cd)$.*

If $Ha = Hc$ then $a \in Hc$, hence $a = h_1 c$ for some $h_1 \in H$. If $Hb = Hd$, then $b \in Hd$, hence $b = h_2 d$ from some $h_2 \in H$. Thus,

$$ab = h_1 ch_2 \, d = h_1(ch_2)d$$

But $ch_2 \in cH = Hc$ (the last equality is true by Theorem 1). Thus, $ch_2 = h_3 c$ for some $h_3 \in H$. Returning to ab,

$$ab = h_1(ch_2)d = h_1(h_3 c)d = (h_1 h_3)(cd)$$

and this last element is clearly in $H(cd)$.

We have shown that $ab \in H(cd)$. Thus, by (*) in Chapter 13, $H(ab) = H(cd)$.

We are now ready to proceed with the construction promised at the beginning of this chapter. Let G be a group and let H be a normal subgroup of G. Think of the set which consists of *all the cosets of H*. This set is conventionally denoted by the symbol G/H. Thus, if Ha, Hb, Hc, ... are cosets of H, then

$$G/H = \{Ha, Hb, Hc, \ldots\}$$

We have just seen that *coset multiplication* is a valid operation on this set. In fact,

Theorem 3 G/H *with coset multiplication is a group.*

Coset multiplication is associative, because

$$Ha \cdot (Hb \cdot Hc) = Ha \cdot H(bc) = Ha(bc) = H(ab)c$$

$$= H(ab) \cdot Hc = (Ha \cdot Hb) \cdot Hc$$

The identity element of G/H is $H = He$, for $Ha \cdot He = Ha$ and $He \cdot Ha = Ha$ for every coset Ha.

Finally, the inverse of any coset Ha is the coset Ha^{-1}, because $Ha \cdot Ha^{-1} = Haa^{-1} = He$ and $Ha^{-1} \cdot Ha = Ha^{-1}a = He$.

The group G/H is called the *factor group*, or *quotient group* of G by H.

And now, the *pièce de résistance*:

Theorem 5 G/H *is a homomorphic image of G.*

The most obvious function from G to G/H is the function f which carries every element to its own coset, that is, the function given by

$$f(x) = Hx$$

This function is a homomorphism, because

$$f(xy) = Hxy = Hx \cdot Hy = f(x)f(y)$$

f is called the *natural homomorphism* from G onto G/H. Since there is a homomorphism from G onto G/H, G/H is a homomorphic image of G.

Thus, when we construct quotient groups of G, we are, in fact, constructing homomorphic images of G. The quotient group construction is useful because it is a way of actually manufacturing homomorphic images of any group G. In fact, as we will soon see, it is a way of manufacturing *all* the homomorphic images of G.

Our first example is intended to clarify the details of quotient group construction. Let $\mathbb{Z}$ be the group of the integers, and let $\langle 6 \rangle$ be the cyclic subgroup of $\mathbb{Z}$ which consists of all the multiples of 6. Since $\mathbb{Z}$ is abelian, and every subgroup of an abelian group is normal, $\langle 6 \rangle$ is a normal subgroup of $\mathbb{Z}$. Therefore, we may form the quotient group $\mathbb{Z}/\langle 6 \rangle$. The elements of this quotient group are all the cosets of the subgroup $\langle 6 \rangle$, namely:

$$\langle 6 \rangle + 0 = \{\ldots, -18, -12, -6, 0, 6, 12, 18, \ldots\}$$

$$\langle 6 \rangle + 1 = \{\ldots, -17, -11, -5, 1, 7, 13, 19, \ldots\}$$

$$\langle 6 \rangle + 2 = \{\ldots, -16, -10, -4, 2, 8, 14, 20, \ldots\}$$

$$\langle 6 \rangle + 3 = \{\ldots, -15, -9, -3, 3, 9, 15, 21, \ldots\}$$

$$\langle 6 \rangle + 4 = \{\ldots, -14, -8, -2, 4, 10, 16, 22, \ldots\}$$

$$\langle 6 \rangle + 5 = \{\ldots, -13, -7, -1, 5, 11, 17, 23, \ldots\}$$

These are *all* the different cosets of $\langle 6 \rangle$, for it is easy to see that $\langle 6 \rangle + 6 = \langle 6 \rangle + 0$, $\langle 6 \rangle + 7 = \langle 6 \rangle + 1$, $\langle 6 \rangle + 8 = \langle 6 \rangle + 2$, and so on.

Now, the operation on $\mathbb{Z}$ is denoted by $+$, and therefore we will call the operation on the cosets *coset addition* rather than coset multiplication. But nothing is changed except the name; for example, the coset $\langle 6 \rangle + 1$ added to the coset $\langle 6 \rangle + 2$ is the coset $\langle 6 \rangle + 3$. The coset $\langle 6 \rangle + 3$ added to the coset $\langle 6 \rangle + 4$ is the coset $\langle 6 \rangle + 7$, which is the same as $\langle 6 \rangle + 1$. To simplify our notation, let us agree to write the cosets in the following shorter form:

$$\bar{0} = \langle 6 \rangle + 0 \qquad \bar{1} = \langle 6 \rangle + 1 \qquad \bar{2} = \langle 6 \rangle + 2$$

$$\bar{3} = \langle 6 \rangle + 3 \qquad \bar{4} = \langle 6 \rangle + 4 \qquad \bar{5} = \langle 6 \rangle + 5$$

Then $\mathbb{Z}/\langle 6 \rangle$ consists of the six elements $\bar{0}, \bar{1}, \bar{2}, \bar{3}, \bar{4}$, and $\bar{5}$, and its operation is summarized in the following table:

$+$	$\bar{0}$	$\bar{1}$	$\bar{2}$	$\bar{3}$	$\bar{4}$	$\bar{5}$
$\bar{0}$	$\bar{0}$	$\bar{1}$	$\bar{2}$	$\bar{3}$	$\bar{4}$	$\bar{5}$
$\bar{1}$	$\bar{1}$	$\bar{2}$	$\bar{3}$	$\bar{4}$	$\bar{5}$	$\bar{0}$
$\bar{2}$	$\bar{2}$	$\bar{3}$	$\bar{4}$	$\bar{5}$	$\bar{0}$	$\bar{1}$
$\bar{3}$	$\bar{3}$	$\bar{4}$	$\bar{5}$	$\bar{0}$	$\bar{1}$	$\bar{2}$
$\bar{4}$	$\bar{4}$	$\bar{5}$	$\bar{0}$	$\bar{1}$	$\bar{2}$	$\bar{3}$
$\bar{5}$	$\bar{5}$	$\bar{0}$	$\bar{1}$	$\bar{2}$	$\bar{3}$	$\bar{4}$

The reader will perceive immediately the similarity between this group and $\mathbb{Z}_6$. As a matter of fact, the quotient group construction of $\mathbb{Z}/\langle 6 \rangle$ is considered to be the rigorous way of constructing $\mathbb{Z}_6$. So from now on, we will consider $\mathbb{Z}_6$ to be the same as $\mathbb{Z}/\langle 6 \rangle$; and, in general, we will consider $\mathbb{Z}_n$ to be the same as $\mathbb{Z}/\langle n \rangle$. In particular, we can see that for any n, $\mathbb{Z}_n$ is a homomorphic image of $\mathbb{Z}$.

Let us repeat: The motive for the quotient group construction is that it gives us a way of actually *producing* all the homomorphic images of any group G. However, what is even more fascinating about the quotient group construction is that in practical instances, we can often choose H so as to "factor out" unwanted properties of G, and preserve in G/H only "desirable" traits. (By "desirable" we mean desirable within the context of some specific application or use.) Let us look at a few examples.

First, we will need two simple properties of cosets, which are given in the next theorem.

Theorem 5 *Let G be a group and H a subgroup of G. Then*
(a) $Ha = Hb$ iff $ab^{-1} \in H$ and
(b) $Ha = H$ iff $a \in H$

If $Ha = Hb$, then $a \in Hb$, so $a = hb$ for some $h \in H$. Thus,

$$ab^{-1} = h \in H$$

If $ab^{-1} \in H$, then $ab^{-1} = h$ for $h \in H$, and therefore $a = hb \in Hb$. It follows by (*) of Chapter 13 that $Ha = Hb$.

This proves (a). It follows that $Ha = He$ iff $ae^{-1} = a \in H$, which proves (b).

For our first example, let G be an abelian group and let H consist of *all the elements of G which have finite order*. It is easy to show that H is a subgroup of G. (The details may be supplied by the reader.) Remember that in an abelian group every subgroup is normal, hence H is a normal subgroup of G, and therefore we may form the quotient group G/H. We will show next that in G/H, *no* element except the neutral element has finite order.

For suppose G/H has an element Hx of finite order. Since the neutral element of G/H is H, this means there is an integer $m \neq 0$ such that $(Hx)^m = H$, that is, $Hx^m = H$. Therefore, by Theorem 5b, $x^m \in H$, so x^m has finite order, say t:

$$(x^m)^t = x^{mt} = e$$

But then x has finite order, so $x \in H$. Thus, by Theorem 5b, $Hx = H$. This proves that in G/H, the only element Hx of finite order is the neutral element H.

Let us recapitulate: If H is the subgroup of G which consists of all the elements of G which have finite order, then in G/H, *no* element (except the neutral element) has finite order. Thus, in a sense, we have *"factored out"* all *the elements of finite order* (*they are all in* H) *and produced a quotient group* G/H *whose elements all have infinite order* (except for the neutral element, which necessarily has order 1).

Our next example may bring out this idea even more clearly. Let G be an arbitrary group; by a *commutator* of G we mean any element of the form $aba^{-1}b^{-1}$ where a and b are in G. The reason such a product is called a commutator is that

$$aba^{-1}b^{-1} = e \qquad \text{iff} \qquad ab = ba$$

In other words, $aba^{-1}b^{-1}$ reduces to the neutral element whenever a and b commute—and *only* in that case! Thus, in an abelian group all the commutators are equal to e. In a group which is not abelian, the number of distinct commutators may be regarded as a measure of the extent to which G departs from being commutative. (The fewer the commutators, the closer the group is to being an abelian group.)

We will see in a moment that if H is a subgroup of G which *contains all the commutators of* G, then G/H is abelian! What this means, in a fairly accurate sense, is that *when we factor out the commutators of* G *we get a quotient group which has no commutators* (except, trivially, the neutral element) *and which is therefore abelian.*

To say that G/H is abelian is to say that for any two elements Hx and Hy in G/H, $HxHy = HyHx$, that is, $Hxy = Hyx$. But by Theorem 5b,

$$Hxy = Hyx \qquad \text{iff} \qquad xy(yx)^{-1} \in H$$

Now $xy(yx)^{-1}$ is the commutator $xyx^{-1}y^{-1}$, so if all the commutators are in H, then G/H is abelian.

EXERCISES

A. Examples of Finite Quotient Groups

In each of the following, G is a group and H is a normal subgroup of G. *List the elements of* G/H *and then write the table of* G/H.

Example $G = \mathbb{Z}_6$ and $H = \{0, 3\}$

The elements of G/H are the three cosets $H = H + 0 = \{0, 3\}$, $H + 1 = \{1, 4\}$, and $H + 2 = \{2, 5\}$. (Note that $H + 3$ is the same as $H + 0$, $H + 4$ is the same as $H + 1$, and $H + 5$ is the same as $H + 2$.) The table of G/H is

+	H	$H+1$	$H+2$
H	H	$H+1$	$H+2$
$H+1$	$H+1$	$H+2$	H
$H+2$	$H+2$	H	$H+1$

1 $G = \mathbb{Z}_{10}$, $H = \{0, 5\}$. (Explain why $G/H \cong \mathbb{Z}_5$.)

2 $G = S_3$, $H = \{\varepsilon, \beta, \delta\}$.

3 $G = D_4$, $H = \{R_0, R_2\}$. (See page 68.)

4 $G = D_4$, $H = \{R_0, R_2, R_4, R_5\}$.

5 $G = \mathbb{Z}_4 \times \mathbb{Z}_2$, $H = \langle(0, 1)\rangle = $ the subgroup of $\mathbb{Z}_4 \times \mathbb{Z}_2$ generated by $(0, 1)$.

6 $G = P_3$, $H = \{\emptyset, \{1\}\}$. (P_3 is the group of subsets of $\{1, 2, 3\}$.)

B. Examples of Quotient Groups of $\mathbb{R} \times \mathbb{R}$

In each of the following, H is a subset of $\mathbb{R} \times \mathbb{R}$.

(a) Prove that H is a normal subgroup of $\mathbb{R} \times \mathbb{R}$. (Remember that every subgroup of an abelian group is normal.)

(b) In geometrical terms, describe the elements of the quotient group G/H.

(c) In geometrical terms or otherwise, describe the operation of G/H.

1 $H = \{(x, 0) : x \in \mathbb{R}\}$

2 $H = \{(x, y) : y = -x\}$

3 $H = \{(x, y) : y = 2x\}$

C. Relating Properties of H to Properties of G/H

In each of the problems below, G is a group and H is a normal subgroup of G. *Prove the following.* (Theorem 5 will play a crucial role.)

1 If $x^2 \in H$ for every $x \in G$, then every element of G/H is its own inverse. Conversely, if every element of G/H is its own inverse, then $x^2 \in H$ for all $x \in G$.

2 Let m be a fixed integer. If $x^m \in H$ for every $x \in G$, then the order of every element in G/H is a divisor of m. Conversely, if the order of every element in G/H is a divisor of m, then $x^m \in H$ for every $x \in G$.

3 Suppose that for every $x \in G$, there is an integer n such that $x^n \in H$; then every element of G/H has finite order. Conversely, if every element of G/H has finite order, then for every $x \in G$ there is an integer n such that $x^n \in H$.

4 Every element of G/H has a square root iff for every $x \in G$, there is some $y \in G$ such that $xy^2 \in H$.

5 G/H is cyclic iff there is an element $a \in G$ with the following property: for every $x \in G$, there is some integer n such that $xa^n \in H$.

6 If G is an abelian group, let H_p be the set of all $x \in G$ whose order is a power of p. Prove that H_p is a subgroup of G. Prove that G/H_p has no elements whose order is a power of p.

7 (a) If G/H is abelian, H contains all the commutators of G.

(b) Let K be a normal subgroup of G, and H a normal subgroup of K. If G/H is abelian, then G/K and K/H are both abelian. [Use (a) and the last paragraph of this chapter.]

D. Properties of G Determined by Properties of G/H and H

There are some group properties which, if they are true in G/H and in H, must be true in G. Here is a sampling. Let G be a group, and H a normal subgroup of G. *Prove*:

1 If every element of G/H has finite order, and every element of H has finite order, then every element of G has finite order.

2 If every element of G/H has a square root, and every element of H has a square root, then every element of G has a square root.

3 Let p be a prime number. A group G is called a *p-group* if the order of every element x in G is a power of p. Prove: If G/H and H are p-groups, then G is a p-group.

4 If G/H and H are finitely generated, then G is finitely generated. (A group is said to be finitely generated if it is generated by a finite subset of its elements.)

E. Order of Elements in Quotient Groups

Let G be a group, and H a normal subgroup of G. *Prove the following*:

1 For each element $a \in G$, the order of the element Ha in G/H is a divisor of the order of a in G. (HINT: Use Chapter 14, Exercise F1.)

2 If $(G : H) = m$, the order of every element of G/H is a divisor of m.

3 If $(G : H) = p$, where p is a prime, then the order of every element $a \notin H$ in G is a multiple of p. [Use (1).]

4 If G has a normal subgroup of index p, where p is a prime, then G has at least one element of order p.

5 If $(G : H) = m$, then $a^m \in H$ for every $a \in G$.

6 In $\mathbb{Q}/\mathbb{Z}$, every element has finite order.

† F. Quotient of a Group by its Center

The *center* of a group G is the normal subgroup C of G consisting of all those elements of G which commute with every element of G. Suppose the quotient group

G/C is a cyclic group; say it is generated by the element Ca of G/C. *Prove each of the following:*

1 For every $x \in G$, there is some integer m such that $Cx = Ca^m$.

2 For every $x \in G$, there is some integer m such that $x = ca^m$, where $c \in C$.

3 For any two elements x and y in G, $xy = yx$. (HINT: Use part 2 to write $x = ca^m$, $y = c'a^n$, and remember that $c, c' \in C$.)

4 Conclude that if G/C is cyclic, then G is abelian.

† G. Using the Class Equation to Determine the Size of the Center

(Prerequisite: Chapter 13, Exercise I.)

Let G be a finite group. Elements a and b in G are called *conjugates* of one another (in symbols, $a \sim b$) iff $a = xbx^{-1}$ for some $x \in G$ (this is the same as $b = x^{-1}ax$). The relation $\sim$ is an equivalence relation in G; the equivalence class of any element a is called its *conjugacy class*, hence G is partitioned into conjugacy classes (as shown in the diagram); the size of each conjugacy class divides the order of G. (For these facts, see Chapter 13, Exercise Set I.)

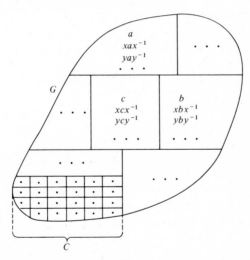

"Each element of the center C is alone in its conjugacy class."

Let $S_1, S_2, \ldots, S_t$ be the distinct conjugacy classes of G, and let $k_1, k_2, \ldots, k_t$ be their sizes. Then $|G| = k_1 + k_2 + \cdots + k_t$. (This is called the *class equation* of G.)

Let G be a group whose order is a power of p, say $|G| = p^k$. Let C denote the center of G. *Prove the following:*

1 The conjugacy class of a contains a (and no other element) iff $a \in C$.

2 Let c be the order of C. Then $|G| = c + k_s + k_{s+1} + \cdots + k_t$, where $k_s, \ldots, k_t$ are the sizes of all the distinct conjugacy classes of elements $x \notin C$.

3 For each $i \in \{s, s + 1, \ldots, t\}$, k_i is equal to a power of p. (See Chapter 13, Exercise I6.)

4 Solving the equation $|G| = c + k_s + \cdots + k_t$ for c, explain why c is a multiple of p.

We may conclude from part 4 that C must contain more than the one element e; in fact, $|C|$ is a multiple of p.

5 If $|G| = p^2$, G must be abelian. (Use the preceding exercise F.)

6 If $|G| = p^2$, then either $G \cong \mathbb{Z}_{p^2}$ or $G \cong \mathbb{Z}_p \times \mathbb{Z}_p$.

† H. Induction on $|G|$: An Example

Many theorems of mathematics are of the form "$P(n)$ is true for every positive integer n." [Here, $P(n)$ is used as a symbol to denote some statement involving n.] Such theorems can be proved by induction as follows:

(a) Show that $P(n)$ is true for $n = 1$.

(b) For any fixed positive integer k, show that, if $P(n)$ is true for every $n < k$, then $P(n)$ must also be true for $n = k$.

If we can show (a) and (b), we may safely conclude that $P(n)$ is true for all positive integers n.

Some theorems of algebra can be proved by induction on the order n of a group. Here is a classical example: Let G be a finite abelian group. We will show that G must contain at least one element of order p, for every prime factor p of $|G|$. If $|G| = 1$, this is true by default, since no prime p can be a factor of 1. Next, let $|G| = k$, and suppose our claim is true for every abelian group whose order is less than k. Let p be a prime factor of k.

Take any element $a \neq e$ in G. If $\text{ord}(a) = p$ or a multiple of p, we are done!

1 If $\text{ord}(a) = tp$ (for some positive integer t), what element of G has order p?

2 Suppose $\text{ord}(a)$ *is not equal to a multiple of* p. Then $G/\langle a \rangle$ is a group having fewer than k elements. (Explain why.) The order of $G/\langle a \rangle$ is a multiple of p. (Explain why.)

3 Why must $G/\langle a \rangle$ have an element of order p?

4 Conclude that G has an element of order p. (HINT: Use Exercise E1.)

SIXTEEN

THE FUNDAMENTAL HOMOMORPHISM THEOREM

Let G be any group. In Chapter 15 we saw that every quotient group of G is a homomorphic image of G. Now we will see that, conversely, every homomorphic image of G is a quotient group of G. More exactly, *every homomorphic image of G is isomorphic to a quotient group of G.*

It will follow that, for any groups G and H, H is a homomorphic image of G iff H is (or is isomorphic to) a quotient group of G. Therefore, the notions of *homomorphic image* and of *quotient group* are interchangeable.

The thread of our reasoning begins with a simple theorem.

Theorem 1 *Let $f : G \to H$ be a homomorphism with kernel K. Then*

$$f(a) = f(b) \qquad iff \qquad Ka = Kb$$

(In other words, any two elements a and b in G have the *same image* under f iff they are in the same coset of K.)

Indeed,

$$f(a) = f(b) \qquad \text{iff} \qquad f(a)[f(b)]^{-1} = e$$
$$\text{iff} \qquad f(ab^{-1}) = e$$
$$\text{iff} \qquad ab^{-1} \in K$$
$$\text{iff} \qquad Ka = Kb \qquad \text{(by Chapter 15, Theorem 5a)}$$

What does this theorem really tell us? It says that if f is a homomorphism from G to H with kernel K, then all the elements in any fixed coset of K have the same image, and, conversely, elements which have the same image are in the same coset of K.

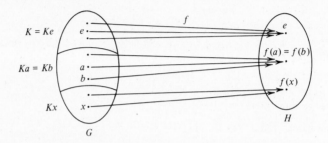

It is therefore clear, already, that there is a one-to-one correspondence matching cosets of K with elements in H. It remains only to show that this correspondence is an isomorphism. But first, how exactly does this correspondence match up specific cosets of K with specific elements of H? Clearly, for each x, the coset Kx is matched with the element $f(x)$. Once this is understood, the next theorem is easy.

Theorem 2 *Let* $f: G \to H$ *be a homomorphism of G* **onto** *H. If K is the kernel of* f, *then*

$$H \cong G/K$$

To show that G/K is isomorphic to H, we must look for an isomorphism from G/K to H. We have just seen that there is a function from G/K to H which matches each coset Kx with the element $f(x)$; call this function ϕ. Thus, ϕ is defined by the identity

$$\phi(Kx) = f(x)$$

This definition does not make it obvious that $\phi(Kx)$ is *uniquely defined*. (If it is not, then we cannot properly call ϕ a function.) We must make sure that if Ka is the same coset as Kb, then $\phi(Ka)$ is the same as $\phi(Kb)$: that is,

$$\text{if} \quad Ka = Kb \quad \text{then} \quad f(a) = f(b)$$

As a matter of fact, this is true by Theorem 1.

Now, let us show that ϕ is an isomorphism:

ϕ *is injective :* If $\phi(Ka) = \phi(Kb)$, then $f(a) = f(b)$, so by Theorem 1, $Ka = Kb$.

ϕ *is surjective,* because every element of H is of the form $f(x) = \phi(Kx)$.

Finally, $\phi(Ka \cdot Kb) = \phi(Kab) = f(ab) = f(a)f(b) = \phi(Ka)\phi(Kb)$.

Thus, ϕ is an isomorphism from G/K onto H.

Theorem 2 is often called the *fundamental homomorphism theorem.* It asserts that every homomorphic image of G is isomorphic to a quotient group of G. Which specific quotient group of G? Well, if f is a homomorphism from G onto H, then H is isomorphic to the quotient group of G by *the kernel of f.*

The fact that f is a homomorphism from G *onto* H may be symbolized by writing

$$f : G \longrightarrow\!\!\!\!\!\rightarrow H$$

Furthermore, the fact that K is the kernel of this homomorphism may be indicated by writing

$$f : G \xrightarrow[K]{} \!\!\!\!\rightarrow H$$

Thus, in capsule form, the fundamental homomorphism theorem says that

$$\text{If} \quad f : G \xrightarrow[K]{} \!\!\!\!\rightarrow H \quad \text{then} \quad H \cong G/K$$

Let us see a few examples:

We saw in the opening paragraph of Chapter 14 that

$$f = \begin{pmatrix} 0 & 1 & 2 & 3 & 4 & 5 \\ 0 & 1 & 2 & 0 & 1 & 2 \end{pmatrix}$$

is a homomorphism from $\mathbb{Z}_6$ onto $\mathbb{Z}_3$. Visibly, the kernel of f is $\{0, 3\}$, which is the subgroup of $\mathbb{Z}_6$ generated by 3, that is, the subgroup $\langle 3 \rangle$. This situation may be symbolized by writing

$$f : \mathbb{Z}_6 \xrightarrow[\langle 3 \rangle]{} \!\!\!\!\rightarrow \mathbb{Z}_3$$

We conclude by Theorem 2 that

$$\mathbb{Z}_3 \cong \mathbb{Z}_6/\langle 3 \rangle$$

For another kind of example, let G and H be any groups and consider their direct product $G \times H$. Remember that $G \times H$ consists of all the ordered pairs (x, y) as x ranges over G and y ranges over H. You multiply ordered pairs by multiplying corresponding components; that is, the operation on $G \times H$ is given by

$$(a, b) \cdot (c, d) = (ac, bd)$$

Now, let f be the function from $G \times H$ onto H given by

$$f(x, y) = y$$

It is easy to check that f is a homomorphism. Furthermore, (x, y) is in the kernel of f iff $f(x, y) = y = e$. This means that the kernel of f consists of all the ordered pairs whose second component is e. Call this kernel G^*; then

$$G^* = \{(x, e) : x \in G\}$$

We symbolize all this by writing

$$f : G \times H \xrightarrow[G^*]{} H$$

By the fundamental homomorphism theorem, we deduce that $H \cong (G \times H)/G^*$. [It is easy to see that G^* is an isomorphic copy of G; thus, identifying G^* with G, we have shown that, roughly speaking, $(G \times H)/G \cong H$.]

Other uses of the fundamental homomorphism theorem are given in the exercises.

EXERCISES

In the exercises which follow, FHT will be used as an abbreviation for fundamental homomorphism theorem.

A. Examples of the FHT Applied to Finite Groups

In each of the following, use the fundamental homomorphism theorem to prove that the two given groups are isomorphic. Then display their tables.

Example $\mathbb{Z}_2$ *and* $\mathbb{Z}_6/\langle 2 \rangle$.

$$f = \begin{pmatrix} 0 & 1 & 2 & 3 & 4 & 5 \\ 0 & 1 & 0 & 1 & 0 & 1 \end{pmatrix}$$

is a homomorphism from $\mathbb{Z}_6$ onto $\mathbb{Z}_2$. (Do not prove that f is a homomorphism.)

The kernel of f is $\{0, 2, 4\} = \langle 2 \rangle$. Thus,

$$\mathbb{Z}_6 \xrightarrow[\langle 2 \rangle]{} \mathbb{Z}_2$$

It follows by the FHT that $\mathbb{Z}_2 \cong \mathbb{Z}_6/\langle 2 \rangle$.

1. $\mathbb{Z}_5$ and $\mathbb{Z}_{20}/\langle 5 \rangle$.
2. $\mathbb{Z}_3$ and $\mathbb{Z}_6/\langle 3 \rangle$.
3. $\mathbb{Z}_2$ and $S_3/\{\varepsilon, \beta, \delta\}$.
4. P_2 and P_3/K, where $K = \{\phi, \{3\}\}$. [HINT: Consider the function $f(C) = C \cap \{1, 2\}$. P_3 is the group of subsets of $\{1, 2, 3\}$, and P_2 of $\{1, 2\}$.]
5. $\mathbb{Z}_3$ and $(\mathbb{Z}_3 \times \mathbb{Z}_3)/K$, where $K = \{(0, 0), (1, 1), (2, 2)\}$. [HINT: Consider the function $f(a, b) = a - b$ from $\mathbb{Z}_3 \times \mathbb{Z}_3$ to $\mathbb{Z}_3$.]

B. Example of the FHT Applied to $\mathscr{F}(\mathbb{R})$

Let $\alpha: \mathscr{F}(\mathbb{R}) \to \mathbb{R}$ be defined by $\alpha(f) = f(1)$ and let $\beta: \mathscr{F}(\mathbb{R}) \to \mathbb{R}$ be defined by $\beta(f) = f(2)$.

1. Prove that α and β are homomorphisms from $\mathscr{F}(\mathbb{R})$ *onto* $\mathbb{R}$.
2 Let J be the set of all the functions from $\mathbb{R}$ to $\mathbb{R}$ whose graph passes through the point $(1, 0)$ and let K be the set of all the functions whose graph passes through $(2, 0)$. Use the FHT to prove that $\mathbb{R} \cong \mathscr{F}(\mathbb{R})/J$ and $\mathbb{R} \cong \mathscr{F}(\mathbb{R})/K$.
3 Conclude that $\mathscr{F}(\mathbb{R})/J \cong \mathscr{F}(\mathbb{R})/K$.

C. Example of the FHT Applied to Abelian Groups

Let G be an abelian group. Let $H = \{x^2 : x \in G\}$ and $K = \{x \in G : x^2 = e\}$.

1 Prove that $f(x) = x^2$ is a homomorphism of G *onto* H.
2 Find the kernel of f.
3 Use the FHT to conclude that $H \cong G/K$.

† D. Group of Inner Automorphisms of a Group G

Let G be a group. By an *automorphism* of G we mean an isomorphism $f : G \to G$.

1 The symbol Aut(G) is used to designate the set of all the automorphisms of G. Prove that the set Aut (G), with the operation $\circ$ of composition, is a group *by proving that Aut(G) is a subgroup of S_G.*

2 By an *inner automorphism* of G we mean any function ϕ_a of the following form:

$$\text{for every } x \in G \qquad \phi_a(x) = axa^{-1}$$

Prove that every inner automorphism of G is an automorphism of G.

3 Prove that, for arbitrary $a, b \in G$,

$$\phi_a \circ \phi_b = \phi_{ab} \quad \text{and} \quad (\phi_a)^{-1} = \phi_{a^{-1}}$$

4 Let $I(G)$ designate the set of all the inner automorphisms of G. That is, $I(G) = \{\phi_a : a \in G\}$. Use part 3 to prove that $I(G)$ is a subgroup of $\text{Aut}(G)$. Conclude that $I(G)$ is a group.

5 By the *center* of G we mean the set of all those elements of G which commute with every element of G, that is, the set C defined by

$$C = \{a \in G : ax = xa \text{ for every } x \in G\}$$

Prove that $a \in C$ if and only if $axa^{-1} = x$ for every $x \in G$.

6 Let $h: G \to I(G)$ be the function defined by $h(a) = \phi_a$. Prove that h is a homomorphism from G onto $I(G)$ and that C is its kernel.

7 Use the FHT to conclude that $I(G)$ is isomorphic with G/C.

† **E. The FHT Applied to Direct Products of Groups**

Let G and H be groups. Suppose J is a normal subgroup of G and K is a normal subgroup of H.

1 Show that the function $f(x, y) = (Jx, Ky)$ is a homomorphism from $G \times H$ *onto* $(G/J) \times (H/K)$.

2 Find the kernel of f.

3 Use the FHT to conclude that $(G \times H)/(J \times K) \cong (G/J) \times (H/K)$.

Let G, M, and N be groups, let $f : G \to M$ be a homomorphism from G onto M, and let $h: G \to N$ be a homomorphism from G onto N.

4 Show that $\phi(x) = (f(x), h(x))$ is a homomorphism from G *onto* $M \times N$.

5 If $H = \ker f$ and $K = \ker h$, prove that $\ker \phi = H \cap K$.

6 Use the FHT to conclude that $G/(H \cap K) \cong M \times N$.

† **F. The First Isomorphism Theorem**

Let G be a group; let H and K be subgroups of G, with H a normal subgroup of G. *Prove the following:*

1 $H \cap K$ is a normal subgroup of K.

2 If $HK = \{xy : x \in H$ and $y \in K\}$, then HK is a subgroup of G.

3 H is a normal subgroup of HK.

4 Every member of the quotient group HK/H may be written in the form Hk for some $k \in K$.

5 The function $f(k) = Hk$ is a homomorphism from K onto HK/H, and its kernel is $H \cap K$.

6 By the FHT, $H/(H \cap K) \cong HK/H$. (*This is referred to as the first isomorphism theorem.*)

† G. A Sharper Cayley Theorem

If H is a subgroup of a group G, let X designate the set of all the cosets of H in G. For each element $a \in G$, define $\rho_a : X \to X$ as follows:

$$\rho_a(Hx) = H(xa)$$

1 Prove that each ρ_a is a permutation of X.

2 Prove that $h: G \to S_X$ defined by $h(a) = \rho_a$ is a homomorphism.

3 Prove that the set $\{a \in H : xax^{-1} \in H$ for every $x \in G\}$, that is, the set of all the elements of H whose conjugates are all in H, is the kernel of h.

4 Prove that if H contains no normal subgroup of G except $\{e\}$, then G is isomorphic to a subgroup of S_X.

† H. Quotient Groups Isomorphic to the Circle Group

Every complex number $a + bi$ may be represented as a point in the complex plane.

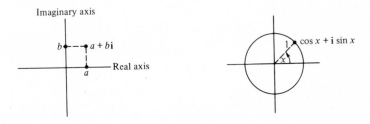

The *unit circle* in the complex plane consists of all the complex numbers whose distance from the origin is 1; thus, clearly, the unit circle consists of all the complex numbers which can be written in the form

$$\cos x + i \sin x$$

for some real number x.

1 For each $x \in \mathbb{R}$, it is conventional to write cis $x = \cos x + i \sin x$. Prove that cis $(x + y) = (\text{cis } x)(\text{cis } y)$.

2 Let T designate the set $\{\text{cis } x : x \in \mathbb{R}\}$, that is, the set of all the complex numbers lying on the unit circle, with the operation of multiplication. Use part 1 to prove that T is a group. (T is called the *circle group*.)

3 Prove that $f(x) = \text{cis } x$ is a homomorphism from $\mathbb{R}$ onto T.

4 Prove that ker $f = \{2n\pi : n \in \mathbb{Z}\} = \langle 2\pi \rangle$.

5 Use the FHT to conclude that $T \cong \mathbb{R}/\langle 2\pi \rangle$.

6 Prove that $g(x) = \text{cis } 2\pi x$ is a homomorphism from $\mathbb{R}$ onto T, with kernel $\mathbb{Z}$.

7 Conclude that $T \cong \mathbb{R}/\mathbb{Z}$.

† I. The Second Isomorphism Theorem

Let H and K be normal subgroups of a group G, with $H \subseteq K$. Define $\phi : G/H \rightarrow G/K$ by: $\phi(Ha) = Ka$. *Prove the following :*

1 ϕ is a well-defined function. [That is, if $Ha = Hb$, then $\phi(Ha) = \phi(Hb)$.]

2 ϕ is a homomorphism.

3 ϕ is surjective.

4 ker $\phi = K/H$.

5 Conclude (using the FHT) that $(G/H)/(K/H) \cong G/K$.

† J. The Correspondence Theorem

Let f be a homomorphism from G onto H with kernel K:

$$G \xrightarrow[K]{} H$$

If S is any subgroup of H, let $S^* = \{x \in G : f(x) \in S\}$. *Prove :*

1 S^* is a subgroup of G.

2 $K \subseteq S^*$.

3 Let g be the restriction of f to S^*. [That is, $g(x) = f(x)$ for every $x \in S^*$, and S^* is the domain of g.] Then g is a homomorphism from S^* *onto* S, and $K = $ ker g.

4 $S \cong S^*/K$.

† K. Cauchy's Theorem

Prerequisites : Chapter 13, Exercise I, and Chapter 15, Exercises G and H.

If G is a group and p is any prime divisor of $|G|$, it will be shown here that G has at least one element of order p. This has already been shown for abelian groups in Chapter 15, Exercise H4. Thus, assume here that G is not abelian. The argument will proceed by induction; thus, let $|G| = k$, and assume our claim is true for any group of order less than k. Let **C** be the center of G, let C_a be the centralizer of a for each $a \in G$, and let $k = c + k_s + \cdots + k_t$ be the class equation of G, as in Chapter 15, Exercise G2. *Prove*:

1 If p is a factor of $| C_a |$ for any $a \in G$, where $a \notin \mathbf{C}$, we are done. (Explain why.)

2 For any $a \notin \mathbf{C}$ in G, if p is not a factor of $| C_a |$, then p is a factor of $(G : C_a)$.

3 Solving the equation $k = c + k_s + \cdots + k_t$ for c, explain why p is a factor of c. We are now done. (Explain why.)

† L. Subgroups of p-Groups (Prelude to Sylow)

Prerequisites: Exercise J; Chapter 15, Exercises G and H.

Let p be a prime number. A *p-group* is any group whose order is a power of p. It will be shown here that if $| G | = p^k$ then G has a normal subgroup of order p^m for every m between 1 and k. The proof is by induction on $| G |$; we therefore assume our result is true for all p-groups smaller than G. *Prove the following*:

1 There is an element a in the center of G such that $\operatorname{ord}(a) = p$. (See Chapter 15, Exercises G and H.)

2 $\langle a \rangle$ is a normal subgroup of G.

3 Explain why it may be assumed that $G/\langle a \rangle$ has a normal subgroup of order p^{m-1}

4 Use Exercise $J(4)$ to prove that G has a normal subgroup of order p^m.

SUPPLEMENTARY PROBLEMS

Exercise Sets M through Q are included as a challenge for the ambitious reader. Two important results of group theory are proved in these exercises: one is called Sylow's theorem, the other is called the Basis theorem of finite abelian groups.

† M. p-Sylow Subgroups

Prerequisites: Exercises J and K, Exercise I1, page 142, and Exercise D3, page 150.

Let p be a prime number. A finite group G is called a *p-group* if the order of every element x in G is a power p. (The orders of different elements may be different powers of p.) If H is a subgroup of any finite group G, and H is a p-group, we call H a *p-subgroup* of G. Finally, if K is a p-subgroup of G, and K is maximal (in the sense that K is not contained in any larger p-subgroup of G), then K is called a *p-Sylow subgroup* of G. *Prove the following*:

1 The order of any p-group is a power of p. (HINT: Use Exercise K.)

2 Every conjugate of a p-Sylow subgroup of G is a p-Sylow subgroup of G.

Let K be a p-Sylow subgroup of G, and $N = N(K)$ the normalizer of K.

3 Let $a \in N$, and suppose the order of Ka in N/K is a power of p. Let $S = \langle Ka \rangle$ be the cyclic subgroup of N/K generated by Ka. Prove that N has a subgroup S^* such that S^*/K is a p-group. (HINT: See Exercise J4.)

4 Prove that S^* is a p-subgroup of G (use Exercise D3, page 150). Then explain why $S^* = K$, and why it follows that $Ka = K$.

5 Use parts 3 and 4 to prove: no element of N/K has order a power of p (except, trivially, the identity element).

6 If $a \in N$ and the order of a is a power of p, then the order of Ka (in N/K) is also a power of p. (*Why?*) Thus, $Ka = K$. (*Why?*)

7 Use part 6 to prove: if $aKa^{-1} = K$ and the order of a is a power of p, then $a \in K$.

† **N. Sylow's Theorem.**

Prerequisites: Exercises K *and* M, *and Exercise* I10, *page* 142.

Let G be a finite group, and K a p-Sylow subgroup of G. Let X be the set of all the conjugates of K. See Exercise M2. If $C_1, C_2 \in X$, let $C_1 \sim C_2$ iff $C_1 = aC_2a^{-1}$ for some $a \in K$.

1 Prove that $\sim$ is an equivalence relation on X.

Thus, $\sim$ partitions X into equivalence classes. If $C \in X$, let the equivalence class of C be denoted by $[C]$.

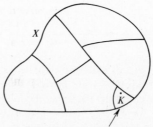

K is the only member of its class

2 For each $C \in X$, prove that the number of elements in $[C]$ is a divisor of $|K|$. (HINT: Use Exercise I10, page 142.) Conclude that for each $C \in X$, the number of elements in $[C]$ is either 1 or a power of p.

3 Use Exercise M7 to prove that the only class with a single element is $[K]$.

4 Use parts 2 and 3 to prove that the number of elements in X is $kp + 1$, for some integer k.

5 Use part 4 to prove that $(G : N)$ is *not* a multiple of p.

6 Prove that $(N : K)$ is *not* a multiple of p. (Use Exercises K and M5.)

7 Use parts 5 and 6 to prove that $(G : K)$ is *not* a multiple of p.

8 Conclude: Let G be a finite group of order $p^k m$, where p is not a factor of m. Every p-Sylow subgroup K of G has order p^k.

Combining part 8 with Exercise L gives:

Let G be a finite group and let p be a prime number. For each n such that p^n divides $|G|$, G has a subgroup of order p^n.

This is known as Sylow's Theorem.

† O. Lifting Elements from Cosets

The purpose of this exercise is to prove a property of cosets which is needed in Exercise Q. Let G be a finite abelian group, and let a be an element of G such that $\mathrm{ord}(a)$ is a multiple of $\mathrm{ord}(x)$ for every $x \in G$. Let $H = \langle a \rangle$. We will prove:

For every $x \in G$, there is some $y \in G$ such that $Hx = Hy$ and $\mathrm{ord}(y) = \mathrm{ord}(Hy)$.

This means that every coset of H contains an element y whose order is the same as the coset's order.

Let x be any element in G, and let $\mathrm{ord}(a) = t$, $\mathrm{ord}(x) = s$, and $\mathrm{ord}(Hx) = r$.

1 Explain why r is the least positive integer such that x^r equals some power of a, say $x^r = a^m$.

2 Deduce from our hypotheses that r divides s, and s divides t.

Thus, we may write $s = ru$ and $t = sv$, so in particular, $t = ruv$.

3 Explain why $a^{muu} = e$, and why it follows that $mu = tz$ for some integer z. Then explain why $m = rvz$.

4 Setting $y = a^{-vz}$, prove that $Hx = Hy$ and $\mathrm{ord}(y) = r$, as required.

† P. Decomposition of a Finite Abelian Group into p-Groups

Let G be an abelian group of order $p^k m$, where p^k and m are relatively prime (that is, p^k and m have no common factors except ± 1). [REMARK: If two integers j and k are relatively prime, then there are integers s and t such that $sj + tk = 1$. This is proved on page 218.]

Let G_{p^k} be the subgroup of G consisting of all elements whose order divides p^k. Let G_m be the subgroup of G consisting of all elements whose order divides m. *Prove:*

1 For any $x \in G$ and integers s and t, $x^{sp^k} \in G_m$ and $x^{tm} \in G_{p^k}$.

2 For every $x \in G$, there are $y \in G_{p^k}$ and $z \in G_m$ such that $x = yz$.

3 $G_{p^k} \cap G_m = \{e\}$.

4 $G \cong G_{p^k} \times G_m$. (See Exercise H, Chapter 14.)

5 Suppose $|G|$ has the following factorization into primes: $|G| = p_1^{k_1}p_2^{k_2} \cdots p_n^{k_n}$. Then $G \cong G_1 \times G_2 \times \cdots \times G_n$ where for each $i = 1, \ldots, n$, G_i is a p_i-group.

Q. Basis Theorem for Finite Abelian Groups

Prerequisite: Exercise P.

As a provisional definition, let us call a finite abelian group "decomposable" if there are elements $a_1, \ldots, a_n \in G$ such that:

(D1)   For every $x \in G$, there are integers $k_1, \ldots, k_n$ such that $x = a_1^{k_1}a_2^{k_2} \cdots a_n^{k_n}$.

(D2)   If there are integers $l_1, \ldots, l_n$ such that $a_1^{l_1}a_2^{l_2} \cdots a_n^{l_n} = e$ then $a_1^{l_1} = a_2^{l_2} = \cdots = a_n^{l_n} = e$.

If (D1) and (D2) hold, we will write $G = [a_1, a_2, \ldots, a_n]$.

1 Let G' be the set of all products $a_2^{l_2} \cdots a_n^{l_n}$, as $l_2, \ldots, l_n$ range over $\mathbb{Z}$. Prove that G' is a subgroup of G, and $G' = [a_2, \ldots, a_n]$.

2 Prove: $G \cong \langle a_1 \rangle \times G'$. Conclude that $G \cong \langle a_1 \rangle \times \langle a_2 \rangle \times \cdots \times \langle a_n \rangle$.

In the remaining exercises of this set, let p be a prime number, and assume G is a finite abelian group such that the order of every element in G is some power of p. Let $a \in G$ be an element whose order is the highest possible in G. We will argue by induction to prove that G is "decomposable." Let $H = \langle a \rangle$.

3 Explain why we may assume that $G/H = [Hb_1, \ldots, Hb_n]$ for some $b_1, \ldots, b_n \in G$.

By Exercise O, we may assume that for each $i = 1, \ldots, n$, $\text{ord}(b_i) = \text{ord}(Hb_i)$. We will show that $G = [a, b_1, \ldots, b_n]$.

4 Prove that for every $x \in G$, there are integers $k_0, k_1, \ldots, k_n$ such that

$$x = a^{k_0}b_1^{k_1} \cdots b_n^{k_n}$$

5 Prove that if $a^{l_0}b_1^{l_1} \cdots b_n^{l_n} = e$, then $a^{l_0} = b_1^{l_1} = \cdots = b_n^{l_n} = e$. Conclude that $G = [a, b_1, \ldots, b_n]$.

6 Use Exercise P5, together with parts 2 and 5 above, to prove: Every finite abelian group G is a direct product of cyclic groups of prime order. (*This is called the Basis Theorem of finite Abelian groups.*)

It can be proved that the above decomposition of a finite Abelian group into cyclic p-groups is unique, except for the order of the factors. We leave it to the ambitious reader to supply the proof of uniqueness.

SEVENTEEN

RINGS: DEFINITIONS AND ELEMENTARY PROPERTIES

In presenting scientific knowledge it is elegant as well as enlightening to begin with the simple and move toward the more complex. If we build upon a knowledge of the simplest things, it is easier to understand the more complex ones. In the first part of this book we dedicated ourselves to the study of groups—surely one of the simplest and most fundamental of all algebraic systems. We will now move on, and, using the knowledge and insights gained in the study of groups, we will begin to examine algebraic systems which have *two* operations instead of just one.

The most basic of the two-operational systems is called a *ring*; it will be defined in a moment. The surprising fact about rings is that, despite their having *two* operations and being more complex than groups, their fundamental properties follow exactly the pattern already laid out for groups. With remarkable, almost compelling ease, we will find two-operational analogs of the notions of subgroup and quotient group, homomorphism and isomorphism—as well as other algebraic notions—and we will discover that rings behave just like groups with respect to these notions.

The two operations of a ring are traditionally called *addition* and *multiplication*, and are denoted as usual by $+$ and $\cdot$, respectively. We must remember, however, that the elements of a ring are not necessarily numbers (for example, there are rings of functions, rings of switching circuits, and so on); and therefore "addition" does not necessarily refer to the conventional addition of numbers, nor does multiplication necessarily refer to the conventional operation of multiplying numbers. In fact, $+$ and $\cdot$ are nothing more than symbols denoting the two operations of a ring.

> *By a ring we mean a set A with operations called addition and multiplication which satisfy the following axioms :*
> (1) *A with addition alone is an abelian group.*
> (2) *Multiplication is associative.*
> (3) *Multiplication is distributive over addition.* That is, *for all a, b, and c in A,*
>
> $$a(b + c) = ab + ac$$
> *and*
> $$(b + c)a = ba + ca$$

Since A with addition alone is an abelian group, there is in A a neutral element for addition: it is called the *zero* element and is written 0. Also, every element has an additive inverse called its *negative*; the negative of a is denoted by $-a$. Subtraction is defined by

$$a - b = a + (-b)$$

The easiest examples of rings are the traditional number systems. The set $\mathbb{Z}$ of the integers, with conventional addition and multiplication, is a ring called the *ring of the integers*. We designate this ring simply with the letter $\mathbb{Z}$. (The context will make it clear whether we are referring to the *ring* of the integers or the additive *group* of the integers.)

Similarly, $\mathbb{Q}$ is the ring of the rational numbers, $\mathbb{R}$ the ring of the real numbers, and $\mathbb{C}$ the ring of the complex numbers. In each case, the operations are conventional addition and multiplication.

Remember that $\mathscr{F}(\mathbb{R})$ represents the set of all the functions from $\mathbb{R}$ to $\mathbb{R}$; that is, the set of all real-valued functions of a real variable. In calculus we learned to add and multiply functions: if f and g are any two functions from $\mathbb{R}$ to $\mathbb{R}$, their sum $f + g$ and their *product* fg are defined as follows:

$$[f + g](x) = f(x) + g(x) \qquad \text{for every real number } x$$
and
$$[fg](x) = f(x)g(x) \qquad \text{for every real number } x$$

$\mathscr{F}(\mathbb{R})$ with these operations for adding and multiplying functions is a ring called the *ring of real functions*. It is written simply as $\mathscr{F}(\mathbb{R})$. On page 47 we saw that $\mathscr{F}(\mathbb{R})$ with only addition of functions is an abelian group. It is left as an exercise for you to verify that multiplication of functions is associative and distributive over addition of functions.

The rings $\mathbb{Z}$, $\mathbb{Q}$, $\mathbb{R}$, $\mathbb{C}$, and $\mathscr{F}(\mathbb{R})$ are all *infinite rings*, that is, rings with infinitely many elements. There are also *finite rings*: rings with a finite number of elements. As an important example, consider the group $\mathbb{Z}_n$, and define an operation of multiplication on $\mathbb{Z}_n$ by allowing the product ab to be the remainder of the usual product of integers a and b after division by n. (For example, in $\mathbb{Z}_5$, $2 \cdot 4 = 3$, $3 \cdot 3 = 4$, and $4 \cdot 3 = 2$.) This operation is called *multiplication modulo n*. $\mathbb{Z}_n$ with addition and multiplication modulo n is a ring: the details are given in Chapter 19.

Let A be any ring. Since A with addition alone is an abelian group, everything we know about abelian groups applies to it. However, it is important to remember that A with addition is an abelian group *in additive notation* and, therefore, before applying theorems about groups to A, these theorems must be translated into additive notation. For example, Theorems 1, 2, and 3 of Chapter 4 read as follows when the notation is additive and the group is Abelian:

$$a + b = a + c \quad \text{implies} \quad b = c \qquad (*)$$

$$a + b = 0 \quad \text{implies} \quad a = -b \quad \text{and} \quad b = -a \qquad (**)$$

$$-(a + b) = (-a) + (-b) \quad \text{and} \quad -(-a) = a \qquad (***)$$

Therefore (*), (**), and (***) are true in every ring.

What happens in a ring when we multiply elements by zero? What happens when we multiply elements by the *negatives* of other elements? The next theorem answers these questions.

Theorem 1 *Let a and b be any elements of a ring A.*
(i) $a0 = 0$ *and* $0a = 0$
(ii) $a(-b) = -(ab)$ *and* $(-a)b = -(ab)$
(iii) $(-a)(-b) = ab$

Part (i) asserts that multiplication by zero always yields zero, and parts (ii) and (iii) state the familiar rules of signs.

To prove (i) we note that

$$
\begin{aligned}
aa + 0 &= aa \\
&= a(a + 0) \quad \text{because } a = a + 0 \\
&= aa + a0 \quad \text{by the distributive law}
\end{aligned}
$$

Thus, $aa + 0 = aa + a0$. By (*) we may eliminate the term aa on both sides of this equation, and therefore $0 = a0$.

To prove (ii), we have

$$a(-b) + ab = a[(-b) + b] \qquad \text{by the distributive law}$$
$$= a0$$
$$= 0 \qquad\qquad \text{by part (i)}$$

Thus, $a(-b) + ab = 0$. By (**) we deduce that $a(-b) = -(ab)$. The twin formula $(-a)b = -(ab)$ is deduced analogously.

We prove part (iii) by using part (ii) twice:

$$(-a)(-b) = -[a(-b)] = -[-(ab)] = ab$$

The general definition of a ring is sparse and simple. However, particular rings may also have "optional features" which make them more versatile and interesting. Some of these options are described next.

By definition, addition is commutative in every ring but multiplication is not. When multiplication *also* is commutative in a ring, we call that ring a *commutative ring*.

A ring A does not necessarily have a neutral element for multiplication. If there *is* in A a neutral element for multiplication, it is called the *unity* of A, and is denoted by the symbol 1. Thus, $a \cdot 1 = a$ and $1 \cdot a = a$ for every a in A. If A has a unity, we call A a *ring with unity*. The rings $\mathbb{Z}$, $\mathbb{Q}$, $\mathbb{R}$, $\mathbb{C}$, and $\mathscr{F}(\mathbb{R})$ are all examples of commutative rings with unity.

Incidentally, a ring whose only element is 0 is called a *trivial ring*; a ring with more than one element is *nontrivial*. In a nontrivial ring with unity, necessarily $1 \neq 0$. This is true because if $1 = 0$ and x is any element of the ring, then

$$x = x1 = x0 = 0$$

In other words, if $1 = 0$ then every element of the ring is equal to 0, hence 0 is the only element of the ring.

If A is a ring with unity, there may be elements in A which *have a multiplicative inverse*. Such elements are said to be *invertible*. Thus, an element a is invertible in a ring if there is some x in the ring such that

$$ax = xa = 1$$

For example, in $\mathbb{R}$ every nonzero element is invertible: its multiplicative inverse is its reciprocal. On the other hand, in $\mathbb{Z}$ the only invertible elements are 1 and -1.

Zero is never an invertible element of a ring except if the ring is trivial; for if zero had a multiplicative inverse x, we would have $0x = 1$, that is, $0 = 1$.

If A is a *commutative ring with unity in which every nonzero element is invertible, A is called a field.* Fields are of the utmost importance in mathematics; for example, $\mathbb{Q}$, $\mathbb{R}$, and $\mathbb{C}$ are fields. There are also *finite* fields, such as $\mathbb{Z}_5$ (it is easy to check that every nonzero element of $\mathbb{Z}_5$ is invertible). Finite fields have beautiful properties and fascinating applications, which will be examined later in this book.

In elementary mathematics we learned the commandment that if the product of two numbers is equal to zero, say

$$ab = 0$$

then one of the two factors, either a or b (or both) must be equal to zero. This is certainly true if the numbers are real (or even complex) numbers, but the rule is *not* inviolable in every ring. For example, in $\mathbb{Z}_6$,

$$2 \cdot 3 = 0$$

even though the factors 2 and 3 are both nonzero. Such numbers, when they exist, are called *divisors of zero*.

> In any ring, a nonzero element a *is called a* **divisor of zero** *if there is a nonzero element* b *in the ring such that the product* ab *or* ba *is equal to zero.*

(Note carefully that *both* factors have to be nonzero.) Thus, 2 and 3 are divisors of zero in $\mathbb{Z}_6$; 4 is also a divisor of zero in $\mathbb{Z}_6$, because $4 \cdot 3 = 0$. For another example, let $\mathcal{M}_2(\mathbb{R})$ designate the set of all 2×2 matrices of real numbers, with addition and multiplication of matrices as described on page 8. The simple task of checking that $\mathcal{M}_2(\mathbb{R})$ satisfies the ring axioms is assigned as Exercise C at the end of this chapter. $\mathcal{M}_2(\mathbb{R})$ is rampant with examples of divisors of zero. For instance

$$\begin{pmatrix} 0 & 1 \\ 0 & 1 \end{pmatrix}\begin{pmatrix} 1 & 1 \\ 0 & 0 \end{pmatrix} = \begin{pmatrix} 0 & 0 \\ 0 & 0 \end{pmatrix}$$

hence

$$\begin{pmatrix} 0 & 1 \\ 0 & 1 \end{pmatrix} \quad \text{and} \quad \begin{pmatrix} 1 & 1 \\ 0 & 0 \end{pmatrix}$$

are both divisors of zero in $\mathcal{M}_2(\mathbb{R})$.

Of course, there are rings which have no divisors of zero at all! For example, $\mathbb{Z}$, $\mathbb{Q}$, $\mathbb{R}$, and $\mathbb{C}$ do not have any divisors of zero. It is important to note carefully what it means for a ring to have *no divisors of zero*: it means that *if the product of two elements in the ring is equal to zero, at least one of the factors is zero.* (Our commandment from elementary mathematics!)

It is also decreed in elementary algebra that a nonzero number a may be canceled in the equation $ax = ay$ to yield $x = y$. While undeniably true in the number systems of mathematics, this rule is not true in every ring. For example, in $\mathbb{Z}_6$,

$$2 \cdot 5 = 2 \cdot 2$$

yet we cannot cancel the common factor 2. A similar example involving 2×2 matrices may be seen on page 9. When cancelation *is* possible, we say the ring has the "cancelation property."

*A ring is said to have the **cancelation property** if*

$$ab = ac \quad or \quad ba = ca \quad implies \quad b = c$$

for any elements a, b, and c in the ring if $a \neq 0$.

There is a surprising and unexpected connection between the cancelation property and divisors of zero:

Theorem 2 *A ring has the cancelation property iff it has no divisors of zero.*

The proof is very straightforward. Let A be a ring, and suppose first that A has the cancelation property. To prove that A has no divisors of zero we begin by letting $ab = 0$, and show that a or b is equal to 0. If $a = 0$, we are done. Otherwise, we have

$$ab = 0 = a0$$

so by the cancelation property (canceling a), $b = 0$.

Conversely, assume A has no divisors of zero. To prove that A has the cancelation property, suppose $ab = ac$ where $a \neq 0$. Then

$$ab - ac = a(b - c) = 0$$

Remember, there are no divisors of zero! Since $a \neq 0$, necessarily $b - c = 0$, so $b = c$.

An *integral domain* is defined to be a commutative ring with unity having the cancelation property. By Theorem 2, an integral domain may also be defined as a commutative ring with unity having no divisors of zero. It is easy to see that every field is an integral domain. The converse, however, is not true: for example, $\mathbb{Z}$ is an integral domain but not a field. We will have a lot to say about integral domains in the following chapters.

EXERCISES

through p.176

A. Examples of Rings

In each of the following, a set A with operations of addition and multiplication is given. *Prove that A satisfies all the axioms to be a commutative ring with unity. Indicate the zero element, the unity, and the negative of an arbitrary a.*

1 A is the set $\mathbb{Z}$ of the integers, with the following "addition" $\oplus$ and "multiplication" $\odot$:

$$a \oplus b = a + b - 1 \qquad a \odot b = ab - (a + b) + 2$$

2 A is the set $\mathbb{Q}$ of the rational numbers, and the operations are $\oplus$ and $\odot$ defined as follows:

$$a \oplus b = a + b + 1 \qquad a \odot b = ab + a + b$$

3 A is the set $\mathbb{Q} \times \mathbb{Q}$ of ordered pairs of rational numbers, and the operations are the following addition $\oplus$ and multiplication $\odot$:

$$(a, b) \oplus (c, d) = (a + c, b + d)$$

$$(a, b) \odot (c, d) = (ac - bd, ad + bc)$$

4 $A = \{x + y\sqrt{2} : x, y \in \mathbb{Z}\}$ with conventional addition and multiplication.
5 Prove that the ring in part 1 is an integral domain.
6 Prove that the ring in part 2 is a field, and indicate the multiplicative inverse of an arbitrary nonzero element.
7 Do the same for the ring in part 3.

B. Ring of Real Functions

1 Verify that $\mathscr{F}(\mathbb{R})$ satisfies all the axioms for being a commutative ring with unity. Indicate the zero and unity, and describe the negative of any f.
2 Describe the divisors of zero in $\mathscr{F}(\mathbb{R})$.
3 Describe the invertible elements in $\mathscr{F}(\mathbb{R})$.
4 Explain why $\mathscr{F}(\mathbb{R})$ is neither a field nor an integral domain.

C. Ring of 2 × 2 Matrices

Let $\mathscr{M}_2(\mathbb{R})$ designate the set of all 2×2 matrices

$$\begin{pmatrix} a & b \\ c & d \end{pmatrix}$$

whose entries are real numbers a, b, c, and d, with the following addition and multiplication:

$$\begin{pmatrix} a & b \\ c & d \end{pmatrix} + \begin{pmatrix} r & s \\ t & u \end{pmatrix} = \begin{pmatrix} a+r & b+s \\ c+t & d+u \end{pmatrix}$$

and

$$\begin{pmatrix} a & b \\ c & d \end{pmatrix} \begin{pmatrix} r & s \\ t & u \end{pmatrix} = \begin{pmatrix} ar+bt & as+bu \\ cr+dt & cs+du \end{pmatrix}$$

1 Verify that $\mathscr{M}_2(\mathbb{R})$ satisfies the ring axioms.
2 Show that $\mathscr{M}_2(\mathbb{R})$ is commutative and has a unity.
3 Explain why $\mathscr{M}_2(\mathbb{R})$ is not an integral domain or a field.

D. Rings of Subsets of a Set

If D is a set, then the power set of D is the set P_D of all the subsets of D. Addition and multiplication are defined as follows: If A and B are elements of P_D (that is, subsets of D), then

$$A + B = (A - B) \cup (B - A) \qquad \text{and} \qquad AB = A \cap B$$

It was shown in Chapter 3, Exercise C, that P_D with addition alone is an abelian group. *Now prove the following:*

1 P_D is a commutative ring with unity. (You may assume $\cap$ is associative; for the distributive law, use the same diagram and approach as was used to prove that addition is associative in Chapter 3, Exercise C.)
2 Describe the divisors of zero in P_D.
3 Describe the invertible elements in P_D.
4 Explain why P_D is neither a field nor an integral domain.
5 Give the tables of P_3, that is, P_D where $D = \{1, 2, 3\}$.

E. Ring of Quaternions

A *quaternion* (in matrix form) is a 2×2 matrix of complex numbers of the form

$$\alpha = \begin{pmatrix} a+bi & c+di \\ -c+di & a-bi \end{pmatrix}$$

1 Prove that the set of all the quaternions, with the matrix addition and multiplication explained on page 8, is a ring with unity. This ring is denoted by the symbol $\mathscr{Q}$. Find an example to show that $\mathscr{Q}$ is not commutative. (You may assume matrix addition and multiplication are associative and obey the distributive law.)

2 Let

$$\mathbf{1} = \begin{pmatrix} 1 & 0 \\ 0 & 1 \end{pmatrix} \qquad \mathbf{i} = \begin{pmatrix} i & 0 \\ 0 & -i \end{pmatrix} \qquad \mathbf{j} = \begin{pmatrix} 0 & 1 \\ -1 & 0 \end{pmatrix} \qquad \mathbf{k} = \begin{pmatrix} 0 & i \\ i & 0 \end{pmatrix}$$

Show that the quaternion α, defined previously, may be written in the form

$$\alpha = a\mathbf{1} + b\mathbf{i} + c\mathbf{j} + d\mathbf{k}$$

(This is the standard notation for quaternions.)

3 Prove the following formulas:

$$\mathbf{i}^2 = \mathbf{j}^2 = \mathbf{k}^2 = -1 \qquad \mathbf{ij} = -\mathbf{ji} = \mathbf{k} \qquad \mathbf{jk} = -\mathbf{kj} = \mathbf{i} \qquad \mathbf{ki} = -\mathbf{ik} = \mathbf{j}$$

4 The *conjugate* of α is

$$\bar{\alpha} = \begin{pmatrix} a - bi & -c - di \\ c - di & a + bi \end{pmatrix}$$

The *norm* of α is $a^2 + b^2 + c^2 + d^2$, and is written $\|\alpha\|$. Show directly (by matrix multiplication) that

$$\bar{\alpha}\alpha = \alpha\bar{\alpha} = \begin{pmatrix} t & 0 \\ 0 & t \end{pmatrix} \qquad \text{where } t = \|\alpha\|$$

Conclude that the multiplicative inverse of α is $(1/t)\bar{\alpha}$.

5 A *skew field* is a (not necessarily commutative) ring with unity in which every nonzero element has a multiplicative inverse. Conclude from parts 1 and 4 that $\mathcal{2}$ is a skew field.

F. Ring of Endomorphisms

Let G be an abelian group in additive notation. An *endomorphism* of G is a homomorphism from G to G. Let End(G) denote the set of all the endomorphisms of G, and define addition and multiplication of endomorphisms as follows:

$$[f + g](x) = f(x) + g(x) \qquad \text{for every } x \text{ in } G$$

$$[fg] = f \circ g \qquad \text{the composite of } f \text{ and } g$$

1 Prove that End(G) with these operations is a ring with unity.

2 List the elements of End($\mathbb{Z}_4$), then give the addition and multiplication tables for End($\mathbb{Z}_4$).

REMARK: The endomorphisms of $\mathbb{Z}_4$ are easy to find. Any endomorphisms of $\mathbb{Z}_4$ will carry 1 to either 0, 1, 2, or 3. For example, take the last case: if

$$1 \xrightarrow{f} 3$$

then necessarily

$$1 + 1 \xrightarrow{f} 3 + 3 = 2 \qquad 1 + 1 + 1 \xrightarrow{f} 3 + 3 + 3 = 1 \qquad \text{and} \qquad 0 \xrightarrow{f} 0$$

hence f is completely determined by the fact that

$$1 \xrightarrow{f} 3$$

G. Direct Product of Rings

If A and B are rings, their *direct product* is a new ring, denoted by $A \times B$, and defined as follows: $A \times B$ consists of all the ordered pairs (x, y) where x is in A and y is in B. Addition in $A \times B$ consists of adding corresponding components:

$$(x_1, y_1) + (x_2, y_2) = (x_1 + x_2, y_1 + y_2)$$

Multiplication in $A \times B$ consists of multiplying corresponding components:

$$(x_1, y_1) \cdot (x_2, y_2) = (x_1 x_2, y_1 y_2)$$

1 If A and B are rings, verify that $A \times B$ is a ring.

2 If A and B are commutative, show that $A \times B$ is commutative. If A and B each has a unity, show that $A \times B$ has a unity.

3 Describe carefully the divisors of zero in $A \times B$.

4 Describe the invertible elements in $A \times B$.

5 Explain why $A \times B$ can never be an integral domain or a field. (Assume $A \times B$ has more than one element.)

H. Elementary Properties of Rings

Prove each of the following:

1 In any ring, $a(b - c) = ab - ac$ and $(b - c)a = ba - ca$.

2 In any ring, if $ab = -ba$, then $(a + b)^2 = (a - b)^2 = a^2 + b^2$.

3 In any integral domain, if $a^2 = b^2$, then $a = \pm b$.

4 In any integral domain, only 1 and -1 are their own multiplicative inverses. (Note that $x = x^{-1}$ iff $x^2 = 1$.)

5 Show that the commutative law for addition need not be assumed in defining a ring with unity: it may be proved from the other axioms. [HINT: Use the distributive law to expand $(a + b)(1 + 1)$ in two different ways.]

6 Let A be any ring. Prove that if the additive group of A is cyclic, then A is a commutative ring.

7 In any integral domain, if $a^n = 0$ for some integer n, then $a = 0$.

I. Properties of Invertible Elements

Prove that each of the following is true in a nontrivial ring with unity.

1 If a is invertible and $ab = ac$, then $b = c$.

2 An element a can have no more than *one* multiplicative inverse.

3 If $a^2 = 0$ then $a + 1$ and $a - 1$ are invertible.

4 If a and b are invertible, their product ab is invertible.

5 The set S of all the invertible elements in a ring is a multiplicative group.

6 By part 5, the set of all the nonzero elements in a field is a multiplicative group. Now use Lagrange's theorem to prove that in a finite field with m elements, $x^{m-1} = 1$ for every $x \neq 0$.

7 If $ax = 1$, x is a *right inverse* of a; if $ya = 1$, y is a *left inverse* of a. Prove that if a has a right inverse x *and* a left inverse y, then a is invertible, and its inverse is equal to x and to y. (First show that $yaxa = 1$.)

8 In a commutative ring, if ab is invertible, then a and b are both invertible.

J. Properties of Divisors of Zero

Prove that each of the following is true in a nontrivial ring.

1 If $a \neq \pm 1$ and $a^2 = 1$, then $a + 1$ and $a - 1$ are divisors of zero.

2 If ab is a divisor of zero, then a or b is a divisor of zero.

3 In a commutative ring with unity, a divisor of zero cannot be invertible.

4 Suppose $ab \neq 0$ in a commutative ring. If either a or b is a divisor of zero, so is ab.

5 Suppose a is neither 0 nor a divisor of zero. If $ab = ac$, then $b = c$.

6 $A \times B$ always has divisors of zero.

K. Boolean Rings

A ring A is a boolean ring if $a^2 = a$ for every $a \in A$. *Prove that each of the following is true in an arbitrary boolean ring A.*

1 For every $a \in A$, $a = -a$. [HINT: Expand $(a + a)^2$.]

2 Use part 1 to prove that A is a commutative ring. [HINT: Expand $(a + b)^2$.]

In parts 3 and 4, assume A has a unity.

3 Every element except 0 and 1 is a divisor of zero. [Consider $x(x - 1)$.]

4 1 is the only invertible element in A.

5 Letting $a \vee b = a + b + ab$, we have the following in A:

$$a \vee bc = (a \vee b)(a \vee c) \qquad a \vee (1 + a) = 1 \qquad a \vee a = a \qquad a(a \vee b) = a$$

L. The Binomial Formula

An important formula in elementary algebra is the binomial expansion formula for an expression $(a + b)^n$. The formula is as follows:

$$(a + b)^n = \sum_{k=0}^{n} \binom{n}{k} a^{n-k} b^k$$

where the binomial coefficient

$$\binom{n}{k} = \frac{n(n-1)(n-2)\cdots(n-k+1)}{k!}$$

This theorem is true in every commutative ring. (If k is any positive integer and a is an element of a ring, ka refers to the sum $a + a + \cdots + a$ with k terms, as in elementary algebra.) The proof of the binomial theorem in a commutative ring is no different from the proof in elementary algebra. We shall review it here.

The proof of the binomial formula is by induction on the exponent n. The formula is trivially true for $n = 1$. In the induction step, we *assume* the expansion for $(a + b)^n$ is as above, and we must prove that

$$(a + b)^{n+1} = \sum_{k=0}^{n+1} \binom{n+1}{k} a^{n+1-k} b^k$$

Now,

$$(a + b)^{n+1} = (a + b)(a + b)^n$$

$$= (a + b) \sum_{k=0}^{n} \binom{n}{k} a^{n-k} b^k$$

$$= \sum_{k=0}^{n} \binom{n}{k} a^{n+1-k} b^k + \sum_{k=0}^{n} \binom{n}{k} a^{n-k} b^{k+1}$$

Collecting terms, we find that the coefficient of $a^{n+1-k} b^k$ is

$$\binom{n}{k} + \binom{n}{k-1}$$

By direct computation, show that

$$\binom{n}{k} + \binom{n}{k-1} = \binom{n+1}{k}$$

It will follow that $(a + b)^{n+1}$ is as claimed, and the proof is complete.

M. Nilpotent and Unipotent Elements

An element a of a ring is *nilpotent* if $a^n = 0$ for some positive integer n.

1 In a ring with unity, prove that if a is nilpotent, then $a + 1$ and $a - 1$ are both invertible. [HINT: Use the factorization

$$1 - a^n = (1 - a)(1 + a + a^2 + \cdots + a^{n-1})$$

for $1 - a$, and a similar formula for $1 + a$.]

2 In a commutative ring, prove that any product xa of a nilpotent element a by any element x is nilpotent.

3 In a commutative ring, prove that the sum of two nilpotent elements is nilpotent. (HINT: You must use the binomial formula; see Exercise L.)

An element a of a ring is *unipotent* iff $1 - a$ is nilpotent.

4 In a commutative ring, prove that the product of two unipotent elements a and b is unipotent. [HINT: Use the binomial formula to expand $1 - ab = (1 + a) + a(1 - b)$ to power $n + m$.]

5 In a commutative ring, prove that every unipotent element is invertible. (HINT: Use the binomial expansion formula.)

EIGHTEEN

IDEALS AND HOMOMORPHISMS

We have already seen several examples of smaller rings contained within larger rings. For example, $\mathbb{Z}$ is a ring inside the larger ring $\mathbb{Q}$, and $\mathbb{Q}$ itself is a ring inside the larger ring $\mathbb{R}$. When a ring B is part of a larger ring A, we call B a *subring* of A. The notion of subring is the precise analog for rings of the notion of subgroup for groups. Here are the relevant definitions:

Let A be a ring, and B a nonempty subset of A. If the sum of any two elements of B is again in B, then B is *closed with respect to addition*. If the negative of every element of B is in B, then B is *closed with respect to negatives*. Finally, if the product of any two elements of B is again in B, then B is *closed with respect to multiplication*. B is called a *subring* of A if B is closed with respect to addition, multiplication, and negatives. Why is B then called a subring of A? Quite elementary:

If a nonempty subset $B \subseteq A$ is closed with respect to addition, multiplication, and negatives, then B with the operations of A is a ring.

This fact is easy to check: If a, b, and c are any three elements of B, then a, b, and c are also elements of A because $B \subseteq A$. But A is a ring, so

$$a + (b + c) = (a + b) + c$$

$$a(bc) = (ab)c$$

$$a(b + c) = ab + ac$$

and
$$(b + c)a = ba + ca$$

Thus, in B addition and multiplication are associative and the distributive law is satisfied. Now, B was assumed to be nonempty, so there is an element $b \in B$; but B is closed with respect to negatives, so $-b$ is also in B. Finally, B is closed with respect to addition, hence $b + (-b) \in B$, that is, 0 is in B. Thus, B satisfies all the requirements for being a ring.

For example, $\mathbb{Q}$ is a subring of $\mathbb{R}$ because the sum of two rational numbers is rational, the product of two rational numbers is rational, and the negative of every rational number is rational.

By the way, if B is a nonempty subset of A, there is a more compact way of checking that B is a subring of A:

> B is a subring of A if and only if B is closed with respect to subtraction and multiplication.

The reason is that *B is closed with respect to subtraction iff B is closed with respect to both addition and negatives.* This last fact is easy to check, and is given as an exercise.

Awhile back, in our study of groups, we singled out certain special subgroups called *normal subgroups*. We will now describe certain special subrings called *ideals* which are the counterpart of normal subgroups: that is, ideals are in rings as normal subgroups are in groups.

Let A be a ring, and B a nonempty subset of A. We will say that B *absorbs products in* A (or, simply, B *absorbs products*) if, whenever we multiply an element in B by an element in A (regardless of whether the latter is inside B or outside B), their product is always in B. In other words,

$$\text{for all } b \in B \text{ and } x \in A, \ xb \text{ and } bx \text{ are in } B.$$

> *A nonempty subset B of a ring A is called an ideal of A if B is closed with respect to addition and negatives, and B absorbs products in A.*

A simple example of an ideal is the set $\mathbb{E}$ of the even integers. $\mathbb{E}$ is an ideal of $\mathbb{Z}$ because the sum of two even integers is even, the negative of any even integer is even, and, finally, the product of an even integer by *any* integer is always even.

In a commutative ring with unity, the simplest example of an ideal is the set of all the multiples of a fixed element a by all the elements in the ring; in other words, the set of all the products

$$xa$$

as a remains fixed and x ranges over all the elements of the ring. This set is obviously an ideal because

$$xa + ya = (x + y)a$$

$$-(xa) = (-x)a$$

and
$$y(xa) = (yx)a$$

This ideal is called the *principal ideal generated by a*, and is donoted by

$$\langle a \rangle$$

As in the case of subrings, if B is a nonempty subset of A, there is a more compact way of checking that B is an ideal of A:

B is an ideal of A if and only if B is closed with respect to subtraction and B absorbs products in A.

We shall see presently that ideals play an important role in connection with homomorphisms.

Homomorphisms are almost the same for rings as for groups.

*A **homomorphism** from a ring A to a ring B is a function $f: A \rightarrow B$ satisfying the identities*

$$f(x_1 + x_2) = f(x_1) + f(x_2)$$

and

$$f(x_1 x_2) = f(x_1)f(x_2)$$

There is a longer but more informative way of writing these two identities:

(1) *If $f(x_1) = y_1$ and $f(x_2) = y_2$, then $f(x_1 + x_2) = y_1 + y_2$*
(2) *If $f(x_1) = y_1$ and $f(x_2) = y_2$, then $f(x_1 x_2) = y_1 y_2$.*

In other words, if f happens to carry x_1 to y_1 and x_2 to y_2, then, necessarily, it must carry $x_1 + x_2$ to $y_1 + y_2$ and $x_1 x_2$ to $y_1 y_2$. Symbolically,

If $x_1 \xrightarrow{f} y_1$ and $x_2 \xrightarrow{f} y_2$, then necessarily

$$x_1 + x_2 \xrightarrow{f} y_1 + y_2 \quad and \quad x_1 x_2 \xrightarrow{f} y_1 y_2$$

One can easily confirm for oneself that a function f with this property will transform the addition and multiplication tables of its domain into the addition and multiplication tables of its range. (We may imagine infinite rings to have "nonterminating" tables.) Thus, a homomorphism from a ring A *onto* a ring B is a function which transforms A into B.

For example, the ring $\mathbb{Z}_6$ is transformed into the ring $\mathbb{Z}_3$ by

$$f = \begin{pmatrix} 0 & 1 & 2 & 3 & 4 & 5 \\ 0 & 1 & 2 & 0 & 1 & 2 \end{pmatrix}$$

as we may verify by comparing their tables. The addition tables are compared on page 132, and we may do the same with their multiplication tables:

·	0	1	2	3	4	5
0	0	0	0	0	0	0
1	0	1	2	3	4	5
2	0	2	4	0	2	4
3	0	3	0	3	0	3
4	0	4	2	0	1	2
5	0	5	4	3	2	1

Replace
x by $f(x)$

$\longrightarrow$

·	0	1	2	0	1	2
0	0	0	0	0	0	0
1	0	1	2	0	1	2
2	0	2	1	0	2	1
0	0	0	0	0	0	0
1	0	1	2	0	1	2
2	0	2	1	0	2	1

Eliminate duplicate
information

$\longrightarrow$

(For example, $2 \cdot 2 = 1$
appears four separate
times in this table.)

·	0	1	2
0	0	0	0
1	0	1	2
2	0	2	1

If there is a homomorphism from A *onto* B, we call B a *homomorphic image* of A. If f is a homomorphism from a ring A to a ring B, not necessarily *onto*, the range of f is a subring of B. (This fact is routine to verify.) Thus, the range of a ring homomorphism is always a ring. And obviously, the range of a homomorphism is always a homomorphic image of its domain.

Intuitively, if B is a homomorphic image of A, this means that certain features of A are faithfully preserved in B while others are deliberately lost. This may be illustrated by developing further an example described in Chapter 14. The *parity ring* P consists of two elements, e and o, with addition and multiplication given by the tables

+	e	o
e	e	o
o	o	e

and

·	e	o
e	e	e
o	e	o

We should think of e as "even" and o as "odd," and the tables as describing the rules for adding and multiplying odd and even integers. For example, even $+$ odd $=$ odd, even *times* odd $=$ even, and so on.

The function $f: \mathbb{Z} \to P$ which carries every even integer to e and every odd integer to o is easily seen to be a homomorphism from $\mathbb{Z}$ to P; this is made clear on page 134. Thus, P is a homomorphic image of $\mathbb{Z}$. Although the ring P is very much smaller that the ring $\mathbb{Z}$, and therefore few of the features of $\mathbb{Z}$ can be expected to reappear in P, nevertheless *one* aspect of the structure of $\mathbb{Z}$ is retained absolutely intact in P, namely the structure of odd and even numbers. As we pass from $\mathbb{Z}$ to P, the *parity* of the integers (their being even or odd), with its arithmetic, is faithfully preserved while all else is lost. Other examples will be given in the exercises.

If f is a homomorphism from a ring A to a ring B, the *kernel* of f is the set of all the elements of A which are carried by f onto the zero element of B. In symbols, the kernel of f is the set

$$K = \{x \in A : f(x) = 0\}$$

It is a very important fact that the *kernel of f is an ideal of A.* (The simple verification of this fact is left as an exercise.)

If A and B are rings, an *isomorphism* from A to B is a homomorphism which is a one-to-one correspondence from A to B. In other words, it is an injective and surjective homomorphism. If there *is* an isomorphism from A to B we say that A is *isomorphic to B*, and this fact is expressed by writing

$$A \cong B$$

EXERCISES

A. Examples of Subrings

Prove that each of the following is a subring of the indicated ring.

1 $\{x + \sqrt{3}y : x, y \in \mathbb{Z}\}$ is a subring of $\mathbb{R}$.
2 $\{x + 2^{1/3}y + 2^{2/3}z : x, y, z \in \mathbb{Z}\}$ is a subring of $\mathbb{R}$.
3 $\{x2^y : x, y \in \mathbb{Z}\}$ is a subring of $\mathbb{R}$.
4 Let $\mathscr{C}(\mathbb{R})$ be the set of all the functions from $\mathbb{R}$ to $\mathbb{R}$ which are continuous on $(-\infty, \infty)$, and let $\mathscr{D}(\mathbb{R})$ be the set of all the functions from $\mathbb{R}$ to $\mathbb{R}$ which are differentiable on $(-\infty, \infty)$. Then $\mathscr{C}(\mathbb{R})$ and $\mathscr{D}(\mathbb{R})$ are subrings of $\mathscr{F}(\mathbb{R})$.
5 Let $\mathscr{U}(\mathbb{R})$ be the set of all functions from $\mathbb{R}$ to $\mathbb{R}$ which are continuous on the interval $[0, 1]$. Then $\mathscr{U}(\mathbb{R})$ is a subring of $\mathscr{F}(\mathbb{R})$, and $\mathscr{C}(\mathbb{R})$ is a subring of $\mathscr{U}(\mathbb{R})$.
6 The subset of $\mathscr{M}_2(\mathbb{R})$ consisting of all matrices of the form

$$\begin{pmatrix} 0 & 0 \\ 0 & x \end{pmatrix}$$

is a subring of $\mathscr{M}_2(\mathbb{R})$.

B. Examples of Ideals

1 Identify which of the following are ideals of $\mathbb{Z} \times \mathbb{Z}$, and explain: $\{(n, n) : n \in \mathbb{Z}\}$; $\{(5n, 0) : n \in \mathbb{Z}\}$; $\{(n, m) : n + m \text{ is even}\}$; $\{(n, m) : nm \text{ is even}\}$; $\{(2n, 3m) : n, m \in \mathbb{Z}\}$.

2 List all the ideals of $\mathbb{Z}_{12}$.

3 Explain why every subring of $\mathbb{Z}_n$ is necessarily an ideal.

4 Explain why the subring of Exercise A6 is not an ideal.

5 Explain why $\mathscr{C}(\mathbb{R})$ is not an ideal of $\mathscr{F}(\mathbb{R})$.

6 Prove that each of the following is an ideal of $\mathscr{F}(\mathbb{R})$:

(a) The set of all f which are constant on the interval $[0, 1]$.
(b) The set of all f such that $f(x) = 0$ for every rational x.
(c) The set of all f such that $f(0) = 0$.

7 List all the ideals of P_3. (P_3 is defined in Chapter 17, Exercise D.)

8 Give an example of a subring of P_3 which is not an ideal.

9 Give an example of a subring of $\mathbb{Z}_3 \times \mathbb{Z}_3$ which is not an ideal.

C. Elementary Properties of Subrings

Prove each of the following:

1 A nonempty subset B of a ring A is closed with respect to addition and negatives iff B is closed with respect to subtraction.

2 Conclude from part 1 that B is a subring of A iff B is closed with respect to subtraction and multiplication.

3 If A is a finite ring and B is a subring of A, then the order of B is a divisor of the order of A.

4 If a subring B of an integral domain A contains 1, then B is an integral domain. (B is then called a *subdomain* of A.)

5 Every subring of a field is an integral domain.

6 If a subring B of a field F is closed with respect to multiplicative inverses, then B is a field. (B is then called a *subfield* of F.)

7 Find subrings of $\mathbb{Z}_8$ which illustrate each of the following:

(i) A is a ring with unity, B is a subring of A, but B is not a ring with unity.
(ii) A and B are rings with unity, B is a subring of A, but the unity of B is not the same as the unity of A.

8 Let A be a ring, $f : A \to A$ a homomorphism, and $B = \{x \in A : f(x) = x\}$. Then B is a subring of A.

9 The *center* of a ring A is the set of all the elements $a \in A$ such that $ax = xa$ for every $x \in A$. Prove that the center of A is a subring of A.

D. Elementary Properties of Ideals

Let A be a ring and J a nonempty subset of A.

1 Using Exercise C1, explain why J is an ideal of A iff J is closed with respect to subtraction and J absorbs products in A.

2 If A is a ring with unity, prove that J is an ideal of A iff J is closed with respect to addition and J absorbs products in A.

3 Prove that the intersection of any two ideals of A is an ideal of A.

4 Prove that if J is an ideal of A and $1 \in J$, then $J = A$.

5 Prove that if J is an ideal of A and J contains an invertible element a of A, then $J = A$.

6 Explain why a field F can have no nontrivial ideals (that is, no ideals except $\{0\}$ and F).

E. Examples of Homomorphisms

Prove that each of the following is a homomorphism. Then describe its kernel and its range.

1 $\phi: \mathscr{F}(\mathbb{R}) \to \mathbb{R}$ given by $\phi(f) = f(0)$.

2 $h: \mathbb{R} \times \mathbb{R} \to \mathbb{R}$ given by $h(x, y) = x$.

3 $h: \mathbb{R} \to \mathscr{M}_2(\mathbb{R})$ given by

$$h(x) = \begin{pmatrix} x & 0 \\ 0 & 0 \end{pmatrix}$$

4 $h: \mathbb{R} \times \mathbb{R} \to \mathscr{M}_2(\mathbb{R})$ given by

$$h(x, y) = \begin{pmatrix} x & 0 \\ 0 & y \end{pmatrix}$$

5 Let A be the set $\mathbb{R} \times \mathbb{R}$ with the usual addition and the following "multiplication":

$$(a, b) \odot (c, d) = (ac, bc)$$

Granting that A is a ring, let $f: A \to \mathscr{M}_2(\mathbb{R})$ be given by

$$f(x, y) = \begin{pmatrix} x & 0 \\ y & 0 \end{pmatrix}$$

6 $h: P_C \to P_C$ given by $h(A) = A \cap D$, where D is a fixed subset of C.

7 List all the homomorphisms from $\mathbb{Z}_2$ to $\mathbb{Z}_4$; from $\mathbb{Z}_3$ to $\mathbb{Z}_6$.

F. Elementary Properties of Homomorphisms

Let A and B be rings, and $f: A \to B$ a homomorphism. *Prove each of the following.*

1 $f(A) = \{f(x) : x \in A\}$ is a subring of B.

2 The kernel of f is an ideal of A.

3 $f(0) = 0$, and for every $a \in A, f(-a) = -f(a)$.

4 f is injective iff its kernel is equal to $\{0\}$.

5 If B is an integral domain, then either $f(1) = 1$ or $f(1) = 0$. If $f(1) = 0$, then $f(x) = 0$ for every $x \in A$. If $f(1) = 1$, the image of every invertible element of A is an invertible element of B.

6 Any homomorphic image of a commutative ring is a commutative ring. Any homomorphic image of a field is a field.

7 If the domain A of the homomorphism f is a field, and if the range of f has more than one element, then f is injective. (HINT: Use Exercise D6.)

G. Examples of Isomorphisms   to 193

1 Let A be the ring of Exercise A2 in Chapter 17. Show that the function $f(x) = x - 1$ is an isomorphism from $\mathbb{Q}$ to A, hence $\mathbb{Q} \cong A$.

2 Let $\mathscr{S}$ be the following subset of $\mathscr{M}_2(\mathbb{R})$:

$$\mathscr{S} = \left\{ \begin{pmatrix} a & b \\ -b & a \end{pmatrix} : a, b \in \mathbb{R} \right\}$$

Prove that the function

$$f(a + b\mathbf{i}) = \begin{pmatrix} a & b \\ -b & a \end{pmatrix}$$

is an isomorphism from $\mathbb{C}$ to $\mathscr{S}$. [REMARK: You must begin by checking that f is a well-defined function; that is, if $a + b\mathbf{i} = c + d\mathbf{i}$, then $f(a + b\mathbf{i}) = f(c + d\mathbf{i})$. To do this, note that if $a + b\mathbf{i} = c + d\mathbf{i}$ then $a - c = (d - b)\mathbf{i}$; this last equation is impossible unless both sides are equal to zero, for otherwise it would assert that a given real number is equal to an imaginary number.]

3 Prove that $\{(x, x) : x \in \mathbb{Z}\}$ is a subring of $\mathbb{Z} \times \mathbb{Z}$, and show $\{(x, x) : x \in \mathbb{Z}\} \cong \mathbb{Z}$.

4 Show that the set of all 2×2 matrices of the form

$$\begin{pmatrix} 0 & 0 \\ 0 & x \end{pmatrix}$$

is a subring of $\mathscr{M}_2(\mathbb{R})$, then prove this subring is isomorphic to $\mathbb{R}$.

For any integer k, let $k\mathbb{Z}$ designate the subring of $\mathbb{Z}$ which consists of all the multiples of k.

5 Prove that $\mathbb{Z} \not\cong 2\mathbb{Z}$; then prove that $2\mathbb{Z} \not\cong 3\mathbb{Z}$. Finally, explain why if $k \neq l$, then $k\mathbb{Z} \not\cong l\mathbb{Z}$.

H. Further Properties of Ideals

Let A be a ring, and let J and K be ideals of A. Prove each of the following. (In parts 3 and 4, assume A is a commutative ring.)

1 If $J \cap K = \{0\}$, then $jk = 0$ for every $j \in J$ and $k \in K$.

2 For any $a \in A, I_a = \{ax + j + k : x \in A, j \in J, k \in K\}$ is an ideal of A.

3 The *radical* of J is the set rad $J = \{a \in A : a^n \in J$ for some $n \in \mathbb{Z}\}$. For any ideal J, rad J is an ideal of A.

4 For any $a \in A$, $\{x \in A : ax = 0\}$ is an ideal (called the *annihilator* of a). Furthermore, $\{x \in A : ax = 0$ for every $a \in A\}$ is an ideal (called the *annihilating ideal* of A). If A is a ring with unity, its annihilating ideal is equal to $\{0\}$.

5 Show that $\{0\}$ and A are ideals of A. (They are *trivial* ideals; every other ideal of A is a *proper* ideal.) A proper ideal J of A is called *maximal* if it is not strictly contained in any strictly larger proper ideal: that is, if $J \subseteq K$, where K is an ideal containing some element not in J, then necessarily $K = A$. Show that the following is an example of a maximal ideal: In $\mathscr{F}(\mathbb{R})$, the ideal $J = \{f : f(0) = 0\}$. [HINT: Use D5. Note that if $g \in K$ and $g(0) \neq 0$ (that is, $g \notin J$), then the function $h(x) = g(x) - g(0)$ is in J, hence $h(x) - g(x) \in K$. Explain why this last function is an invertible element of $\mathscr{F}(\mathbb{R})$.]

I. Further Properties of Homomorphisms

Let A and B be rings. *Prove each of the following*:

1 If $f : A \to B$ is a homomorphism from A onto B with kernel K, and J is an ideal of A such that $K \subseteq J$, then $f(J)$ is an ideal of B.

2 If $f : A \to B$ is a homomorphism from A onto B, and B is a *field*, then the kernel of f is a maximal ideal. (HINT: Use part 1, with D6. Maximal ideals are defined in Exercise H5.)

3 There are no nontrivial homomorphisms from $\mathbb{Z}$ to $\mathbb{Z}$. [The trivial homomorphisms are $f(x) = 0$ and $f(x) = x$.]

4. If n is a multiple of m, then $\mathbb{Z}_m$ is a homomorphic image of $\mathbb{Z}_n$.

5 If n is odd, there is an injective homomorphism from $\mathbb{Z}_2$ into $\mathbb{Z}_{2n}$.

† J. A Ring of Endomorphisms

Let A be a commutative ring. *Prove each of the following*:

1 For each element a in A, the function π_a defined by $\pi_a(x) = ax$ satisfies the identity $\pi_a(x + y) = \pi_a(x) + \pi_a(y)$. (In other words, π_a is an endomorphism of the additive group of A.)

2 π_a is injective iff a is not a divisor of zero.

3 π_a is surjective iff a is invertible.

4 Let $\mathscr{A}$ denote the set $\{\pi_a : a \in A\}$ with the two operations

$$[\pi_a + \pi_b](x) = \pi_a(x) + \pi_b(x) \qquad \text{and} \qquad \pi_a \pi_b = \pi_a \circ \pi_b$$

Verify that $\mathscr{A}$ is a ring.

5 If $\phi : A \to \mathscr{A}$ is given by $\phi(a) = \pi_a$, then ϕ is a homomorphism.

6 If A has a unity, then ϕ is an isomorphism. Similarly, if A has no divisors of zero then ϕ is an isomorphism.

NINETEEN
QUOTIENT RINGS

We continue our journey into the elementary theory of rings, traveling a road which runs parallel to the familiar landscape of groups. In our study of groups we discovered a way of actually *constructing* all the homomorphic images of any group G. We constructed quotient groups of G, and showed that every quotient group of G is a homomorphic image of G. We will now imitate this procedure and construct *quotient rings*.

We begin by defining cosets of rings:

Let A be a ring, and J an ideal of A. For any element $a \in A$, the symbol $J + a$ denotes the set of all sums $j + a$, as a remains fixed and j ranges over J. That is,

$$J + a = \{j + a : j \in J\}$$

*$J + a$ is called a **coset** of J in A.*

It is important to note that, if we provisionally ignore multiplication, A with addition alone is an abelian group and J is a subgroup of A. Thus, the cosets we have just defined are (if we ignore multiplication) *precisely the cosets of the subgroup J in the group A, with the notation being additive.* Consequently, everything we already know about group cosets continues to apply in the present case—only, care must be taken to translate known facts about group cosets into *additive notation*. For example, Property (*) of Chapter 13, with Theorem 5 of Chapter 15, reads as follows in additive notation:

$$a \in J + b \qquad iff \qquad J + a = J + b \qquad\qquad (*)$$

$$J + a = J + b \qquad iff \qquad a - b \in J \qquad\qquad (**)$$

$$J + a = J \qquad iff \qquad a \in J \qquad\qquad (***)$$

We also know, by the reasoning which leads up to Lagrange's theorem, that the family of all the cosets $J + a$, as a ranges over A, is a partition of A.

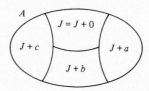

There is a way of *adding* and *multiplying cosets* which works as follows:

$$(J + a) + (J + b) = J + (a + b)$$

$$(J + a)(J + b) = J + ab$$

In other words, the sum of the coset of a and the coset of b is the coset of $a + b$; the product of the coset of a and the coset of b is the coset of ab.

It is important to know that the sum and product of cosets, defined in this fashion, are determined without ambiguity. Remember that $J + a$ may be the same coset as $J + c$ [by (*) this happens iff c is an element of $J + a$], and, likewise, $J + b$ may be the same coset as $J + d$. Therefore, we have the equations

$$
\begin{array}{ccc}
(J + a) + (J + b) = J + (a + b) & & (J + a)(J + b) = J + ab \\
\parallel \qquad\qquad \parallel & \text{and} & \parallel \qquad\qquad \parallel \\
(J + c) + (J + d) = J + (c + d) & & (J + c)(J + d) = J + cd
\end{array}
$$

Obviously we must be absolutely certain that $J + (a + b) = J + (c + d)$ and $J + ab = J + cd$. The next theorem provides us with this important guarantee.

Theorem 1 *Let J be an ideal of A. If $J + a = J + c$ and $J + b = J + d$, then*
(i) $J + (a + b) = J + (c + d)$, *and*
(ii) $J + ab = J + cd$.

We are given that $J + a = J + c$ and $J + b = J + d$, hence by (**),

$$a - c \in J \qquad \text{and} \qquad b - d \in J$$

Since J is closed with respect to addition, $(a - c) + (b - d) = (a + b) -$

$(c + d)$ is in J. It follows by (**) that $J + (a + b) = J + (c + d)$, which proves (i). On the other hand, since J absorbs products in A,

$$\underbrace{(a - c)b}_{ab - cb} \in J \quad \text{and} \quad \underbrace{c(b - d)}_{cb - cd} \in J$$

and therefore $(ab - cb) + (cb - cd) = ab - cd$ is in J. It follows by (**) that $J + ab = J + cd$. This proves (ii).

Now, think of the set which consists of *all the cosets of J* in A. This set is conventionally denoted by the symbol A/J. For example, if $J + a, J + b, J + c, \ldots$ are cosets of J, then

$$A/J = \{J + a, J + b, J + c, \ldots\}$$

We have just seen that coset addition and multiplication are valid operations on this set. In fact,

Theorem 2 *A/J with coset addition and multiplication is a ring.*

Coset addition and multiplication are associative, and multiplication is distributive over addition. (These facts may be routinely checked.) The zero element of A/J is the coset $J = J + 0$, for if $J + a$ is any coset,

$$(J + a) + (J + 0) = J + (a + 0) = J + a$$

Finally, the negative of $J + a$ is $J + (-a)$, because

$$(J + a) + (J + (-a)) = J + (a + (-a)) = J + 0$$

The ring A/J is called the *quotient ring* of A by J.

And now, the crucial connection between quotient rings and homomorphisms:

Theorem 3 *A/J is a homomorphic image of A.*

Following the plan already laid out for groups, the *natural homomorphism* from A onto A/J is the function f which carries every element to its own coset, that is, the function f given by

$$f(x) = J + x$$

This function is very easily seen to be a homomorphism.

Thus, when we construct quotient rings of A, we are, in fact, constructing homomorphic images of A. The quotient ring construction is useful because it is a way of actually manufacturing homomorphic images of any ring A.

The quotient ring construction is now illustrated with an important example. Let $\mathbb{Z}$ be the ring of the integers, and let $\langle 6 \rangle$ be the ideal of $\mathbb{Z}$ which consists of all the multiples of the number 6. The elements of the quotient ring $\mathbb{Z}/\langle 6 \rangle$ are all the cosets of the ideal $\langle 6 \rangle$, namely:

$$\langle 6 \rangle + 0 = \{\ldots, -18, -12, -6, 0, \quad 6, 12, 18, \ldots\} = \bar{0}$$

$$\langle 6 \rangle + 1 = \{\ldots, -17, -11, -5, 1, \quad 7, 13, 19, \ldots\} = \bar{1}$$

$$\langle 6 \rangle + 2 = \{\ldots, -16, -10, -4, 2, \quad 8, 14, 20, \ldots\} = \bar{2}$$

$$\langle 6 \rangle + 3 = \{\ldots, -15, -9, \quad -3, 3, \quad 9, 15, 21, \ldots\} = \bar{3}$$

$$\langle 6 \rangle + 4 = \{\ldots, -14, -8, \quad -2, 4, 10, 16, 22, \ldots\} = \bar{4}$$

$$\langle 6 \rangle + 5 = \{\ldots, -13, -7, \quad -1, 5, 11, 17, 23, \ldots\} = \bar{5}$$

We will represent these cosets by means of the simplified notation $\bar{0}, \bar{1}, \bar{2}, \bar{3}, \bar{4}, \bar{5}$. The rules for adding and multiplying cosets give us the following tables:

$+$	$\bar{0}$	$\bar{1}$	$\bar{2}$	$\bar{3}$	$\bar{4}$	$\bar{5}$
$\bar{0}$	$\bar{0}$	$\bar{1}$	$\bar{2}$	$\bar{3}$	$\bar{4}$	$\bar{5}$
$\bar{1}$	$\bar{1}$	$\bar{2}$	$\bar{3}$	$\bar{4}$	$\bar{5}$	$\bar{0}$
$\bar{2}$	$\bar{2}$	$\bar{3}$	$\bar{4}$	$\bar{5}$	$\bar{0}$	$\bar{1}$
$\bar{3}$	$\bar{3}$	$\bar{4}$	$\bar{5}$	$\bar{0}$	$\bar{1}$	$\bar{2}$
$\bar{4}$	$\bar{4}$	$\bar{5}$	$\bar{0}$	$\bar{1}$	$\bar{2}$	$\bar{3}$
$\bar{5}$	$\bar{5}$	$\bar{0}$	$\bar{1}$	$\bar{2}$	$\bar{3}$	$\bar{4}$

$\cdot$	$\bar{0}$	$\bar{1}$	$\bar{2}$	$\bar{3}$	$\bar{4}$	$\bar{5}$
$\bar{0}$	$\bar{0}$	$\bar{0}$	$\bar{0}$	$\bar{0}$	$\bar{0}$	$\bar{0}$
$\bar{1}$	$\bar{0}$	$\bar{1}$	$\bar{2}$	$\bar{3}$	$\bar{4}$	$\bar{5}$
$\bar{2}$	$\bar{0}$	$\bar{2}$	$\bar{4}$	$\bar{0}$	$\bar{2}$	$\bar{4}$
$\bar{3}$	$\bar{0}$	$\bar{3}$	$\bar{0}$	$\bar{3}$	$\bar{0}$	$\bar{3}$
$\bar{4}$	$\bar{0}$	$\bar{4}$	$\bar{2}$	$\bar{0}$	$\bar{4}$	$\bar{2}$
$\bar{5}$	$\bar{0}$	$\bar{5}$	$\bar{4}$	$\bar{3}$	$\bar{2}$	$\bar{1}$

One cannot fail to notice the analogy between the quotient ring $\mathbb{Z}/\langle 6 \rangle$ and the ring $\mathbb{Z}_6$. In fact, we will regard them as one and the same. More generally, for every positive integer n, we consider $\mathbb{Z}_n$ to be the same as $\mathbb{Z}/\langle n \rangle$. In particular, this makes it clear that $\mathbb{Z}_n$ is a homomorphic image of $\mathbb{Z}$.

By Theorem 3, any quotient ring A/J is a homomorphic image of A. Therefore the quotient ring construction is a way of actually producing homomorphic images of any ring A. In fact, as we will now see, it is a way of producing *all* the homomorphic images of A.

Theorem 4 *Let $f: A \to B$ be a homomorphism from a ring A **onto** a ring B, and let K be the kernel of f. Then $B \cong A/K$.*

To show that A/K is isomorphic with B, we must look for an isomorphism from A/K to B. Mimicking the procedure which worked successfully for groups, we let ϕ be the function from A/K to B which matches each coset $K + x$ with the element $f(x)$; that is,

$$\phi(K + x) = f(x)$$

Remember that if we ignore multiplication for just a moment, A and B are groups and f is a group homomorphism from A onto B, with kernel K. Therefore we may apply Theorem 2 of Chapter 16: *ϕ is a well-defined, bijective function from A/K to B.* Finally,

$$\phi((K + a) + (K + b)) = \phi(K + (a + b)) = f(a + b)$$

$$= f(a) + f(b) = \phi(K + a) + \phi(K + b)$$

and $\quad \phi((K + a)(K + b)) = \phi(K + ab) = f(ab)$

$$= f(a)f(b) = \phi(K + a)\phi(K + b)$$

Thus, ϕ is an isomorphism from A/K onto B.

Theorem 4 is called the *fundamental homomorphism theorem* for rings. Theorems 3 and 4 together assert that every quotient ring of A is a homomorphic image of A, and, conversely, every homomorphic image of A is isomorphic to a quotient ring of A. Thus, for all practical purposes, quotients and homomorphic images of a ring are the same.

As in the case of groups, there are many practical instances in which it is possible to select an ideal J of A so as to "factor out" unwanted traits of A, and obtain a quotient ring A/J with "desirable" features.

As a simple example, let A be a ring, not necessarily commutative, and let J be an ideal of A which contains all the differences

$$ab - ba$$

as a and b range over A. It is quite easy to show that the quotient ring A/J is then commutative. Indeed, to say that A/J is commutative is to say that for any two cosets $J + a$ and $J + b$,

$$(J + a)(J + b) = (J + b)(J + a) \qquad \text{that is} \qquad J + ab = J + ba$$

By (**), this last equation is true iff $ab - ba \in J$. Thus, if every difference $ab - ba$ is in J, then any two cosets commute.

A number of important quotient ring constructions, similar in principle to this one, are given in the exercises.

An ideal J of a commutative ring is said to be a *prime ideal* if for any two elements a and b in the ring,

$$\text{If} \qquad ab \in J \qquad \text{then} \qquad a \in J \qquad \text{or} \qquad b \in J$$

Whenever J is a prime ideal of a commutative ring with unity A, the quotient ring A/J is an integral domain. (The details are left as an exercise.)

An ideal of a ring is called *proper* if it is not equal to the whole ring. A proper ideal J of a ring A is called a *maximal ideal* if there exists no proper ideal K of A such that $J \subseteq K$ with $J \neq K$ (in other words, J is not contained in any strictly larger proper ideal). It is an important fact that if A is a commutative ring with unity, then J *is a maximal ideal of A iff A/J is a field*.

To prove this assertion, let J be a maximal ideal of A. If A is a commutative ring with unity, it is easy to see that A/J is one also. In fact, it should be noted that the unity of A/J is the coset $J + 1$, because if $J + a$ is any coset, $(J + a)(J + 1) = J + a1 = J + a$. Thus, to prove that A/J is a field, it remains only to show that if $J + a$ is any nonzero coset, there is a coset $J + x$ such that $(J + a)(J + x) = J + 1$.

The zero coset is J. Thus, by (***), to say that $J + a$ is *not* zero, is to say that $a \notin J$. Now, let K be the set of all the sums

$$xa + j$$

as x ranges over A and j ranges over J. It is easy to check that K is an ideal. Furthermore, K contains a because $a = 1a + 0$, and K contains every element $j \in J$ because j can be written as $0a + j$. Thus, *K is an ideal which contains J and is strictly larger than J* (for remember that $a \in K$ but $a \notin J$). But J is a *maximal* ideal! Thus, K must be the whole ring A.

It follows that $1 \in K$, so $1 = xa + j$ for some $x \in A$ and $j \in J$. Thus, $1 - xa = j \in J$, so by (**), $J + 1 = J + xa = (J + x)(J + a)$. In the quotient ring A/J, $J + x$ is therefore the multiplicative inverse of $J + a$.

The converse proof consists, essentially, of "unraveling" the preceding argument; it is left as an entertaining exercise.

EXERCISES

A. Examples of Quotient Rings

In each of the following, A is a ring and J is an ideal of A. List the elements of A/J, and then write the addition and multiplication tables of A/J.

Example $A = \mathbb{Z}_6$, $J = \{0, 3\}$.

The elements of A/J are the three cosets $J = J + 0 = \{0, 3\}$, $J + 1 = \{1, 4\}$, and $J + 2 = \{2, 5\}$. The tables for A/J are as follows:

	J	J + 1	J + 2
J	J	J + 1	J + 2
J + 1	J + 1	J + 2	J
J + 2	J + 2	J	J + 1

	J	J + 1	J + 2
J	J	J	J
J + 1	J	J + 1	J + 2
J + 2	J	J + 2	J + 1

1 $A = \mathbb{Z}_{10}$, $J = \{0, 5\}$.
2 $A = P_3$, $J = \{\emptyset, \{1\}\}$. (P_3 is defined in Chapter 17, Exercise D.)
3 $A = \mathbb{Z}_2 \times \mathbb{Z}_6$; $J = \{(0, 0), (0, 2), (0, 4)\}$.

B. Examples of the Use of the FHT

In each of the following, use the FHT (fundamental homomorphism theorem) to prove that the two given rings are isomorphic. Then display their tables.

Example $\mathbb{Z}_2$ *and* $\mathbb{Z}_6/\langle 2 \rangle$.

The following function is a homomorphism from $\mathbb{Z}_6$ onto $\mathbb{Z}_2$:

$$f = \begin{pmatrix} 0 & 1 & 2 & 3 & 4 & 5 \\ 0 & 1 & 0 & 1 & 0 & 1 \end{pmatrix}$$

(Do not prove that f is a homomorphism.)
The kernel of f is $\{0, 2, 4\} = \langle 2 \rangle$. Thus:

$$\mathbb{Z}_6 \xrightarrow[\langle 2 \rangle]{} \mathbb{Z}_2$$

It follows by the FHT that $\mathbb{Z}_2 \cong \mathbb{Z}_6/\langle 2 \rangle$.

1 $\mathbb{Z}_5$ and $\mathbb{Z}_{20}/\langle 5 \rangle$.
2 $\mathbb{Z}_3$ and $\mathbb{Z}_6/\langle 3 \rangle$.
3 P_2 and P_3/K, where $K = \{\emptyset, \{3\}\}$. [HINT: See Chapter 18, Exercise E6. Consider the function $f(X) = X \cap \{1, 2\}$.]
4 $\mathbb{Z}_2$ and $\mathbb{Z}_2 \times \mathbb{Z}_2/K$, where $K = \{(0, 0), (0, 1)\}$.

C. Quotient Rings and Homomorphic Images in $\mathscr{F}(\mathbb{R})$

1 Let ϕ be the function from $\mathscr{F}(\mathbb{R})$ to $\mathbb{R} \times \mathbb{R}$ defined by $\phi(f) = (f(0), f(1))$. Prove that ϕ is a homomorphism from $\mathscr{F}(\mathbb{R})$ *onto* $\mathbb{R} \times \mathbb{R}$, and describe its kernel.
2 Let J be the subset of $\mathscr{F}(\mathbb{R})$ consisting of all f whose graph passes through the points $(0, 0)$ and $(1, 0)$. Referring to part 1, explain why J is an ideal of $\mathscr{F}(\mathbb{R})$, and $\mathscr{F}(\mathbb{R})/J \cong \mathbb{R} \times \mathbb{R}$.
3 Let ϕ be the function from $\mathscr{F}(\mathbb{R})$ to $\mathscr{F}(\mathbb{Q})$ defined as follows:

$$\phi(f) = f_\mathbb{Q} = \text{the restriction of } f \text{ to } \mathbb{Q}$$

(NOTE: The domain of f_Q is Q and on this domain f_Q is the same function as f.) Prove that ϕ is a homomorphism from $\mathscr{F}(\mathbb{R})$ *onto* $\mathscr{F}(Q)$, and describe the kernel of ϕ.

4 Let J be the subset of $\mathscr{F}(\mathbb{R})$ consisting of all f such that $f(x) = 0$ for every rational x. Referring to part 3, explain why J is an ideal of $\mathscr{F}(\mathbb{R})$ and $\mathscr{F}(\mathbb{R})/J \cong \mathscr{F}(Q)$.

D. Elementary Applications of the Fundamental Homomorphism Theorem

In each of the following let A be a commutative ring. If $a \in A$ and n is a positive integer, the notation na will stand for

$$a + a + \cdots + a \qquad (n \text{ terms})$$

1 Suppose $2x = 0$ for every $x \in A$. Prove that $(x + y)^2 = x^2 + y^2$ for all x and y in A. Conclude that the function $h(x) = x^2$ is a homomorphism from A to A. If $J = \{x \in A : x^2 = 0\}$ and $B = \{x^2 : x \in A\}$, explain why J is an ideal of A, B is a subring of A, and $A/J \cong B$.

2 Suppose $6x = 0$ for every $x \in A$. Prove that the function $h(x) = 3x$ is a homomorphism from A to A. If $J = \{x : 3x = 0\}$ and $B = \{3x : x \in A\}$, explain why J is an ideal of A, B is a subring of A, and $A/J \cong B$.

3 If a is an idempotent element of A (that is, $a^2 = a$), prove that the function $\pi_a(x) = ax$ is a homomorphism from A into A. Show that the kernel of π_a is I_a, the annihilator of a (defined in Exercise H4 of Chapter 18). Show that the range of π_a is $\langle a \rangle$. Conclude by the FHT that $A/I_a \cong \langle a \rangle$.

4 For each $a \in A$, let π_a be the function given by $\pi_a(x) = ax$. Define the following addition and multiplication on $\bar{A} = \{\pi_a : a \in A\}$:

$$\pi_a + \pi_b = \pi_{a+b} \qquad \text{and} \qquad \pi_a \pi_b = \pi_{ab}$$

($\bar{A}$ is a ring; however, do not prove this.) Show that the function $\phi(a) = \pi_a$ is a homomorphism from A onto $\bar{A}$. Let I designate the annihilating ideal of A (defined in Exercise H4 of Chapter 18). Use the FHT to show that $A/I \cong \bar{A}$.

E. Properties of Quotient Rings A/J in Relation to Properties of J

Let A be a ring and J an ideal of A. Use (*), (**), and (***) of this chapter. *Prove each of the following*:

1 Every element of A/J has a square root iff for every $x \in A$, there is some $y \in A$ such that $x - y^2 \in J$.

2 Every element of A/J is its own negative iff $x + x \in J$ for every $x \in A$.

3 A/J is a boolean ring iff $x^2 - x \in J$ for every $x \in A$. (S is called a boolean ring iff $s^2 = s$ for every $s \in S$.)

4 If J is the ideal of all the nilpotent elements of a commutative ring A, then A/J

has no nilpotent elements (except zero). (Nilpotent elements are defined in Chapter 17, Exercise M; by M2 and M3 they form an ideal.)

5 Every element of A/J is nilpotent iff J has the following property: for every $x \in A$, there is a positive integer n such that $x^n \in J$.

6 A/J has a unity element iff there exists an element $a \in A$ such that $ax - x \in J$ and $xa - x \in J$ for every $x \in A$.

F. Prime and Maximal Ideals

Let A be a commutative ring with unity, and J an ideal of A. *Prove each of the following.*

1 A/J is a commutative ring with unity.

2 J is a prime ideal iff A/J is an integral domain.

3 Every maximal ideal of A is a prime ideal. (HINT: Use the fact, proved in this chapter, that if J is a maximal ideal then A/J is a field. Also, use part 2.)

4 If A/J is a field, then J is a maximal ideal. (HINT: Use Theorem 3 of this chapter and Exercise I2 of Chapter 18.)

G. Further Properties of Quotient Rings in Relation to their Ideals

Let A be a ring and J an ideal of A. *Prove the following.* (In parts 1 to 3, assume that A is a commutative ring with unity.)

1 A/J is a field iff for every element $a \in A$, where $a \notin J$, there is some $b \in A$ such that $ab - 1 \in J$.

2 Every nonzero element of A/J is either invertible or a divisor of zero iff the following property holds, where $a, x \in A$: For every $a \notin J$, there is some $x \notin J$ such that either $ax \in J$ or $ax - 1 \in J$.

3 An ideal J of a ring A is called *primary* iff for all $a, b \in A$, if $ab \in J$, then either $a \in J$ or $b^n \in J$ for some positive integer n. Prove that every zero divisor in A/J is nilpotent iff J is primary.

4 An ideal J of a ring A is called *semiprime* iff it has the following property: For every $a \in A$, if $a^n \in J$ for some positive integer n, then necessarily $a \in J$. Prove that J is semiprime iff A/J has no nilpotent elements (except zero).

5 Prove that an integral domain can have no nonzero nilpotent elements. Then use part 4, together with E2, to prove that every prime ideal is semiprime.

H. Z_n as a Homomorphic Image of Z

Recall that the function

$$f(a) = \bar{a}$$

is the natural homomorphism from $\mathbb{Z}$ onto $\mathbb{Z}_n$. If a polynomial equation $p = 0$ is satisfied in $\mathbb{Z}$, necessarily $f(p) = f(0)$ is true in $\mathbb{Z}_n$. Let us take a specific example; there are integers x and y satisfying $11x^2 - 8y^2 + 29 = 0$ (we may take $x = 3$ and $y = 4$). It follows that there must be elements $\bar{x}$ and $\bar{y}$ in $\mathbb{Z}_6$ which satisfy $\overline{11}\,\bar{x}^2 - \overline{8}\bar{y}^2 + \overline{29} = \bar{0}$ in $\mathbb{Z}_6$, that is, $\bar{5}\,\bar{x}^2 - \bar{2}\,\bar{y}^2 + \bar{5} = \bar{0}$. (We take $\bar{x} = \bar{3}$ and $\bar{y} = \bar{4}$.) The problems which follow are based on this observation.

1 Prove that the equation $x^2 - 7y^2 - 24 = 0$ has no integer solutions. (HINT: If there are integers x and y satisfying this equation, what equation will $\bar{x}$ and $\bar{y}$ satisfy in $\mathbb{Z}_7$?)

2 Prove that $x^2 + (x + 1)^2 + (x + 2)^2 = y^2$ has no integer solutions.

3 Prove that $x^2 + 10y^2 = n$ (where n is an integer) has no integer solutions if the last digit of n is 2, 3, 7, or 8.

4 Prove that the sequence 3, 8, 13, 18, 23, ... does not include the square of any integer. (HINT: The image of each number on this list, under the natural homomorphism from $\mathbb{Z}$ to $\mathbb{Z}_5$, is $\bar{3}$.)

5 Prove that the sequence 2, 10, 18, 26, ... does not include the cube of any integer.

6 Prove that the sequence 3, 11, 19, 27, ... does not include the sum of two squares of integers.

7 Prove that if n is a product of two consecutive integers, its units digit must be 0, 2, or 6.

8 Prove that if n is the product of three consecutive integers, its units digit must be 0, 2, 4, or 6.

TWENTY

INTEGRAL DOMAINS

Let us recall that an integral domain is a commutative ring with unity having the cancelation property, that is,

$$if \quad a \neq 0 \quad and \quad ab = ac \quad then \quad b = c \qquad (1)$$

At the end of Chapter 17 we saw that an integral domain may also be defined as a commutative ring with unity having no divisors of zero, which is to say that

$$if \quad ab = 0 \quad then \quad a = 0 \quad or \quad b = 0 \qquad (2)$$

for as we saw, (1) and (2) are equivalent properties in any commutative ring.

The system $\mathbb{Z}$ of the integers is the exemplar and prototype of integral domains. In fact, the term "integral domain" means a system of algebra ("domain") having *integerlike* properties. However, $\mathbb{Z}$ is not the only integral domain: there are a great many integral domains different from $\mathbb{Z}$.

Our first few comments will apply to rings generally. To begin with, we introduce a convenient notation for multiples, which parallels the *exponent notation* for powers. Additively, the sum

$$a + a + \cdots + a$$

of n equal terms is written as n $\cdot$ a. We also define $0 \cdot a$ to be 0, and let $(-n) \cdot a = -(n \cdot a)$ for all positive integers n. Then

$$m \cdot a + n \cdot a = (m + n) \cdot a \qquad and \qquad m \cdot (n \cdot a) = (mn) \cdot a$$

for every element a of a ring and all integers m and n. These formulas are the translations into additive notation of the *laws of exponents* given in Chapter 10.

If A is a ring, A with addition alone is a group. Remember that in additive notation the *order* of an element a in A is the least positive integer n such that $n \cdot a = 0$. If there is no such positive integer n, then a is said to have order infinity. To emphasize the fact that we are referring to the order of a in terms of *addition*, we will call it the *additive order* of a. ,

In a ring with unity, if 1 has additive order n, we say the ring has "characteristic n." In other words, if A is a ring with unity,

*the **characteristic** of A is the least positive integer n such that*

$$\underbrace{1 + 1 + \cdots + 1}_{\text{n times}} = 0$$

*If there is no such positive integer n, A has **characteristic** 0.*

These concepts are especially simple in an integral domain. Indeed,

Theorem 1 *All the nonzero elements in an integral domain have the same additive order.*

That is, every $a \neq 0$ has the same additive order as the additive order of 1. The truth of this statement becomes transparently clear as soon as we observe that

$$n \cdot a = a + a + \cdots + a = 1a + \cdots + 1a = (1 + \cdots + 1)a = (n \cdot 1)a$$

hence $n \cdot a = 0$ iff $n \cdot 1 = 0$. (Remember that in an integral domain, if the product of two factors is equal to 0, at least one factor must be 0.)

It follows, in particular, that if the characteristic of an integral domain is a positive integer n, then

$$n \cdot x = 0$$

for every element x in the domain.

Furthermore,

Theorem 2 *In an integral domain with nonzero characteristic, the characteristic is a prime number.*

For if the characteristic were a composite number mn, then by the distributive law,

$$(m \cdot 1)(n \cdot 1) = \underbrace{(1 + \cdots + 1)}_{m \text{ terms}}\underbrace{(1 + \cdots + 1)}_{n \text{ terms}} = \underbrace{1 + 1 + \cdots + 1}_{mn \text{ terms}} = (mn) \cdot 1 = 0$$

Thus, either $m \cdot 1 = 0$ or $n \cdot 1 = 0$, which is impossible because mn was chosen to be the *least* positive integer such that $(mn) \cdot 1 = 0$.

A very interesting rule of arithmetic is valid in integral domains whose characteristic is not zero.

Theorem 3 *In any integral domain of characteristic* p,

$$(a + b)^p = a^p + b^p \qquad \textit{for all elements a and b}$$

This formula becomes clear when we look at the binomial expansion of $(a + b)^p$. Remember that by the binomial formula,

$$(a + b)^p = a^p + \binom{p}{1} \cdot a^{p-1}b + \cdots + \binom{p}{p-1} \cdot ab^{p-1} + b^p$$

where the binomial coefficient

$$\binom{p}{k} = \frac{p(p-1)(p-2) \cdots (p-k+1)}{k!}$$

It is demonstrated in Exercise L of Chapter 17 that the binomial formula is correct in every commutative ring.

Note that if p is a prime number and $0 < k < p$, then

$$\binom{p}{k} \textit{ is a multiple of } p$$

because every factor of the denominator is less than p, hence p does not cancel out. Thus, each term of the binomial expansion above, except for the first and last terms, is of the form px, which is equal to 0 because the domain has characteristic p. Thus, $(a + b)^p = a^p + b^p$.

It is obvious that every field is an integral domain: for if $a \neq 0$ and $ax = ay$ in a field, we can multiply both sides of this equation by the multiplicative inverse of a to cancel a. However, not every integral domain is a field: for example, $\mathbb{Z}$ is not a field. Nevertheless,

Theorem 4 *Every **finite** integral domain is a field.*

List the elements of the integral domain in the following manner:

$$0, 1, a_1, a_2, \ldots, a_n$$

In this manner of listing, there are n + 2 elements in the domain. Take any a_i, and show that it is invertible: to begin with, note that the products

$$a_i 0, \ a_i 1, \ a_i a_1, \ a_i a_2, \ldots, a_i a_n$$

are all distinct: for if $a_i x = a_i y$, then $x = y$. Thus, there are n + 2 *distinct* products $a_i x$; but there are exactly n + 2 elements in the domain, so every element in the domain is equal to one of these products. In particular, $1 = a_i x$ for some x, hence a_i is invertible.

OPTIONAL

The integral domain $\mathbb{Z}$ is not a field because it does not contain the quotients m/n of integers. However, $\mathbb{Z}$ can be *enlarged* to a field by adding to it all the quotients of integers; the resulting field, of course, is $\mathbb{Q}$ the field of the rational numbers. $\mathbb{Q}$ consists of all quotients of integers, and it contains $\mathbb{Z}$ (or rather, an isomorphic copy of $\mathbb{Z}$) when we identify each integer n with the quotient n/1. We say that $\mathbb{Q}$ is the *field of quotients* of $\mathbb{Z}$.

It is a fascinating fact that the method for constructing $\mathbb{Q}$ from $\mathbb{Z}$ can be applied to any integral domain. Starting from any integral domain A, it is possible to construct a field which contains A: a field of quotients of A. This is not merely a mathematical curiosity, but a valuable addition to our knowledge. In applications it often happens that a system of algebra we are dealing with lacks a needed property, but is contained in a larger system which *has* that property—and that is almost as good! In the present case, A is not a field but may be enlarged to one.

Thus, if A is any integral domain, we will proceed to construct a field A^* consisting of all the quotients of elements in A; and A^* will contain A, or rather an isomorphic copy of A, when we identify each element a of A with the quotient $a/1$. The construction will be carefully outlined and the busy work left as an exercise.

Given A, let S denote the set of all ordered pairs (a, b) of elements of A, where $b \neq 0$. That is,

$$S = \{(a, b) : a, b \in A \quad \text{and} \quad b \neq 0\}$$

In order to understand the next step, we should think of (a, b) as a/b. [It is too early in the proof to introduce fractional notation, but nevertheless each ordered pair (a, b) should be *thought of* as a fraction a/b.] Now, a problem of representation arises here, because it is obvious that the quotient xa/xb is equal to the quotient a/b; to put the same fact differently, the quotients a/b and c/d are equal whenever $ad = bc$. That is, if $ad = bc$, then a/b and c/d are two different ways of writing the *same* quotient. Motivated by this observation, we *define* $(a, b) \sim (c, d)$ *to mean that* $ad = bc$, and easily verify that $\sim$ is an equivalence relation on the set S. (Equivalence relations are explained in Chapter 12.) Then we let $[a, b]$ denote the *equivalence class* of (a, b), that is,

$$[a, b] = \{(c, d) \in S : (c, d) \sim (a, b)\}$$

Intuitively, all the pairs which represent a given quotient are lumped together in one equivalence class; thus, *each quotient is represented by exactly one equivalence class.*

Let us recapitulate the formal details of our construction up to this point: Given the set S of ordered pairs of elements in A, we define an equivalence relation $\sim$ in S by letting $(a, b) \sim (c, d)$ iff $ad = bc$. We let $[a, b]$ designate the equivalence class of (a, b), and finally, *we let A^* denote the set of all the equivalence classes $[a, b]$.* The elements of A^* will be called *quotients.*

Before going on, observe carefully that

$$[a, b] = [r, s] \qquad \text{iff} \qquad (a, b) \sim (r, s) \qquad \text{iff} \qquad as = br \qquad (*)$$

As our next step, we define operations of addition and multiplication in A^*:

$$[a, b] + [c, d] = [ad + bc, bd]$$

and
$$[a, b] \cdot [c, d] = [ac, bd]$$

To understand these definitions, simply remember the formulas

$$\frac{a}{b} + \frac{c}{d} = \frac{ad + bc}{bd} \qquad \text{and} \qquad \frac{a}{b} \cdot \frac{c}{d} = \frac{ac}{bd}$$

We must make certain these definitions are unambiguous; that is, if $[a, b] = [r, s]$ and $[c, d] = [t, u]$, we have the equations

$$
\begin{array}{ccc}
[a, b] + [c, d] = [ad + bc, bd] & & [a, b] \cdot [c, d] = [ac, bd] \\
\| \qquad \| & \text{and} & \| \qquad \| \\
[r, s] + [t, u] = [ru + st, su] & & [r, s] \cdot [t, u] = [rt, su]
\end{array}
$$

and we must therefore verify that $[ad + bc, bd] = [ru + st, su]$ and $[ac, bd] = [rt, su]$. This is left as an exercise. It is also left for the student to verify that addition and multiplication are associative and commutative and the distributive law is satisfied.

The zero element is $[0, 1]$, because $[a, b] + [0, 1] = [a, b]$. The negative of $[a, b]$ is $[-a, b]$, for $[a, b] + [-a, b] = [0, b^2] = [0, 1]$. [The last equation is true because of (*).] The unity is $[1, 1]$, and the multiplicative inverse of $[a, b]$ is $[b, a]$, for $[a, b] \cdot [b, a] = [ab, ab] = [1, 1]$. Thus, A^* is a field!

Finally, if A' is the subset of A^* which contains every $[a, 1]$, we let ϕ be the function from A to A' defined by $\phi(a) = [a, 1]$. This function is injective because, by (*), if $[a, 1] = [b, 1]$ then $a = b$. It is obviously surjective and is easily shown to be a homomorphism. Thus, ϕ is an isomorphism from A to A', so A^* contains an isomorphic copy A' of A.

EXERCISES

A. Characteristic of an Integral Domain

Let A be an integral domain. *Prove each of the following.*

1 Let a be any nonzero element of A. If $\text{n} \cdot a = 0$, where $n \neq 0$, then n is a multiple of the characteristic of A.

2 If A has characteristic zero, $\text{n} \neq 0$, and $\text{n} \cdot a = 0$, then $a = 0$.

3 If A has characteristic 3, and $5 \cdot a = 0$, then $a = 0$.

4 If there is a nonzero element a in A such that $256 \cdot a = 0$, then A has characteristic 2.

5 If there are distinct nonzero elements a and b in A such that $125 \cdot a = 125 \cdot b$, then A has characteristic 5.

6 If there are nonzero elements a and b in A such that $(a + b)^2 = a^2 + b^2$ then A has characteristic 2.

7 If there are nonzero elements a and b in A such that $10a = 0$ and $14b = 0$ then A has characteristic 2.

B. Characteristic of a Finite Integral Domain

Let A be a finite integral domain. *Prove each of the following.*

1 If A has characteristic q, then q is a divisor of the order of A.

2 If the order of A is a prime number p, then the characteristic of A must be equal to p.

3 If the order of A is p^m, where p is a prime, the characteristic of A must be equal to p.

4 If A has 81 elements, its characteristic is 3.

5 If A, with addition alone, is a cyclic group, the order of A is a prime number.

C. Finite Rings

Let A be a finite commutative ring with unity. *Prove each of the following.*

1 Every nonzero element of A is either a divisor of zero or invertible. (HINT: Use an argument analogous to the proof of Theorem 4.)
2 If $a \neq 0$ is not a divisor of zero, then some positive power of a is equal to 1. (HINT: Consider $a, a^2, a^3, \ldots$. Since A is finite, there must be positive integers n < m such that $a^n = a^m$.)
3 Use part 2 to prove: If a is invertible, then a^{-1} is equal to a positive power of a.

D. Field of Quotients of an Integral Domain

The following questions refer to the construction of a field of quotients of A, as outlined on pages 200 to 202.

1 If $[a, b] = [r, s]$ and $[c, d] = [t, u]$, prove that $[a, b] + [c, d] = [r, s] + [t, u]$.
2 If $[a, b] = [r, s]$ and $[c, d] = [t, u]$, prove that $[a, b][c, d] = [r, s][t, u]$.
3 If $(a, b) \sim (c, d)$ means $ad = bc$, prove that $\sim$ is an equivalence relation on S.
4 Prove that addition in A^* is associative and commutative.
5 Prove that multiplication in A^* is associative and commutative.
6 Prove the distributive law in A^*.
7 Verify that $\phi : A \rightarrow A'$ is a homomorphism.

E. Further Properties of the Characteristic of an Integral Domain

Let A be an integral domain. *Prove each of the following.*

1 Let $a \in A$. If A has characteristic p, and $n \cdot a = 0$ where n is *not* a multiple of p, then $a = 0$
2 If p is a prime, and there is a nonzero element $a \in A$ such that $p \cdot a = 0$, then A has characteristic p.
3 If p is a prime, and there is a nonzero element $a \in A$ such that $p^m \cdot a = 0$ for some integer m, then A has characteristic p.
4 If A has characteristic p, then the function $f(a) = a^p$ is a homomorphism from A to A.
5 Let A have order p, where p is a prime. Explain why

$$A = \{0, 1, 2 \cdot 1, 3 \cdot 1, \ldots, (p-1) \cdot 1\}$$

Prove that $A \cong \mathbb{Z}_p$.
6 If A has characteristic p, then for any positive integer n,
  (i) $(a + b)^{p^n} = a^{p^n} + b^{p^n}$; and (ii) $(a_1 + a_2 + \cdots + a_r)^{p^n} = a_1^{p^n} + \cdots + a_r^{p^n}$.

7 Let $A \subseteq B$ where A and B are integral domains. A has characteristic p iff B has characteristic p.

F. Finite Fields

By Theorem 4, "finite integral domain" and "finite field" are the same. *Prove the following:*

1 Every finite field has nonzero characteristic.

2 If A is a finite field of characteristic p, the function $f(a) = a^p$ is an automorphism of A, that is, an isomorphism from A to A. (HINT: Use Exercise E4 above and Exercise F7 of Chapter 18. To show that f is surjective, compare the number of elements in the domain and in the range of f.)

*The function $f(a) = a^p$ is called the **Froebenius automorphism**.*

3 Use part 2 to prove: In a finite field of characteristic p, every element has a p-th root.

TWENTY-ONE
THE INTEGERS

There are two possible ways of describing the system of the integers.

On the one hand, we may attempt to describe it concretely.

On the other hand, we may find a list of axioms from which it is possible to deduce all the properties of the integers, so the *only* system which has all these properties is the system of the integers.

The second of these two ways is the way of mathematics. It is elegant, economical, and simple. We select as axioms only those particular properties of the integers which are absolutely necessary in order to prove further properties of the integers. And we select a sufficiently *complete* list of axioms so that, using them, one can prove all the properties of the integers needed in mathematics.

We have already seen that the integers are an integral domain. However, there are numerous examples of integral domains which bear little resemblance to the set of the integers. For example, there are finite integral domains such as $\mathbb{Z}_5$, fields (remember that every field is an integral domain) such as $\mathbb{Q}$ and $\mathbb{R}$, and others. Thus, in order to pin down the integers—that is, in order to find a list of axioms which applies to the integers and *only* the integers—we must select some additional axioms and add them to the axioms of integral domains. This we will now proceed to do.

Most of the traditional number systems have two aspects. One aspect is their algebraic structure: they are integral domains or fields. The other aspect—which we have not yet touched upon—is that their elements can be *ordered*. That is, if a and b are distinct elements, we can say that a is less than b or b is less than a. This second aspect—the ordering of elements—will now be formalized.

> An **ordered integral domain** is an integral domain A with a relation, symbolized by $<$, having the following properties:
> (1) For any a and b in A, exactly one of the following is true:
>
> $$a = b \qquad a < b \qquad \text{or} \qquad b < a$$
>
> Furthermore, for any a, b, and c in A,
> (2) If $a < b$ and $b < c$, then $a < c$.
> (3) If $a < b$, then $a + c < b + c$.
> (4) If $a < b$, then $ac < bc$ on the condition that $0 < c$.

The relation $<$ is called an *order relation* on A. The four conditions which an order relation must fulfill are familiar to everyone. (1) and (2) require no comment. (3) asserts that we are allowed to add any given c to both sides of an inequality. (4) asserts that we may multiply both sides of an inequality by any c, on the condition that c is greater than zero.

As usual, $a > b$ has the same meaning as $b < a$. Furthermore, $a \leq b$ means "$a < b$ or $a = b$," and $b \geq a$ means the same as $a \leq b$.

In an ordered integral domain A, an element a is called *positive* if $a > 0$. If $a < 0$ we call a *negative*. Note that if a is positive then $-a$ is negative (proof: add $-a$ to both sides of the inequality $a > 0$). Similarly, if a is negative, then $-a$ is positive.

In any ordered integral domain, *the square of every nonzero element is positive*. Indeed, if c is nonzero, then either $c > 0$ or $c < 0$. If $c > 0$, then, multiplying both sides of the inequality $c > 0$ by c,

$$cc > c0 = 0$$

so $c^2 > 0$. On the other hand, if $c < 0$, then

$$(-c) > 0$$

hence
$$(-c)(-c) > 0(-c) = 0$$

But $(-c)(-c) = c^2$, so once again, $c^2 > 0$.

In particular, since $1 = 1^2$, 1 is always positive.

From the fact that $1 > 0$, we immediately deduce that $1 + 1 > 1$, $1 + 1 + 1 > 1 + 1$, and so on. In general, for any positive integer n,

$$(n + 1) \cdot 1 > n \cdot 1$$

where n $\cdot$ 1 designates the unity element of the ring A added to itself n times. Thus, in any ordered integral domain A, *the set of all the multiples of*

1 is ordered as in $\mathbb{Z}$: namely

$$\cdots < (-2) \cdot 1 < (-1) \cdot 1 < 0 < 1 < 2 \cdot 1 < 3 \cdot 1 < \cdots$$

The set of all the positive elements of A is denoted by A^+. An ordered integral domain A is called an *integral system* if every nonempty subset of A^+ has a least element; in other words, if *every nonempty set of positive elements of A has a least element.* This property is called the *well-ordering property* for A^+.

It is obvious that $\mathbb{Z}$ is an integral system, for every nonempty set of positive integers contains a least number. For example, the smallest element of the set of all the positive even integers is 2. Note that $\mathbb{Q}$ and $\mathbb{R}$ are *not* integral systems. For although both are ordered integral domains, they contain sets of positive numbers, such as $\{x : 0 < x < 1\}$, which have no least element.

In any integral system, *there is no element between 0 and 1.* For suppose A is an integral system in which there are elements x between 0 and *1*. Then the set $\{x \in A : 0 < x < 1\}$ is a nonempty set of positive members of A, so by the well-ordering property it has a least element c. That is,

$$0 < c < 1$$

and c is the least element of A with this property. But then (multiplying by c),

$$0 < c^2 < c$$

Thus, c^2 is between 0 and 1 and is less than c, which is impossible.

Thus, there is no element of A between 0 and *1*.

Finally, in any integral system, *every element is a multiple of 1.* If that were not the case, we could use the well-ordering principle to pick the least positive element of A which is *not* a multiple of *1*: call it b. Now, $b > 0$ and there are no elements of A between 0 and 1, so $b > 1$. (Remember that b cannot be equal to *1* because b is not a multiple of *1*.) Since $b > 1$, it follows that $b - 1 > 0$. But $b - 1 < b$ and b is the *least* positive element which is not a multiple of 1, so $b - 1$ *is* a multiple of *1*. Say

$$b - 1 = n \cdot 1$$

But then $b = n \cdot 1 + 1 = (n + 1) \cdot 1$, which is impossible.

Thus, in any integral system, all the elements are multiples of 1 and these are ordered exactly as in $\mathbb{Z}$. It is now a mere formality to prove that *every integral system is isomorphic to* $\mathbb{Z}$. This is left as Exercise D at the end of this chapter.

Since every integral system is isomorphic to $\mathbb{Z}$, any two integral systems are isomorphic to each other. Thus $\mathbb{Z}$ is, up to isomorphism, the *only* integral system. We have therefore succeeded in giving a complete axiomatic characterization of $\mathbb{Z}$.

Henceforward we consider $\mathbb{Z}$ to be defined by the fact that it is an integral system.

The theorem which follows is the basis of proofs by mathematical induction. It is intuitively clear and easy to prove.

Theorem 1 *Let K represent a set of positive integers. Consider the following two conditions :*

(i) 1 *is in K.*
(ii) *For any positive integer* k *if* k $\in$ K, *then also* k $+ 1 \in$ K.

If K is any set of positive integers satisfying these two conditions, then K consists of all the positive integers.

Indeed, if K does not contain all the positive integers, then by the well-ordering principle, the set of all the positive integers which are *not* in K has a least element. Call it b; b is the least positive integer *not* in K. By Condition (i), $b \neq 1$, hence $b > 1$.

Thus, $b - 1 > 0$, and $b - 1 \in K$. But then, by Condition (ii), $b \in K$, which is impossible.

Let the symbol S_n represent any statement about the positive integer n. For example, S_n might stand for "n is odd," or "n is a prime," or it might represent an equation such as $(n - 1)(n + 1) = n^2 - 1$ or an inequality such as $n \leq n^2$. If, let us say, S_n stands for $n \leq n^2$, then S_1 asserts that $1 \leq 1^2$, S_2 asserts that $2 \leq 2^2$, S_3 asserts that $3 \leq 3^2$, and so on.

Theorem 2: Principle of mathematical induction *Consider the following conditions :*

(i) S_1 *is true.*
(ii) *For any positive integer* k, *if* S_k *is true, then also* S_{k+1} *is true.*

If Conditions (i) and (ii) are satisfied, then S_n *is true for every positive integer* n.

Indeed, if K is the set of all the positive integers k such that S_k is true, then K complies with the conditions of Theorem 1. Thus, K contains all the positive integers. This means that S_n is true for every n.

As a simple illustration of how the principle of mathematical induction is applied, let S_n be the statement that

$$1 + 2 + \cdots + n = \frac{n(n + 1)}{2}$$

that is, the sum of the first n positive integers is equal to n(n + 1)/2. Then S_1 is simply

$$1 = \frac{1 \cdot 2}{2}$$

which is clearly true. Suppose, next, that k is any positive integer and that S_k is true. In other words,

$$1 + 2 + \cdots + k = \frac{k(k + 1)}{2}$$

Then, by adding k + 1 to both sides of this equation, we obtain

$$1 + 2 + \cdots + k + (k + 1) = \frac{k(k + 1)}{2} + (k + 1)$$

that is, $$1 + 2 + \cdots + (k + 1) = \frac{(k + 1)(k + 2)}{2}$$

However, this last equation is exactly S_{k+1}. We have therefore verified that whenever S_k is true, S_{k+1} also is true. Now, the principle of mathematical induction allows us to conclude that

$$1 + 2 + \cdots + n = \frac{n(n + 1)}{2}$$

for every positive integer n.

A variant of the principle of mathematical induction, called the *principle of strong induction*, asserts that S_n is true for every positive integer n on the conditions that

(i) S_1 is true, and
(ii) For any positive integer k, if S_i is true for every i < k, then S_k is true.

The details are outlined in Exercise H at the end of this chapter.

One of the most important facts about the integers is that any integer m may be divided by any positive integer n to yield a quotient q and a positive remainder r. (The remainder is less than the divisor n.) For example, 25 may be divided by 8 to give a quotient of 3 and a remainder of 1:

$$25 = 8 \times 3 + 1$$

$$m \quad n \quad q \quad r$$

This process is known as the *division algorithm*. It is stated in a precise manner as follows:

Theorem 3: Division algorithm *If* m *and* n *are integers and* n *is positive, there exist unique integers* q *and* r *such that*

$$m = nq + r \quad and \quad 0 \leq r < n$$

We call q *the quotient, and* r *the remainder, in the division of* m *by* n.

We begin by showing a simple fact:

There exists an integer x *such that* $xn \leq m$.   (*)

Remember that n is positive, hence $n \geq 1$. As for m, since $m \neq 0$, either $m > 0$ or $m < 0$. We consider these two cases separately:

Suppose $m > 0$. Then

$$0 \leq m \quad \text{hence} \quad (0)n \leq m$$
$$\phantom{0 \leq m \quad \text{hence} \quad} x$$

Suppose $m < 0$. We may multiply both sides of $n \geq 1$ by the positive integer $-m$ to get $(-m)n \geq -m$. Adding $mn + m$ to both sides yields

$$mn \leq m$$
$$ x$$

Thus, regardless of whether m is positive or negative, there is some integer x such that $xn \leq m$.

Let W be the subset of $\mathbb{Z}$ consisting of all the nonnegative integers which are expressible in the form $m - xn$, where x is any integer. By (*), W is not empty; hence by the well-ordering property, W contains a least integer r. Because $r \in W$, r is nonnegative and is expressible in the form $m - nq$ for some integer q. That is,

$$r \geq 0$$

and

$$r = m - nq$$

Thus, we already have $m = nq + r$ and $0 \leq r$. It remains only to verify that $r < n$. Suppose not: suppose $n \le r$, that is, $r - n \ge 0$. But

$$r - n = (m - nq) - n = m - n(q + 1)$$

and clearly $r - n < r$. This means that $m - n(q + 1)$ is an element of W *less than* r, which is impossible because r is the least element of W. We conclude that $n \le r$ is impossible, hence $r < n$.

The verification that q and r are unique is left as an exercise.

EXERCISES

A. Properties of Order Relations in Integral Domains

Let A be an ordered integral domain. *Prove the following, for all a, b, and c in A.*

1 If $a \le b$ and $b \le c$, then $a \le c$.
2 If $a \le b$, then $a + c \le b + c$.
3 If $a \le b$ and $c \ge 0$, then $ac \le bc$.
4 If $a < b$ and $c < 0$, then $bc < ac$.
5 $a < b$ iff $-b < -a$.
6 If $a + c < b + c$, then $a < b$.
7 If $ac < bc$ and $c > 0$, then $a < b$.
8 If $a < b$ and $c < d$, then $a + c < b + d$.

B. Further Properties of Ordered Integral Domains

Let A be an ordered integral domain. *Prove the following, for all a, b, and c in A.*

1 $a^2 - 2ab + b^2 \ge 0$
2 $a^2 + b^2 \ge 2ab$
3 $a^2 + b^2 \ge ab$
4 $a^2 + b^2 \ge -ab$
5 $a^2 + b^2 + c^2 \ge ab + bc + ac$
6 $a^2 + b^2 > ab$, if $a \ne b$
7 $a + b \le ab + 1$, if $a, b \ge 1$
8 $ab + ac + bc + 1 < a + b + c + abc$, if $a, b, c > 1$

C. Uses of Induction

Prove each of the following, using the principle of mathematical induction

1 $1 + 3 + 5 + \cdots + (2n - 1) = n^2$ (The sum of the first n odd integers is n^2.)

2 $1^3 + 2^3 + \cdots + n^3 = (1 + 2 + \cdots + n)^2$

3 $1^2 + 2^2 + \cdots + (n - 1)^2 < \dfrac{n^3}{3} < 1^2 + 2^2 + \cdots + n^2$

4 $1^3 + 2^3 + \cdots + (n - 1)^3 < \dfrac{n^4}{4} < 1^3 + 2^3 + \cdots + n^3$

5 $1^2 + 2^2 + \cdots + n^2 = \frac{1}{6}n(n + 1)(2n + 1)$

6 $1^3 + 2^3 + \cdots + n^3 = \frac{1}{4}n^2(n + 1)^2$

7 $\dfrac{1}{2!} + \dfrac{2}{3!} + \cdots \dfrac{n}{(n + 1)!} = \dfrac{n! - 1}{n!}$

8 The *Fibonacci sequence* is the sequence of integers $F_1, F_2, F_3, \ldots$ defined as follows: $F_1 = 1$; $F_2 = 1$; $F_{n+2} = F_{n+1} + F_n$ for all positive integers n. (That is, every number, after the second one, is the sum of the two preceding ones.) Use induction to prove that for all $n > 0$,

$$F_{n+1}F_{n+2} - F_n F_{n+3} = (-1)^n$$

D. Every Integral System Is Isomorphic to $\mathbb{Z}$

Let A be an integral system. Let $h : \mathbb{Z} \to A$ be defined by: $h(n) = n \cdot 1$. The purpose of this exercise is to prove that h is an isomorphism, from which it follows that $A \cong \mathbb{Z}$. *Prove the following*:

1 For every positive integer n, $n \cdot 1 > 0$. From this, deduce that A has nonzero characteristic.

2 Prove that h is injective and surjective.

3 Prove that h is an isomorphism.

E. Absolute Values

In any ordered integral domain, define $|a|$ by

$$|a| = \begin{cases} a & \text{if} & a \geq 0 \\ -a & \text{if} & a < 0 \end{cases}$$

Using this definition, prove the following:

1 $|-a| = |a|$

2 $a \leq |a|$

3 $a \geq -|a|$

4 If $b > 0$, $|a| \leq b$ iff $-b \leq a \leq b$

5 $|a + b| \leq |a| + |b|$
6 $|a - b| \leq |a| + |b|$
7 $|ab| = |a| \cdot |b|$
8 $|a| - |b| \leq |a - b|$
9 $||a| - |b|| \leq |a - b|$

F. Problems on the Division Algorithm

Prove the following, where k, m, n, q, *and* r *designate integers.*

1 Let $n > 0$ and $k > 0$. If q is the quotient and r is the remainder when m is divided by n, then q is the quotient and kr is the remainder when km is divided by kn.

2 Let $n > 0$ and $k > 0$. If q is the quotient when m is divided by n, and q_1 is the quotient when q is divided by k, then q_1 is the quotient when m is divided by nk.

3 If $n \neq 0$, there exist q and r such that $m = nq + r$ and $0 \leq r < |n|$. (Use Theorem 3, and consider the case where $n < 0$.)

4 In Theorem 3, suppose $m = nq_1 + r_1 = nq_2 + r_2$ where $0 \leq r_1, r_2 < n$. Prove that $r_1 - r_2 = 0$. [HINT: Consider the difference $(nq_1 + r_1) - (nq_2 + r_2)$.]

5 Use part 4 to prove that $q_1 - q_2 = 0$. Conclude that the quotient and remainder, in the division algorithm, are unique.

6 If r is the remainder when m is divided by n, then $\bar{m} = \bar{r}$ in $\mathbb{Z}_n$; and conversely.

G. Laws of Multiples

The purpose of this exercise is to give rigorous proofs (using induction) of the basic identities involved in the use of exponents or multiples. If A is a ring and $a \in A$, we define $n \cdot a$ (where n is any positive integer) by the pair of conditions:

$$(i) \quad 1 \cdot a = a, \quad \text{and} \quad (ii) \quad (n + 1) \cdot a = n \cdot a + a$$

Use mathematical induction (with the above definition) to prove that the following are true for all positive integers n and all elements $a, b \in A$:

1 $n \cdot (a + b) = n \cdot a + n \cdot b$
2 $(n + m) \cdot a = n \cdot a + m \cdot a$
3 $(n \cdot a)b = a(n \cdot b) = n \cdot (ab)$
4 $m \cdot (n \cdot a) = (mn) \cdot a$
5 $n \cdot a = (n \cdot I)a$ where I is the unity element of A
6 $(n \cdot a)(m \cdot b) = (nm) \cdot ab$ (Use parts 3 and 4.)

H. Principle of Strong Induction

Prove the following in $\mathbb{Z}$:

1 Let K denote a set of positive integers. Consider the following conditions:

(i) $1 \in K$.

(ii) For any positive integer k, if every positive integer less than k is in K, then $k \in K$.

If K satisfies these two conditions, prove that K contains all the positive integers.

2 Let S_n represent any statement about the positive integer n. Consider the following conditions:

(i) S_1 is true.

(ii) For any positive integer k, if S_i is true for every $i < k$, S_k is true.

If Conditions (i) and (ii) are satisfied, prove that S_n is true for every positive integer n.

TWENTY-TWO
FACTORING INTO PRIMES

It has been said that the two events most decisive in shaping the course of man's development were the invention of the wheel and the discovery of numbers. From the time—ages ago—when man first learned the use of numbers for counting, they have been a source of endless fascination for him. Alchemists and astrologers extolled the virtues of "mystic" numbers and found in them the key to potent magic. Others, more down to earth, found delight in observing the many regularities and unexpected properties of numbers. The integers have been a seemingly inexhaustible source of problems great and small on which mathematics has fed and continues to draw in our day.

The properties of prime numbers alone have occupied the energies of mathematicians from the time of Euclid. New questions relating to the primes continue to appear, and many continue to resist the best efforts to solve them. Most importantly, a large part of number theory starts out from a few basic facts about prime numbers. They will be outlined in this chapter.

Modern number theory is the oldest as well as one of the newest parts of mathematics. It rests upon some basic data regarding the structure of the domain $\mathbb{Z}$ of the integers. An understanding of this structure is a fundamental part of any mathematical education.

An important feature of any ring is the structure of its ideals. We therefore begin by asking: What are the ideals of $\mathbb{Z}$? We have already made use of *principal* ideals of $\mathbb{Z}$, such as the ideal

$$\langle 6 \rangle = \{\ldots, -18, -12, -6, 0, 6, 12, 18, \ldots\}$$

which consist of all the multiples of one fixed integer. It is natural to inquire whether $\mathbb{Z}$ has any ideals which are not principal, and what they might look like. The answer is far from self-evident.

Theorem 1 *Every ideal of $\mathbb{Z}$ is principal.*

Let J be any ideal of $\mathbb{Z}$. If 0 is the only integer in J, then $J = \langle 0 \rangle$, the principal ideal generated by 0. If there are *nonzero* integers in J, then for each x in J, $-x$ is also in J; thus there are *positive* integers in J. By the well-ordering property we may pick the *least* positive integer in J, and call it n.

We will prove that $J = \langle n \rangle$, which is to say that *every element of J is some multiple of n*. Well, let m be any element of J. By the division algorithm we may write $m = nq + r$ where $0 \leq r < n$. Now m was chosen in J, and $n \in J$, hence nq is in J. Thus,

$$r = m - nq \in J$$

Remember that r is either 0 or else positive and less than n. The second case is impossible, because n is the *least* positive integer in J. Thus, $r = 0$, and therefore $m = nq$, which is a multiple of n.

We have proved that every element of J is a multiple of n, which is to say that $J = \langle n \rangle$.

It is useful to note that by the preceding proof, any ideal J is generated *by the least positive integer in J*.

If r and s are integers, we say that *s is a multiple of r* if there is an integer k such that

$$s = rk$$

If this is the case, we also say that *r is a factor of s*, or *r divides s*, and we symbolize this by writing

$$r \mid s$$

Note that 1 and -1 divide every integer. On the other hand, *an integer r divides 1 iff r is invertible.* In $\mathbb{Z}$ there are very few invertible elements. As a matter of fact,

Theorem 2 *The only invertible elements of $\mathbb{Z}$ are 1 and -1.*

If s is invertible, this means there is an integer r such that

$$rs = 1$$

Clearly $r \neq 0$ and $s \neq 0$ (otherwise their product would be 0). Furthermore, r and s are either both positive or both negative (otherwise their product would be negative).

If r and s are both positive, then $r = 1$ or $r > 1$. In case $r > 1$, we may multiply both sides of $1 < r$ by s to get $s < rs = 1$; this is impossible because s cannot be positive and < 1. Thus, it must be that $r = 1$, hence $1 = rs = 1s = s$, so also $s = 1$.

If r and s are both negative, then $-r$ and $-s$ are positive. Thus,

$$1 = rs = (-r)(-s)$$

and by the preceding case, $-r = -s = 1$. Thus, $r = s = -1$.

A pair of integers r and s are called *associates* if they divide each other, that is, if $r \mid s$ and $s \mid r$. If r and s are associates, this means there are integers k and l such that $r = ks$ and $s = lr$. Thus, $r = ks = klr$, hence $kl = 1$. By Theorem 2, k and l are ± 1, and therefore $r = \pm s$. Thus, we have shown that

$$\text{If } r \text{ and } s \text{ are associates in } \mathbb{Z}, \text{ then } r = \pm s. \qquad (*)$$

An integer t is called a *common divisor* of integers r and s if $t \mid r$ and $t \mid s$. A *greatest common divisor* of r and s is an integer t such that

(i) $t \mid r$ and $t \mid s$, and
(ii) For any integer u, if $u \mid r$ and $u \mid s$, then $u \mid t$.

In other words, t is a greatest common divisor of r and s if t is a common divisor of r and s, and every other common divisor of r and s divides t. Note that the adjective "greatest" in this definition does not mean primarily that t is greater in *magnitude* than any other common divisor, but, rather, that it is a *multiple* of any other common divisor.

The words "greatest common divisor" are familiarly abbreviated by gcd. As an example, 2 is a gcd of 8 and 10; but -2 also is a gcd of 8 and 10. According to the definition, two different gcd's must divide each other; hence by (*), they differ only in sign. Of the two possible gcd's $\pm t$ for r and s, we select the positive one, call it *the* gcd of r and s, and denote it by

$$\gcd(r, s)$$

Does every pair r, s of integers have a gcd? Our experience with the integers tells us that the answer is "yes." We can easily prove this, and more:

Theorem 3 *Any two nonzero integers r and s have a greatest common divisor t. Furthermore, t is equal to a "linear combination" of r and s. That is,*

$$t = kr + ls$$

for some integers k and l.

Let J be the set of all the linear combinations of r and s, that is, the set of all $ur + vs$ as u and v range over $\mathbb{Z}$. J is closed with respect to addition and negatives and absorbs products because

$$(u_1 r + v_1 s) + (u_2 r + v_2 s) = (u_1 + u_2)r + (v_1 + v_2)s$$

$$-(ur + vs) = (-u)r + (-v)s$$

and $$w(ur + vs) = (wu)r + (wv)s$$

Thus, J is an ideal of $\mathbb{Z}$. By Theorem 1, J is a *principal* ideal of $\mathbb{Z}$, say $J = \langle t \rangle$. (J consists of all the multiples of t.)

Now t is in J, which means that t is a linear combination of r and s:

$$t = kr + ls$$

Furthermore, $r = 1r + 0s$ and $s = 0r + 1s$, so r and s are linear combinations of r and s; thus r and s are in J. But all the elements of J are multiples of t, so r and s are multiples of t. That is,

$$t \mid r \quad \text{and} \quad t \mid s$$

Now, if u is any common divisor of r and s, this means that $r = xu$ and $s = yu$ for some integers x and y. Thus,

$$t = kr + ls = kxu + lyu = u(kx + ly)$$

It follows that $u \mid t$. This confirms that t is the gcd of r and s.

A word of warning: the fact that an integer m is a linear combination of r and s does not necessarily imply that m is the gcd of r and s. For example, $3 = (1)15 + (-2)6$, and 3 is the gcd of 15 and 6. On the other hand, $27 = (1)15 + (2)6$, yet 27 is *not* a gcd of 15 and 6.

A pair of integers r and s are said to be *relatively prime* if they have no common divisors except ± 1. For example, 4 and 15 are relatively prime. If r and s are relatively prime, their gcd is equal to 1; so by Theorem 3, there are integers k and l such that $kr + ls = 1$. Actually, the converse of this statement is true too: if some linear combination of r and s is equal to 1 (that is, if there are integers k and l such that $kr + ls = 1$), then r and s are relatively prime. The simple proof of this fact is left as an exercise.

If m is any integer, it is obvious that ± 1 and $\pm m$ are factors of m. We call these the *trivial factors* of m. If m has any *other* factors, we call them *proper* factors of m. For example, ± 1 and ± 6 are the trivial factors of 6, whereas ± 2 and ± 3 are proper factors of 6.

If an integer m *has* proper factors, m is called *composite*. If an integer $p \neq 0$, 1 has *no* proper factors (that is, if all its factors are trivial), then we call p a *prime*. For example, 6 is composite, whereas 7 is a prime.

Composite number lemma *If a positive integer m is composite, then $m = rs$ where*

$$1 < r < m \qquad and \qquad 1 < s < m$$

If m is composite, this means that $m = rs$ for integers r and s which are not equal either to 1 or to m. We may take r and s to be positive, hence $1 < r$ and $1 < s$. Multiplying both sides of $1 < r$ by s gives $s < rs = m$. Analogously, we get $r < m$.

What happens when a composite number is divided by a prime? The next lemma provides an answer to that question.

Euclid's lemma *Let m and n be integers, and let p be a prime.*

$$If \ p \,|\, (mn), \ then \ either \ p \,|\, m \ or \ p \,|\, n.$$

If $p \,|\, m$ we are done. So let us assume that p does not divide m. What integers are common divisors of p and m?

Well, the only divisors of p are ± 1 and $\pm p$. Since we assumed that p does *not* divide m, p and $-p$ are ruled out as *common* divisors of p and m, hence their only common divisors are 1 and -1.

It follows that $\gcd(p, m) = 1$, so by Theorem 3,

$$kp + lm = 1$$

for some integer coefficients k and l. Thus,

$$kpn + lmn = n$$

But $p \,|\, (mn)$, so there is an integer h such that $mn = ph$. Therefore,

$$kpn + lph = n$$

that is, $p(kn + lh) = n$. Thus, $p \,|\, n$.

Corollary 1 *Let $m_1, \ldots, m_t$ be integers, and let p be a prime. If $p \mid (m_1 \cdots m_t)$, then $p \mid m_i$ for one of the factors m_i among $m_1, \ldots, m_t$.*

We may write the product $m_1 \cdots m_t$ as $m_1(m_2 \cdots m_t)$, so by Euclid's lemma, $p \mid m_1$ or $p \mid m_2 \cdots m_t$. In the first case we are done, and in the second case we may apply Euclid's lemma once again, and repeat this up to t times.

Corollary 2 *Let $q_1, \ldots, q_t$ and p be positive primes. If $p \mid (q_1 \cdots q_t)$, then p is equal to one of the factors $q_1, \ldots, q_t$.*

By Corollary 1, p *divides* one of the factors $q_1, \ldots, q_t$, say $p \mid q_i$. But q_i is a prime, so its only divisors are ± 1 and $\pm q_i$; p is positive and not equal to 1, so if $p \mid q_i$, necessarily $p = q_i$.

Theorem 4: Factorization into primes *Every integer $n > 1$ can be expressed as a product of positive primes. That is, there are one or more primes $p_1, \ldots, p_r$ such that*

$$n = p_1 p_2 \cdots p_r$$

Let K represent the set of all the positive integers greater than 1 which *cannot* be written as a product of one or more primes. We will assume there *are* such integers, and derive a contradiction.

By the well-ordering principle, K contains a least integer m; m cannot be a prime, because if it *were* a prime it would not be in K. Thus, m is composite; so by the compositive number lemma,

$$m = rs$$

for positive integers r and s *less* than m and greater than 1; r and s are not in K because m is the least integer in K. This means that r and s *can* be expressed as products of primes, hence so can $m = rs$. This contradiction proves that K is empty; hence every $n > 1$ can be expressed as a product of primes.

Theorem 5: Unique factorization *Suppose n can be factored into positive primes in two ways, namely*

$$n = p_1 \cdots p_r = q_1 \cdots q_t$$

Then $r = t$, and the p_i are the same numbers as the q_j except, possibly, for the order in which they appear.

In the equation $p_1 \cdots p_r = q_1 \cdots q_t$, let us cancel common factors from each side, one by one, until we can do no more canceling. If all the factors are canceled on both sides, this proves the theorem. Otherwise, we are left with some factors on each side, say

$$p_i \cdots p_k = q_j \cdots q_m$$

Now, p_i is a factor of $p_i \cdots p_k$, so $p_i \,|\, q_j \cdots q_m$. Thus, by Corollary 2 to Euclid's lemma, p_i is *equal* to one of the factors $q_j, \ldots, q_m$, which is impossible because we assumed we can do no more canceling.

It follows from Theorems 4 and 5 that every integer m can be factored into primes, and that the prime factors of m are unique (except for the order in which we happen to list them).

EXERCISES

A. Properties of the Relation "*a* Divides *b*"

Prove the following, for any integers a, b, and c.

1 If $a \,|\, b$ and $b \,|\, c$, then $a \,|\, c$.
2 $a \,|\, b$ iff $a \,|\, (-b)$ iff $(-a) \,|\, b$.
3 $1 \,|\, a$ and $(-1) \,|\, a$.
4 $a \,|\, 0$.
5 If $c \,|\, a$ and $c \,|\, b$, then $c \,|\, (ax + by)$ for all $x, y \in \mathbb{Z}$.
6 If $a > 0$ and $b > 0$ and $a \,|\, b$, then $a \le b$.
7 $a \,|\, b$ iff $ac \,|\, bc$, when $c \ne 0$.
8 If $a \,|\, b$ and $c \,|\, d$, then $ac \,|\, bd$.
9 Let p be a prime. If $p \,|\, a^n$ for some $n > 0$, then $p \,|\, a$.

B. Properties of the gcd

Prove the following, for any integers a, b, and c. For each of these problems, you will need only the definition of the gcd.

1 $\gcd(a, 0) = a$, if $a > 0$.
2 $\gcd(a, b) = \gcd(a, b + xa)$ for any $x \in \mathbb{Z}$.
3 Let p be a prime. Then $\gcd(a, p) = 1$ or p. (Explain.)
4 Suppose every common divisor of a and b is a common divisor of c and d, and vice versa. Then $\gcd(a, b) = \gcd(c, d)$.

5 If $\gcd(ab, c) = 1$, then $\gcd(a, c) = 1$ and $\gcd(b, c) = 1$.
6 Let $\gcd(a, b) = c$. Write $a = ca'$ and $b = cb'$. Then

$$\gcd(a', b') = \gcd(a, b') = \gcd(a', b) = 1.$$

C. Properties of Relatively Prime Integers

Prove the following, for all integers a, b, c, d, r, and s. (Theorem 3 will be helpful.)

1 If there are integers r and s such that $ra + sb = 1$, then a and b are relatively prime.
2 If $\gcd(a, c) = 1$ and $c \mid ab$, then $c \mid b$. (Reason as in the proof of Euclid's lemma.)
3 If $a \mid d$ and $c \mid d$ and $\gcd(a, c) = 1$, then $ac \mid d$.
4 If $d \mid ab$ and $d \mid cb$, where $\gcd(a, c) = 1$, then $d \mid b$.
5 If $d = \gcd(a, b)$ where $a = dr$ and $b = ds$, then $\gcd(r, s) = 1$.
6 If $\gcd(a, c) = 1$ and $\gcd(b, c) = 1$, then $\gcd(ab, c) = 1$.

D. Further Properties of gcd's and Relatively Prime Integers

Prove the following, for all integers a, b, c, d, r, and s.

1 Suppose $a \mid b$ and $c \mid b$ and $\gcd(a, c) = d$. Then $ac \mid bd$.
2 If $ac \mid b$ and $ad \mid b$ and $\gcd(c, d) = 1$, then $acd \mid b$.
3 Let $d = \gcd(a, b)$. For any integer x, $d \mid x$ iff x is a linear combination of a and b.
4 Suppose that for all integers x, $x \mid a$ and $x \mid b$ iff $x \mid c$. Then $c = \gcd(a, b)$.
5 Prove by induction: For all $n > 0$, if $\gcd(a, b) = 1$, then $\gcd(a, b^n) = 1$.
6 Suppose $\gcd(a, b) = 1$ and $c \mid ab$. Then there exist integers r and s such that $c = rs$, $r \mid a$, $s \mid b$, and $\gcd(r, s) = 1$.

E. A Property of the gcd

Let a and b be integers. *Prove the following :*

1 Suppose a is odd and b is even, or vice versa. Then $\gcd(a, b) = \gcd(a + b, a - b)$.
2 Suppose a and b are both odd. Then $2\gcd(a, b) = \gcd(a + b, a - b)$.
3 If a and b are both even, explain why either of the two previous conclusions are possible.

F. Least Common Multiples

A *least common multiple* of two integers a and b is a positive integer c such that (i) $a \mid c$ and $b \mid c$; (ii) if $a \mid x$ and $b \mid x$, then $c \mid x$. *Prove the following:*

1 The set of all the common multiples of a and b is an ideal of $\mathbb{Z}$.

2 Every pair of integers a and b has a least common multiple. (HINT: Use part 1.)

The least common multiple of a and b is denoted by lcm(a, b).

3 $a \cdot \text{lcm}(b, c) = \text{lcm}(ab, ac)$.

4 If $a = a_1 c$ and $b = b_1 c$ where $c = \gcd(a, b)$, then $\text{lcm}(a, b) = a_1 b_1 c$.

5 $\text{lcm}(a, ab) = ab$.

6 If $\gcd(a, b) = 1$, then $\text{lcm}(a, b) = ab$.

7 If $\text{lcm}(a, b) = ab$, then $\gcd(a, b) = 1$.

8 Let $\gcd(a, b) = c$. Then $\text{lcm}(a, b) = ab/c$.

9 Let $\gcd(a, b) = c$ and $\text{lcm}(a, b) = d$. Then $cd = ab$.

G. Ideals in $\mathbb{Z}$

Prove the following :

1 $\langle n \rangle$ is a prime ideal iff n is a prime number.

2 Every prime ideal of $\mathbb{Z}$ is a maximal ideal. [HINT: If $\langle p \rangle \subseteq \langle a \rangle$, but $\langle p \rangle \neq \langle a \rangle$, explain why $\gcd(p, a) = 1$ and conclude that $1 \in \langle a \rangle$.]

3 For every prime number p, $\mathbb{Z}_p$ is a field. (HINT: Remember $\mathbb{Z}_p = \mathbb{Z}/\langle p \rangle$. Use the last page of Chapter 19.)

4 If $c = \text{lcm}(a, b)$, then $\langle a \rangle \cap \langle b \rangle = \langle c \rangle$.

5 Every homomorphic image of $\mathbb{Z}$ is isomorphic to $\mathbb{Z}_n$ for some n.

6 Let G be a group and let $a, b \in G$. Then $S = \{n \in \mathbb{Z} : ab^n = b^n a\}$ is an ideal of $\mathbb{Z}$.

7 Let G be a group, H a subgroup of G, and $a \in G$. Prove that

$$S = \{n \in \mathbb{Z} : a^n \in H\}$$

is an ideal of $\mathbb{Z}$.

8 If $\gcd(a, b) = d$, then $\langle a \rangle + \langle b \rangle = \langle d \rangle$. (NOTE: If J and K are ideals of a ring A, then $J + K = \{x + y : x \in J \text{ and } y \in K\}$.)

H. The gcd and the lcm as Operations on $\mathbb{Z}$.

For any two integers a and b, let $a * b = \gcd(a, b)$ and $a \circ b = \text{lcm}(a, b)$. *Prove the following properties of these operations.*

1 $*$ and $\circ$ are associative.

2 There is an identity element for $\circ$, but not for $*$.

3 Which integers have inverses with respect to $\circ$?

4 $a * (b \circ c) = (a * b) \circ (a * c)$.

TWENTY-THREE

ELEMENTS OF NUMBER THEORY

Almost as soon as children are able to count, they learn to distinguish between even numbers and odd numbers. The distinction between even and odd is the most elemental of all concepts relating to numbers. It is also the starting point of the modern science of number theory.

From a sophisticated standpoint, a number is even if the remainder, after dividing the number by 2, is 0. The number is odd if that remainder is 1.

This notion may be generalized in an obvious way. Let n be any positive integer: a number is said to be *congruent to* 0, *modulo n* if the remainder, when the number is divided by n, is 0. The number is said to be *congruent to* 1, *modulo n* if the remainder, when the number is divided by n, is 1. Similarly, the number is *congruent to* 2, *modulo n* if the remainder after division by n is 2; and so on. This is the natural way of generalizing the distinction between odd and even.

Note that "even" is the same as "congruent to 0, modulo 2"; and "odd" is the same as "congruent to 1, modulo 2."

In short, the distinction between odd and even is only one special case of a more general notion. We shall now define this notion formally:

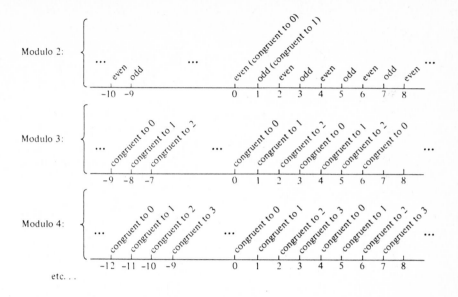

Let n be any positive integer. If a and b are any two integers, we shall say that a *is congruent to* b, *modulo* n if a and b, when they are divided by n, leave the same remainder r. That is, if we use the division algorithm to divide a and b by n, then

$$a = nq_1 + r \qquad \text{and} \qquad b = nq_2 + r$$

where the remainder r is the same in both equations.

Subtracting these two equations, we see that

$$a - b = (nq_1 + r) - (nq_2 + r) = n(q_1 - q_2)$$

Therefore we get the following important fact:

$$a \text{ is congruent to } b, \text{ modulo } n \qquad iff \qquad n \text{ divides } a - b \qquad (1)$$

If a is congruent to b, modulo n, we express this fact in symbols by writing

$$a \equiv b \ (\text{mod } n)$$

which should be read "a is congruent to b, modulo n." We refer to this relation as *congruence modulo n*.

By using (1), it is easily verified that congruence modulo n is a reflexive, symmetric, and transitive relation on $\mathbb{Z}$. It is also easy to check that for any $n > 0$ and any integers a, b, and c,

$$a \equiv b \ (\text{mod } n) \qquad \text{implies} \qquad a + c \equiv b + c \ (\text{mod } n)$$

and $\qquad a \equiv b \ (\text{mod } n) \qquad \text{implies} \qquad ac \equiv bc \ (\text{mod } n)$

(The proofs, which are exceedingly easy, are assigned as Exercise C at the end of this chapter.)

Recall that

$$\langle n \rangle = \{ \ldots, -3n, -2n, -n, 0, n, 2n, 3n, \ldots \}$$

is the ideal of $\mathbb{Z}$ which consists of all the multiples of n. The quotient ring $\mathbb{Z}/\langle n \rangle$ is usually denoted by $\mathbb{Z}_n$, and its elements are denoted by $\bar{0}, \bar{1}, \bar{2}, \ldots, \overline{n-1}$. These elements are cosets:

$$\bar{0} = \langle n \rangle + 0 = \{ \ldots, -2n, -n, 0, n, 2n, \ldots \}$$

$$\bar{1} = \langle n \rangle + 1 = \{ \ldots, -2n + 1, -n + 1, 1, n + 1, 2n + 1, \ldots \}$$

$$\bar{2} = \langle n \rangle + 2 = \{ \ldots, -2n + 2, -n + 2, 2, n + 2, 2n + 2, \ldots \}$$

and so on

It is clear by inspection that different integers are in the same coset iff they differ from each other by a multiple of n. That is,

a and b are in the same coset $\qquad$ *iff* $\qquad$ *n divides a − b*

$$\textit{iff} \qquad a \equiv b \ (\textit{mod } n) \qquad [\text{by (1)}] \qquad (2)$$

If a is any integer, the coset (in $\mathbb{Z}_n$) which contains a will be denoted by $\bar{a}$. For example, in $\mathbb{Z}_6$,

$$\bar{0} = \bar{6} = -\bar{6} = \overline{12} = \overline{18} = \cdots \qquad \bar{1} = \bar{7} = -\bar{5} = \overline{13} = \cdots$$

$$\bar{2} = \bar{8} = -\bar{4} = \overline{14} = \cdots \qquad \text{etc.}$$

In particular, $\bar{a} = \bar{b}$ means that a and b are in the same coset. It follows by (2) that

$$\bar{a} = \bar{b} \qquad \text{in } \mathbb{Z}_n \qquad \textit{iff} \qquad a \equiv b \ (\text{mod } n) \qquad (3)$$

On account of this fundamental connection between congruence modulo n and equality in $\mathbb{Z}_n$, most facts about congruence can be discovered by examining the rings $\mathbb{Z}_n$. These rings have very simple properties, which are presented next. From these properties we will then be able to deduce all we need to know about congruences.

Let n be a positive integer. It is an important fact that for any integer a,

$\bar{a}$ is invertible in $\mathbb{Z}_n$ iff a and n are relatively prime. $\qquad (4)$

Indeed, if a and n are relatively prime, their *gcd* is equal to 1. Therefore, by Theorem 3 of Chapter 22, there are integers s and t such that $sa + tn = 1$. It follows that

$$1 - sa = tn \in \langle n \rangle$$

so by (**) on page 188, 1 and sa belong to the same coset in $\mathbb{Z}/\langle n \rangle$. This is the same as saying that $\bar{1} = \overline{sa} = \bar{s}\,\bar{a}$, hence $\bar{s}$ is the multiplicative inverse of $\bar{a}$ in $\mathbb{Z}_n$. The converse is proved by reversing the steps of this argument.

It follows from (4) that if n is a prime number, every nonzero element of $\mathbb{Z}_n$ is invertible! Thus,

$$\mathbb{Z}_p \text{ is a field, for every prime number } p. \tag{5}$$

In any field, the set of all the nonzero elements, with multiplication as the only operation (ignore addition), is a group. Indeed, the product of any two nonzero elements in nonzero, and the multiplicative inverse of any nonzero element is nonzero. Thus, in $\mathbb{Z}_p$, the set

$$\mathbb{Z}_p^* = \{\bar{1}, \bar{2}, \ldots, \overline{p-1}\}$$

with multiplication as its only operation, *is a group of order $p - 1$.*

Remember that if G is a group whose order is, let us say, m, then $x^m = e$ for every x in G. (This is true by Theorem 5 of Chapter 13.) Now, $\mathbb{Z}_p^*$ has order $p - 1$ and its identity element is $\bar{1}$, so $\bar{a}^{p-1} = \bar{1}$ for every $\bar{a} \neq \bar{0}$ in $\mathbb{Z}_p$. If we use (3) to translate this equality into a congruence, we get a classical result of number theory:

Little theorem of Fermat Let p be a prime. Then,

$$a^{p-1} \equiv 1 \pmod{p} \qquad \textit{for every} \qquad a \not\equiv 0 \pmod{p}$$

Corollary $a^p \equiv a \pmod{p}$ *for every integer a.*

Actually, a version of this theorem is true in $\mathbb{Z}_n$ even where n is *not* a prime number. In this case, let V_n denote the *set of all the invertible elements* in $\mathbb{Z}_n$. Clearly, V_n is a group with respect to multiplication. (Reason: The product of two invertible elements is invertible, and, if a is invertible, so is its inverse.) For any positive integer n, let $\phi(n)$ denote *the number of positive integers, less than n, which are relatively prime to n.* For example, 1, 3, 5, and 7 are relatively prime to 8, hence $\phi(8) = 4$. ϕ is called *Euler's phi-function.*

It follows immediately from (4) that the number of elements in V_n is $\phi(n)$.

Thus, V_n is a group of order $\phi(n)$, and its identity elements is $\bar{1}$. Consequently, for any $\bar{a}$ in V_n, $\bar{a}^{\phi(n)} = \bar{1}$. If we use (3) to translate this equation into a congruence, we get:

Euler's theorem *If a and n are relatively prime,* $a^{\phi(n)} \equiv 1 \pmod{n}$.

OPTIONAL

Congruences are more important in number theory than we might expect. This is because a vast range of problems in number theory—problems which have nothing to do with congruences at first sight—can be transformed into problems involving congruences, and are most easily solved in that form. An example is given next:

A *Diophantine equation* is any polynomial equation (in one or more unknowns) whose coefficients are integers. To solve a Diophantine equation is to find *integer values* of the unknowns which satisfy the equation. We might be inclined to think that the restriction to integer values makes it easier to solve equations; in fact, the very opposite is true. For instance, even in the case of an equation as simple as $4x + 2y = 5$, it is not obvious whether we can find an integer solution consisting of x and y in $\mathbb{Z}$. (As a matter of fact, there is *no* integer solution; try to explain why not.)

Solving Diophantine equations is one of the oldest and most important problems in number theory. Even the problem of solving Diophantine *linear* equations is difficult and has many applications. Therefore, it is a very important fact that *solving linear Diophantine equations is equivalent to solving linear congruences*. Indeed,

$$ax + by = c \quad \text{iff} \quad by = c - ax \quad \text{iff} \quad ax \equiv c \pmod{b}$$

Thus, any solution of $ax \equiv c \pmod{b}$ yields a solution in integers of $ax + by = c$.

Finding solutions of linear congruences is therefore an important matter, and we will turn our attention to it now.

A congruence such as $ax \equiv b \pmod{n}$ may look very easy to solve, but appearances can be deceptive. In fact, many such congruences have no solutions at all! For example, $4x \equiv 5 \pmod 2$ cannot have a solution, because $4x$ is always even [hence, congruent to $0 \pmod 2$], whereas 5 is odd [hence congruent to $1 \pmod 2$]. Our first item of business, therefore, is to find a way of recognizing whether or not a linear congruence has a solution:

Theorem 1 *The congruence* $ax \equiv b$ (mod n) *has a solution iff* $\gcd(a, n) \,|\, b$.

Indeed,

$$ax \equiv b \text{ (mod } n) \qquad \text{iff} \qquad n \,|\, (ax - b) \qquad \text{iff} \qquad ax - b = yn$$

$$\text{iff} \qquad ax - yn = b$$

Next, by the proof of Theorem 3 in Chapter 22, if J is the ideal of all the linear combinations of a and n, then $\gcd(a, n)$ is the least positive integer in J. Furthermore, every integer in J is a multiple of $\gcd(a, n)$. Thus, b is a linear combination of a and n iff $b \in J$ iff b is a multiple of $\gcd(a, n)$. This completes the proof of our theorem.

Now that we are able to recognize when a congruence *has* a solution, let us see what such a solution looks like.

Consider the congruence $ax \equiv b$ (mod n). By a *solution modulo n* of this congruence, we mean a congruence

$$x \equiv c \ (mod \ n)$$

such that any integer x satisfies $x \equiv c$ (mod n) iff it satisfies $ax \equiv b$ (mod n). [That is, the solutions of $ax \equiv b$ (mod n) are all the integers congruent to c, modulo n.] Does every congruence $ax \equiv b$ (mod n) (supposing that it *has* a solution) have a solution modulo n? Unfortunately not! Nevertheless, as a starter, we have the following:

Lemma *If* $\gcd(a, n) = 1$, *then* $ax \equiv b$ (mod n) *has a solution modulo n*.

Indeed, by (3), $ax \equiv b$ (mod n) is equivalent to the equality $\bar{a}\,\bar{x} = \bar{b}$ in $\mathbb{Z}_n$. But by (4), $\bar{a}$ has a multiplicative inverse in $\mathbb{Z}_n$, hence from $\bar{a}\,\bar{x} = \bar{b}$ we get $\bar{x} = \bar{a}^{-1}\bar{b}$. Setting $\bar{a}^{-1}\bar{b} = \bar{c}$, we get $\bar{x} = \bar{c}$ in $\mathbb{Z}_n$, that is, $x \equiv c$ (mod n).

Thus, if a and n are relatively prime, $ax \equiv b$ (mod n) has a solution modulo n. If a and n are *not* relatively prime, we have no solution modulo n; nevertheless, we have a very nice result:

Theorem 2 *If the congruence* $ax \equiv b$ (mod n) *has a solution, then it has a solution modulo m, where*

$$m = \frac{n}{\gcd(a, n)}$$

This means that the solution of $ax \equiv b$ (mod n) is of the form $x \equiv c$ (mod m); it consists of all the integers which are congruent to c, modulo m.

To prove this, let gcd(a, n) = d, and note the following:

$ax \equiv b$ (mod n)　　iff　　$n | (ax - b)$　　iff　　$ax - b = ny$

$$\text{iff}\quad \frac{a}{d}x - \frac{b}{d} = \frac{n}{d}y \quad \text{iff} \quad \frac{a}{d}x \equiv \frac{b}{d}\ (\text{mod}\ \frac{n}{d}) \qquad (*)$$

But a/d and n/d are relatively prime (because we are dividing a and n by d, which is their gcd); hence by the lemma,

$$\frac{a}{d}x \equiv \frac{b}{d}\ (\text{mod}\ \frac{n}{d})$$

has a solution x mod n/d. By (*), this is also a solution of $ax \equiv b$ (mod n).

As an example, let us solve $6x \equiv 4$ (mod 10). Gcd(6, 10) = 2 and $2 | 4$, so by Theorem 1, this congruence has a solution. By (*) in the proof of Theorem 2, this solution is the same as the solution of

$$\frac{6}{2}x \equiv \frac{4}{2}\ (\text{mod}\ \frac{10}{2}) \quad \text{that is} \quad 3x \equiv 2\ (\text{mod } 5)$$

This is equivalent to the equation $\bar{3}\bar{x} = \bar{2}$ in $\mathbb{Z}_5$, and its solution is $\bar{x} = \bar{4}$. So finally, the solution of $6x \equiv 4$ (mod 10) is $x \equiv 4$ (mod 5).

How do we go about solving several linear congruences simultaneously? Well, suppose we are given k congruences,

$$a_1 x \equiv b_1\ (\text{mod } n_1), \qquad a_2 x \equiv b_2\ (\text{mod } n_2), \qquad \dots, \qquad a_k x \equiv b_k\ (\text{mod } n_k)$$

If each of these congruences has a solution, we solve each one individually by means of Theorem 2. This gives us

$$x \equiv c_1\ (\text{mod } m_1), \qquad x \equiv c_2\ (\text{mod } m_2), \qquad \dots, \qquad x \equiv c_k\ (\text{mod } m_k)$$

We are left with the problem of solving this last set of congruences simultaneously.

Is there any integer x which satisfies all k of these congruences? The answer for *two* simultaneous congruences is as follows:

Theorem 3 *Consider $x \equiv a$ (mod n) and $x \equiv b$ (mod m). There is an integer x satisfying both simultaneously iff $a \equiv b$ (mod d), where d = gcd(m, n).*

If x is a simultaneous solution, then $n | (x - a)$ and $m | (x - b)$. Thus,

$$x - a = nq_1 \quad \text{and} \quad x - b = mq_2$$

Subtracting the first equation from the second gives

$$a - b = mq_2 - nq_1$$

But $d \mid m$ and $d \mid n$, so $d \mid (a - b)$; thus, $a \equiv b \pmod{d}$.

Conversely, if $a \equiv b \pmod{d}$, then $d \mid (a - b)$, so $a - b = dq$. By Theorem 3 of Chapter 22, $d = rn + tm$ for some integers r and t. Thus, $a - b = rqn + tqm$. From this equation, we get

$$a - rqn = b + tqm$$

Set $x = a - rqn = b + tqm$; then $x - a = -rqn$ and $x - b = tqm$, hence $n \mid (x - a)$ and $m \mid (x - b)$, so

$$x \equiv a \pmod{n} \qquad \text{and} \qquad x \equiv b \pmod{m}$$

Now that we are able to determine whether or not a pair of congruences *has* a simultaneous solution, let us see what such a solution looks like.

Theorem 4 *If a pair of congruences $x \equiv a \pmod{n}$ and $x \equiv b \pmod{m}$ has a simultaneous solution, then it has a simultaneous solution of the form*

$$x \equiv c \pmod{t}$$

where t is the least common multiple of m and n.

Before proving the theorem, let us observe that the least common multiple (lcm) of any two integers m and n has the following property: let t be the least common multiple of m and n. *Every common multiple of m and n is a multiple of t, and conversely.* That is, for all integers x,

$$m \mid x \qquad \text{and} \qquad n \mid x \qquad \text{iff} \qquad t \mid x$$

(See Exercise F at the end of Chapter 22.) In particular,

$$m \mid (x - c) \qquad \text{and} \qquad n \mid (x - c) \qquad \text{iff} \qquad t \mid (x - c)$$

hence

$$x \equiv c \pmod{m} \qquad \text{and} \qquad x \equiv c \pmod{n} \qquad \text{iff} \qquad x \equiv c \pmod{t} \quad (6)$$

Returning to our theorem, let c be any solution of the given pair of congruences (remember, we are assuming there *is* a simultaneous solution). Then $c \equiv a \pmod{n}$ and $c \equiv b \pmod{m}$. Any other integer x is a simultaneous solution iff $x \equiv c \pmod{n}$ and $x \equiv c \pmod{m}$. But by (6), this is true iff $x \equiv c \pmod{t}$. The proof is complete.

A special case of Theorems 3 and 4 is very important in practice: it is the case where m and n are relatively prime. Note that, in this case, $\gcd(m, n) = 1$ and $\operatorname{lcm}(m, n) = mn$. Thus, by Theorems 3 and 4,

(7) *If m and n are relatively prime, the pair of congruences* $x \equiv a$ (mod *n*) *and* $x \equiv b$ (mod *m*) ***always*** *has a solution. This solution is of the form* $x \equiv c$ (mod *mn*).

(7) can easily be extended to the case of more than two linear congruences. The result is known as the

Chinese remainder theorem *Let* $m_1, m_2, \ldots, m_k$ *be pairwise relatively prime. Then the system of simultaneous linear congruences*

$$x \equiv c_1 \text{ (mod } m_1), \qquad x \equiv c_2 \text{ (mod } m_2), \qquad \ldots, \qquad x \equiv c_k \text{ (mod } m_k)$$

always has a solution, which is of the form $x \equiv c$ (mod $m_1 m_2 \ldots m_k$).

Use Theorem 4 to solve $x \equiv c_1$ (mod m_1) and $x \equiv c_2$ (mod m_2) simultaneously. The solution is of the form $x \equiv d$ (mod $m_1 m_2$). Solve the latter simultaneously with $x \equiv c_3$ (mod m_3), to get a solution mod $m_1 m_2 m_3$. Repeat this process k times.

EXERCISES

A. Solving Single Congruences

1 For each of the following congruences, find m such that the congruence has a unique solution modulo m. If there is no solution, write "none."

    (a) $60x \equiv 12$ (mod 24)   (b) $42x \equiv 24$ (mod 30)   (c) $49x \equiv 30$ (mod 25)
    (d) $39x \equiv 14$ (mod 52)   (e) $147x \equiv 47$ (mod 98)   (f) $39x \equiv 26$ (mod 52)

2 Solve the following linear congruences:

    (a) $12x \equiv 7$ (mod 25)   (b) $35x \equiv 8$ (mod 12)   (c) $15x \equiv 9$ (mod 6)
    (d) $42x \equiv 12$ (mod 30)   (e) $147x \equiv 49$ (mod 98)   (f) $39x \equiv 26$ (mod 52)

3 (a) Explain why $2x^2 \equiv 8$ (mod 10) has the same solutions as $x^2 \equiv 4$ (mod 5).

    (b) Explain why $x \equiv 2$ (mod 5) and $x \equiv 3$ (mod 5) are all the solutions of $2x^2 \equiv 8$ (mod 10).

4 Solve the following quadratic congruences (if there is no solution, write "none"):

    (a) $6x^2 \equiv 9$ (mod 15)   (b) $60x^2 \equiv 18$ (mod 24)   (c) $30x^2 \equiv 18$ (mod 24)
    (d) $4(x + 1)^2 \equiv 14$ (mod 10)   (e) $4x^2 - 2x + 2 \equiv 0$ (mod 6)
    (f) $3x^2 - 6x + 6 \equiv 0$ (mod 15)

5 Solve the following congruences:

    (a) $x^4 \equiv 4$ (mod 6)   (b) $2(x - 1)^4 \equiv 0$ (mod 8)
    (c) $x^3 + 3x^2 + 3x + 1 \equiv 0$ (mod 8)
    (d) $x^4 + 2x^2 + 1 \equiv 4$ (mod 5)

6 Solve the following Diophantine equations (if there is no solution, write "none"):

    (a) $14x + 15y = 11$ (b) $4x + 5y = 1$ (c) $21x + 10y = 9$
    (d) $30x^2 + 24y = 18$

B. Solving Sets of Congruences

Example *Solve the pair of simultaneous congruences* $x \equiv 5 \pmod 6$, $x \equiv 7 \pmod{10}$.

By Theorems 3 and 4, this pair of congruences has a solution modulo 30. From $x \equiv 5 \pmod 6$, we get $x = 6q + 5$. Introducing this into $x \equiv 7 \pmod{10}$ yields $6q + 5 \equiv 7 \pmod{10}$. Thus, successively: $6q \equiv 2 \pmod{10}$, $3q \equiv 1 \pmod 5$, $q \equiv 2 \pmod 5$, $q = 5r + 2$. Introducing this into $x = 6q + 5$ gives $x = 6(5r + 2) + 5 = 30r + 17$. Thus, $x \equiv 17 \pmod{30}$. This is our solution.

1 Solve each of the following pairs of simultaneous congruences:
   (a) $x \equiv 7 \pmod 8$; $x \equiv 11 \pmod{12}$ (b) $x \equiv 12 \pmod{18}$; $x \equiv 30 \pmod{45}$
   (c) $x \equiv 8 \pmod{15}$; $x \equiv 11 \pmod{14}$

2 Solve each of the following pairs of simultaneous congruences:
   (a) $10x \equiv 2 \pmod{12}$; $6x \equiv 14 \pmod{20}$ (b) $4x \equiv 2 \pmod 6$; $9x \equiv 3 \pmod{12}$
   (c) $6x \equiv 2 \pmod 8$; $10x \equiv 2 \pmod{12}$

3 Use Theorem 3 to prove the following: Suppose we are given k congruences

$$x_1 \equiv c_1 \pmod{m_1}, \qquad x_2 \equiv c_2 \pmod{m_2} \qquad \dots, \qquad x_k \equiv c_k \pmod{m_k}$$

There is an x satisfying all k congruences simultaneously iff for all $i, j \in \{1, \dots, k\}$, $c_i \equiv c_j \pmod{d_{ij}}$, where $d_{ij} = \gcd(m_i, m_j)$.

4 Use Theorem 4 to prove the following: If the system of congruences $x_1 \equiv c_1 \pmod{m_1}, \dots, x_k \equiv c_k \pmod{m_k}$ has a simultaneous solution, then it has a simultaneous solution of the form $x \equiv c \pmod t$, where $t = \mathrm{lcm}(m_1, m_2, \dots, m_k)$.

5 Solve each of the following systems of simultaneous linear congruences; if there is no solution, write "none."
   (a) $x \equiv 2 \pmod 3$; $x \equiv 3 \pmod 4$; $x \equiv 1 \pmod 5$; $x \equiv 4 \pmod 7$
   (b) $6x \equiv 4 \pmod 8$; $10x \equiv 4 \pmod{12}$; $3x \equiv 8 \pmod{10}$
   (c) $5x \equiv 3 \pmod 6$; $4x \equiv 2 \pmod 6$; $6x \equiv 6 \pmod 8$

6 Solve the following systems of simultaneous Diophantine equations:
   (a) $4x + 6y = 2$; $9x + 12y = 3$ (b) $3x + 4y = 2$; $5x + 6y = 2$; $3x + 10y = 8$

C. Elementary Properties of Congruence

Prove the following for all integers a, b, c, d and all positive integers m and n.

1 If $a \equiv b \pmod n$ and $b \equiv c \pmod n$, then $a \equiv c \pmod n$.

2 If $a \equiv b \pmod n$, then $a + c \equiv b + c \pmod n$.

3 If $a \equiv b \pmod n$, then $ac \equiv bc \pmod n$.

4 $a \equiv b \pmod 1$.

5 If $ab \equiv 0 \pmod p$, where p is a prime, then $a \equiv 0 \pmod p$ or $b \equiv 0 \pmod p$.

6 If $a^2 \equiv b^2 \pmod p$, where p is a prime, then $a \equiv \pm b \pmod p$.

7 If $a \equiv b \pmod m$, then $a + km \equiv b \pmod m$, for any integer k. In particular, $a + km \equiv a \pmod m$.

8 If $ac \equiv bc \pmod{n}$ and $\gcd(c, n) = 1$, then $a \equiv b \pmod{n}$.
9 If $a \equiv b \pmod{n}$, then $a \equiv b \pmod{m}$ for any m which is a factor of n.

D. Further Properties of Congruence

Prove the following for all integers a, b, c and all positive integers m and n.

1 If $ac \equiv bc \pmod{n}$, and $\gcd(c, n) = d$, then $a \equiv b \pmod{n/d}$.
2 If $a \equiv b \pmod{n}$, then $\gcd(a, n) = \gcd(b, n)$.
3 If $a \equiv b \pmod{p}$ for every prime p, then $a = b$.
4 If $a \equiv b \pmod{n}$, then $a^m \equiv b^m \pmod{n}$ for every positive integer m.
5 If $a \equiv b \pmod{m}$ and $a \equiv b \pmod{n}$ where $\gcd(m, n) = 1$, then $a \equiv b \pmod{mn}$.
6 If $ab \equiv 1 \pmod{c}$, $ac \equiv 1 \pmod{b}$ and $bc \equiv 1 \pmod{a}$, then $ab + bc + ac \equiv 1 \pmod{abc}$. (Assume $a, b, c > 0$.)
7 If $a^2 \equiv 1 \pmod{2}$, then $a^2 \equiv 1 \pmod{4}$.
8 If $a \equiv b \pmod{n}$, then $a^2 + b^2 \equiv 2ab \pmod{n^2}$; and conversely.
9 If $a \equiv 1 \pmod{m}$, then a and m are relatively prime; and conversely.

E. Consequences of Fermat's Theorem

Prove the following.

1 If p is a prime, find $\phi(p)$. Use this to deduce Fermat's theorem from Euler's theorem.
2 If $p > 2$ is a prime and $a \not\equiv 0 \pmod{p}$, then

$$a^{(p-1)/2} \equiv \pm 1 \pmod{p}$$

3 (*a*) Let p be a prime > 2. If $p \equiv 3 \pmod{4}$, then $(p - 1)/2$ is odd.
  (*b*) Let $p > 2$ be a prime such that $p \equiv 3 \pmod{4}$. Then there is *no* solution to the congruence $x^2 + 1 \equiv 0 \pmod{p}$. [HINT: Raise both sides of $x^2 \equiv -1 \pmod{p}$ to the power $(p - 1)/2$, and use Fermat's little theorem.]
4 Let p and q be distinct primes. Then $p^{q-1} + q^{p-1} \equiv 1 \pmod{pq}$.
5 Let p be a prime.
  (*a*) If $m \mid (p - 1)$, then $a^m \equiv 1 \pmod{p}$ provided that $p \nmid a$.
  (*b*) If $m \mid (p - 1)$, then $a^{m+1} \equiv a \pmod{p}$ for all integers a.
6 Let p and q be distinct primes.
  (*a*) If $m \mid (p - 1)$ and $m \mid (q - 1)$, then $a^m \equiv 1 \pmod{pq}$ for any a such that $p \nmid a$ and $q \nmid a$.
  (*b*) If $m \mid (p - 1)$ and $m \mid (q - 1)$, then $a^{m+1} \equiv a \pmod{pq}$ for all integers a.
7 Generalize the result of part 6 to n distinct primes, $p_1, \ldots, p_n$. (State your result, but do not prove it.)

8 Use part 6 to explain why the following are true:

(i) $a^{19} \equiv 1 \pmod{133}$

(ii) $a^{10} \equiv 1 \pmod{66}$, provided a is not a multiple of 2, 3, or 11.

(iii) $a^{13} \equiv a \pmod{105}$

(iv) $a^{49} \equiv a \pmod{1547}$ (HINT: $1547 = 7 \times 13 \times 17$.)

9 Find the following integers x:

(a) $x \equiv 8^{38} \pmod{210}$ (b) $x \equiv 7^{57} \pmod{133}$ (a) $x \equiv 5^{73} \pmod{66}$

F. Consequences of Euler's Theorem

Prove each of the following :

1 If gcd $(a, n) = 1$, the solution modulo n of $ax \equiv b \pmod{n}$ is $x \equiv a^{\phi(n)-1} b \pmod{n}$.

2 If gcd $(a, n) = 1$, then $a^{m\phi(n)} \equiv 1 \pmod{n}$ for all values of m.

3 If gcd $(m, n) = $ gcd $(a, mn) = 1$, then $a^{\phi(m)\phi(n)} \equiv 1 \pmod{mn}$.

4 If p is a prime, $\phi(p^n) = p^n - p^{n-1} = p^{n-1}(p - 1)$. (HINT: For any integer a, a and p^n have a common divisor $\neq \pm 1$ iff a is a multiple of p. There are exactly p^{n-1} multiples of p between 1 and p^n.)

5 For every $a \not\equiv 0 \pmod{p}$, $a^{p^n(p-1)} \equiv 1 \pmod{p^{n+1}}$, where p is a prime.

6 Under the conditions of part 3, if t is a common multiple of $\phi(m)$ and $\phi(n)$, then $a^t \equiv 1 \pmod{mn}$. Generalize to three integers l, m, and n.

7 Use parts 4 and 6 to explain why the following are true:

(i) $a^{12} \equiv 1 \pmod{180}$ for every a such that $\gcd(a, 180) = 1$.

(ii) $a^{42} \equiv 1 \pmod{1764}$ if gcd $(a, 1764) = 1$. (REMARK: $1764 = 4 \times 9 \times 49$.)

(iii) $a^{20} \equiv 1 \pmod{1800}$ if gcd $(a, 1800) = 1$.

8 If gcd $(m, n) = 1$, then $n^{\phi(m)} + m^{\phi(n)} \equiv 1 \pmod{mn}$.

9 If l, m, n are relatively prime in pairs, then $(mn)^{\phi(l)} + (ln)^{\phi(m)} + (lm)^{\phi(n)} \equiv 1 \pmod{mn}$.

G. Wilson's Theorem, and Some Consequences

In any integral domain, if $x^2 = 1$, then $x^2 - 1 = (x + 1)(x - 1) = 0$, hence $x = \pm 1$. Thus, an element $x \neq \pm 1$ cannot be its own multiplicative inverse. As a consequence, in $\mathbb{Z}_p$ the integers $\bar{2}, \bar{3}, \dots, \overline{p - 2}$ may be arranged in pairs, each one being paired off with its multiplicative inverse. *Prove the following:*

1 In $\mathbb{Z}_p$, $\bar{2} \cdot \bar{3} \cdots \overline{p - 2} = \bar{1}$.

2 $(p - 2)! \equiv 1 \pmod{p}$ for any prime number p.

3 $(p - 1)! + 1 \equiv 0 \pmod{p}$ for any prime number p. This is known as *Wilson's Theorem*.

4 For any composite number $n \neq 4$, $(n - 1)! \equiv 0 \pmod{n}$. [HINT: If p is any prime factor of n, then p is a factor of $(n - 1)!$ Why?]

Before going on to the remaining exercises, we make the following observations: Let $p > 2$ be a prime. Then

$$(p - 1)! = 1 \cdot 2 \cdots \frac{p - 1}{2} \cdot \frac{p + 1}{2} \cdot \cdots \cdot (p - 2) \cdot (p - 1)$$

Consequently,

$$(p - 1)! \equiv (-1)^{(p - 1)/2} \left(1 \cdot 2 \cdots \frac{p - 1}{2} \right)^2 \pmod{p}$$

REASON: $p - 1 \equiv -1 \pmod{p}$, $p - 2 \equiv -2 \pmod{p}$, $\cdots$, $(p + 1)/2 \equiv -(p - 1)/2$ $\pmod{p}$.

With this result, prove the following:

5 $[(p - 1)/2]!^2 \equiv (-1)^{(p + 1)/2} \pmod{p}$, for any prime $p > 2$. (HINT: Use Wilson's theorem.)

6 If $p \equiv 1 \pmod 4$, then $\dfrac{p + 1}{2}$ is odd. (Why?) Conclude that

$$\left(\frac{p - 1}{2} \right)!^2 \equiv -1 \pmod{p}$$

7 If $p \equiv 3 \pmod 4$, then $\dfrac{p + 1}{2}$ is even. (Why?) Conclude that

$$\left(\frac{p - 1}{2} \right)!^2 \equiv 1 \pmod{p}$$

8 When $p > 2$ is a prime, the congruence $x^2 + 1 \equiv 0 \pmod{p}$ has a solution if $p \equiv 1$ $\pmod 4$. (HINT: Use part 6.)

9 For any prime $p > 2$, $x^2 \equiv -1 \pmod{p}$ has a solution iff $p \not\equiv 3 \pmod 4$. (HINT: Use part 8 and E3.)

H. Quadratic Residues

An integer a is called a *quadratic residue* modulo m if there is an integer x such that $x^2 \equiv a \pmod{m}$. This is the same as saying that $\bar{a}$ is a square in $\mathbb{Z}_m$. If a is not a quadratic residue modulo m, then a is called a *quadratic nonresidue* modulo m. Quadratic residues are important for solving quadratic congruences, for studying sums of squares, etc. Here, we will examine quadratic residues modulo an arbitrary prime $p > 2$.

Let $h : \mathbb{Z}_p^* \to \mathbb{Z}_p^*$ be defined by $h(\bar{a}) = \bar{a}^2$. Prove:

1 h is a homomorphism. Its kernel is $\{\pm \bar{1}\}$.

2 The range of h has $(p - 1)/2$ elements. If ran $h = R$, R is a subgroup of $\mathbb{Z}_p^*$ having two cosets. One contains all the residues, the other all the nonresidues.

The *Legendre symbol* is defined as follows:

$$\left(\frac{a}{p}\right) = \begin{cases} +1 & \text{if } p \nmid a \text{ and } a \text{ is a residue mod } p. \\ -1 & \text{if } p \nmid a \text{ and } a \text{ is a nonresidue mod } p. \\ 0 & \text{if } p \mid a. \end{cases}$$

3 Referring to part 2, let the two cosets of R called 1 and -1. Then $\mathbb{Z}_p^*/R = \{1, -1\}$. Explain why

$$\left(\frac{a}{p}\right) = h(\bar{a})$$

for every integer a which is not a multiple of p.

4 Evaluate: $\left(\dfrac{17}{23}\right); \left(\dfrac{3}{29}\right); \left(\dfrac{5}{11}\right); \left(\dfrac{8}{13}\right); \left(\dfrac{2}{23}\right).$

5 Prove: if $a \equiv b \pmod{p}$, then $\left(\dfrac{a}{p}\right) = \left(\dfrac{b}{p}\right)$. In particular, $\left(\dfrac{a + kp}{p}\right) = \left(\dfrac{a}{p}\right).$

6 Prove: (i) $\left(\dfrac{a}{p}\right)\left(\dfrac{b}{p}\right) = \left(\dfrac{ab}{p}\right)$ (ii) $\left(\dfrac{a^2}{p}\right) = 1$ if $p \nmid a$

7 $\left(\dfrac{-1}{p}\right) = \begin{cases} 1 \text{ if } p \equiv 1 \pmod{4} \\ -1 \text{ if } p \equiv 3 \pmod{4} \end{cases}$ (HINT: Use Exercises G6 and 7.)

The most important rule for computing

$$\left(\frac{a}{p}\right)$$

is the *law of quadratic reciprocity*, which asserts that for distinct primes $p, q > 2$,

$$\left(\frac{p}{q}\right) = \begin{cases} -\left(\dfrac{q}{p}\right) & \text{if } p, q \text{ are both} \equiv 3 \pmod{4} \\ \left(\dfrac{q}{p}\right) & \text{otherwise} \end{cases}$$

(The proof may be found in any textbook on number theory, for example, *Fundamentals of Number Theory*, by W. J. LeVeque.)

8 Use parts 5 to 7 and the law of quadratic reciprocity to find:

$$\left(\frac{30}{101}\right); \quad \left(\frac{10}{151}\right); \quad \left(\frac{15}{41}\right); \quad \left(\frac{14}{59}\right); \quad \left(\frac{379}{401}\right)$$

Is 14 a quadratic residue, modulo 59?

9 Which of the following congruences is solvable?

(a) $x^2 \equiv 30 \pmod{101}$ (b) $x^2 \equiv 6 \pmod{103}$ (c) $2x^2 \equiv 70 \pmod{106}$

[NOTE: $x^2 \equiv a \pmod{p}$ is solvable iff a is a quadratic residue modulo p iff

$$\left(\frac{a}{p}\right) = 1.]$$

I. Primitive Roots

Recall that V_n is the multiplicative group of all the invertible elements in $\mathbb{Z}_n$. If V_n happens to be cyclic, say $V_n = \langle m \rangle$, then any integer $a \equiv m \pmod{n}$ is called a *primitive root* of n. Prove the following:

1 a is a primitive root of n iff the order of $\bar{a}$ in V_n is $\phi(n)$.

2 Every prime number p has a primitive root. (HINT: For every prime p, $\mathbb{Z}_p{}^*$ is a cyclic group. The simple proof of this fact is given as Theorem 1 in Chapter 33.)

3 Find primitive roots of the following integers (if there are none, say so): 6, 10, 12, 14, 15.

4 Suppose a is a primitive root of m. If b is any integer which is relatively prime to m, then $b \equiv a^k \pmod{m}$ for some $k \geq 1$.

5 Suppose m has a primitive root, and let n be relatively prime to $\phi(m)$. (Suppose $n > 0$.) Prove that if a is relatively prime to m, then $x^n \equiv a \pmod{m}$ has a solution.

6 Let $p > 2$ be a prime. Every primitive root of p is a quadratic nonresidue, modulo p. (HINT: Suppose a primitive root a is a residue; then every power of a is a residue.)

7 A prime p of the form $p = 2^m + 1$ is called a *Fermat prime*. Let p be a Fermat prime. Every quadratic nonresidue mod p is a primitive root of p. (HINT: How many primitive roots are there? How many residues? Compare.)

TWENTY-FOUR
RINGS OF POLYNOMIALS

In elementary algebra an important role is played by polynomials in an unknown x. These are expressions such as

$$2x^3 - \tfrac{1}{2}x^2 + 3$$

whose terms are grouped in powers of x. The exponents, of course, are positive integers and the coefficients are real or complex numbers.

Polynomials are involved in countless applications—applications of every kind and description. For example, polynomial functions are the easiest functions to compute, and therefore one commonly attempts to approximate arbitrary functions by polynomial functions. A great deal of effort has been expended by mathematicians to find ways of achieving this.

Aside from their uses in science and computation, polynomials come up very naturally in the general study of rings, as the following example will show:

Suppose we wish to enlarge the ring $\mathbb{Z}$ by adding to it the number π. It is easy to see that we will have to adjoin to $\mathbb{Z}$ other new numbers besides just π; for the enlarged ring (containing π as well as all the integers) will also contain such things as $-\pi$, $\pi + 7$, $6\pi^2 - 11$, and so on.

As a matter of fact, any ring which contains $\mathbb{Z}$ as a subring and which also contains the number π will have to contain every number of the form

$$a\pi^n + b\pi^{n-1} + \cdots + k\pi + l$$

where a, b, ..., k, l are integers. In other words, it will contain *all the polynomial expressions in π with integer coefficients.*

But the set of all the polynomial expressions in π with integer coefficients is a ring. (It is a subring of $\mathbb{R}$ because it is obvious that the sum and product of any two polynomials in π is again a polynomial in π.) This ring contains $\mathbb{Z}$ because every integer a is a polynomial with a constant term only, and it also contains π.

Thus, if we wish to enlarge the ring $\mathbb{Z}$ by adjoining to it the new number π, it turns out that the "next largest" ring after $\mathbb{Z}$ which contains $\mathbb{Z}$ as a subring and includes π, is exactly the ring of all the polynomials in π with coefficients in $\mathbb{Z}$.

As this example shows, aside from their practical applications, polynomials play an important role in the scheme of ring theory because they are precisely what we need when we wish to enlarge a ring by adding new elements to it.

In elementary algebra one considers polynomials whose coefficients are real numbers, or in some cases, complex numbers. As a matter of fact, the properties of polynomials are pretty much independent of the exact nature of their coefficients. All we need to know is that the coefficients are contained in some ring. For convenience, we will assume this ring is a commutative ring with unity.

Let A be a commutative ring with unity. Up to now we have used letters to denote elements or sets, but now we will use the letter x in a different way. In a polynomial expression such as $ax^2 + bx + c$, where a, b, $c \in A$, we do not consider x to be an element of A, but rather x is a symbol which we use in an entirely formal way. Later we will allow the substitution of other things for x, but at present x *is simply a placeholder.*

Notationally, the terms of a polynomial may be listed in either ascending or descending order. For example, $4x^3 - 3x^2 + x + 1$ and $1 + x - 3x^2 + 4x^3$ denote the same polynomial. In elementary algebra descending order is preferred, but for our purposes ascending order is more convenient.

Let A be a commutative ring with unity, and x an arbitrary symbol. Every expression of the form

$$a_0 + a_1 x + a_2 x^2 + \cdots + a_n x^n$$

*is called a **polynomial in x with coefficients in A**, or more simply, a **polynomial in x over A**. The expressions $a_k x^k$, for $k \in \{1, ..., n\}$, are called the **terms** of the polynomial.*

Polynomials in x are designated by symbols such as $a(x)$, $b(x)$, $q(x)$, and so on. If $a(x) = a_0 + a_1 x + \cdots + a_n x^n$ is any polynomial and $a_k x^k$ is any one of its terms, a_k is called the *coefficient* of x^k. By the *degree* of a polynomial $a(x)$ we mean the *greatest n such that the coefficient of* x^n *is not zero*. In other words, if $a(x)$ has degree n, this means that $a_n \neq 0$ but $a_m = 0$ for every $m > n$. The degree of $a(x)$ is symbolized by

$$\deg a(x)$$

For example, $1 + 2x - 3x^2 + x^3$ is a polynomial degree 3.

The polynomial $0 + 0x + 0x^2 + \cdots$ all of whose coefficients are equal to zero is called the *zero polynomial*, and is symbolized by 0. It is the only polynomial whose *degree is not defined* (because it has no nonzero coefficient).

If a nonzero polynomial $a(x) = a_0 + a_1 x + \cdots + a_n x^n$ has degree n, then a_n is called its *leading coefficient*: it is the last nonzero coefficient of $a(x)$. The term $a_n x^n$ is then called its *leading term*, while a_0 is called its *constant term*.

If a polynomial $a(x)$ has degree zero, this means that its constant term a_0 is its only nonzero term: $a(x)$ is a *constant polynomial*. Beware of confusing a polynomial of degree zero with the zero polynomial.

Two polynomials $a(x)$ and $b(x)$ are *equal* if they have the same degree and corresponding coefficients are equal. Thus, if $a(x) = a_0 + \cdots + a_n x^n$ is of degree n, and $b(x) = b_0 + \cdots + b_m x^m$ is of degree m, then $a(x) = b(x)$ iff $n = m$ and $a_k = b_k$ for each k from 0 to n.

The familiar sigma notation for sums is useful for polynomials. Thus,

$$a(x) = a_0 + a_1 x + \cdots + a_n x^n = \sum_{k=0}^{n} a_k x^k$$

with the understanding that $x^0 = 1$.

Addition and multiplication of polynomials is familiar from elementary algebra. We will now define these operations formally. Throughout these definitions we let $a(x)$ and $b(x)$ stand for the following polynomials:

$$a(x) = a_0 + a_1 x + \cdots + a_n x^n$$

$$b(x) = b_0 + b_1 x + \cdots + b_n x^n$$

Here we do *not* assume that $a(x)$ and $b(x)$ have the same degree, but allow ourselves to insert zero coefficients if necessary to achieve uniformity of appearance.

We add polynomials by adding corresponding coefficients. Thus,

$$a(x) + b(x) = (a_0 + b_0) + (a_1 + b_1)x + \cdots + (a_n + b_n)x^n$$

Note that the degree of $a(x) + b(x)$ is less than or equal to the *higher* of the two degrees, deg $a(x)$ and deg $b(x)$.

Multiplication is more difficult, but quite familiar:

$$a(x)b(x)$$
$$= a_0 b_0 + (a_0 b_1 + b_0 a_1)x + (a_0 b_2 + a_1 b_1 + a_2 b_0)x^2 + \cdots + a_n b_n x^{2n}$$

In other words, the product of $a(x)$ and $b(x)$ is the polynomial

$$c(x) = c_0 + c_1 x + \cdots + c_{2n} x^{2n}$$

whose kth coeffient (for any k from 0 to $2n$) is

$$c_k = \sum_{i+j=k} a_i b_j$$

This is the sum of all the $a_i b_j$ for which $i + j = k$. Note that deg $[a(x)b(x)]$ $\leq$ deg $a(x)$ + deg $b(x)$.

If A is any ring, the symbol

$$A[x]$$

designates the set of all the polynomials in x whose coefficients are in A, with addition and multiplication of polynomials as we have just defined them.

Theorem 1 *Let A be a commutative ring with unity. Then $A[x]$ is a commutative ring with unity.*

To prove this theorem, we must show systematically that $A[x]$ satisfies all the axioms of a commutative ring with unity. Throughout the proof, let $a(x)$, $b(x)$, and $c(x)$ stand for the following polynomials:

$$a(x) = a_0 + a_1 x + \cdots + a_n x^n$$
$$b(x) = b_0 + b_1 x + \cdots + b_n x^n$$

and
$$c(x) = c_0 + c_1 x + \cdots + c_n x^n$$

The axioms which involve only addition are easy to check: for example, addition is commutative because

$$a(x) + b(x) = (a_0 + b_0) + (a_1 + b_1)x + \cdots + (a_n + b_n)x^n$$
$$= (b_0 + a_0) + (b_1 + a_1)x + \cdots + (b_n + a_n)x^n = b(x) + a(x)$$

The associative law of addition is proved similarly, and is left as an exercise. The zero polynomial has already been described, and the negative of $a(x)$ is

$$-a(x) = (-a_0) + (-a_1)x + \cdots + (-a_n)x^n$$

To prove that multiplication is associative requires some care. Let $b(x)c(x) = d(x)$, where $d(x) = d_0 + d_1x + \cdots + d_{2n}x^{2n}$. By the definition of polynomial multiplication, the kth coefficient of $b(x)c(x)$ is

$$d_k = \sum_{i+j=k} b_i c_j$$

Then $a(x)[b(x)c(x)] = a(x)d(x) = e(x)$, where $e(x) = e_0 + e_1x + \cdots + e_{3n}x^{3n}$. Now, the lth coefficient of $a(x)d(x)$ is

$$e_l = \sum_{h+k=l} a_h d_k = \sum_{h+k=l} a_h \left(\sum_{i+j=k} b_i c_j \right)$$

It is easy to see that the sum on the right consists of all the terms $a_h b_i c_j$ such that $h + i + j = l$. Thus,

$$e_l = \sum_{h+i+j=l} a_h b_i c_j$$

For each l from 0 to $3n$, e_l is the lth coefficient of $a(x)[b(x)c(x)]$.

If we repeat this process to find the lth coefficient of $[a(x)b(x)]c(x)$, we discover that it, too, is e_l. Thus,

$$a(x)[b(x)c(x)] = [a(x)b(x)]c(x)$$

To prove the distributive law, let $a(x)[b(x) + c(x)] = d(x)$ where $d(x) = d_0 + d_1x + \cdots + d_{2n}x^{2n}$. By the definitions of polynomial addition and multiplication, the kth coefficient $a(x)[b(x) + c(x)]$ is

$$d_k = \sum_{i+j=k} a_i(b_j + c_j) = \sum_{i+j=k} (a_i b_j + a_i c_j)$$

$$= \sum_{i+j=k} a_i b_j + \sum_{i+j=k} a_i c_j$$

But $\sum_{i+j=k} a_i b_j$ is exactly the kth coefficient of $a(x)b(x)$, and $\sum_{i+j=k} a_i c_j$ is the kth coefficient of $a(x)c(x)$, hence d_k is equal to the kth coefficient of $a(x)b(x) + a(x)c(x)$. This proves that

$$a(x)[b(x) + c(x)] = a(x)b(x) + a(x)c(x)$$

The commutative law of multiplication is simple to verify and is left to the student. Finally, the unity polynomial is the constant polynomial 1.

Theorem 2 *If A is an integral domain, then $A[x]$ is an integral domain.*

If $a(x)$ and $b(x)$ are nonzero polynomials, we must show that their product $a(x)b(x)$ is not zero. Let a_n be the leading coefficient of $a(x)$, and b_m the leading coefficient of $b(x)$. By definition, $a_n \neq 0$, and $b_m \neq 0$. Thus $a_n b_m \neq 0$ because A is an integral domain. It follows that $a(x)b(x)$ has a nonzero coefficient (namely $a_n b_m$), so it is not the zero polynomial.

If A is an integral domain, we refer to $A[x]$ as a *domain of polynomials*, because $A[x]$ is an integral domain. Note that by the preceding proof, if a_n and b_m are the leading coefficients of $a(x)$ and $b(x)$, then $a_n b_m$ is the leading coefficient of $a(x)b(x)$. Thus, deg $a(x)b(x) = n + m$: *In a domain of polynomials $A[x]$, where A is an integral domain,*

$$\deg[a(x) \cdot b(x)] = \deg a(x) + \deg b(x)$$

In the remainder of this chapter we will look at a property of polynomials which is of special interest when all the coefficients lie in a field. Thus, from this point forward, let F be a field, and let us consider polynomials belonging to $F[x]$.

It would be tempting to believe that if F is a field then $F[x]$ also is a field. However, this is not so, for one can easily see that the multiplicative inverse of a polynomial is not generally a polynomial. Nevertheless, by Theorem 2, $F[x]$ *is an integral domain.*

Domains of polynomials over a *field* do, however, have a very special property: any polynomial $a(x)$ may be divided by any nonzero polynomial $b(x)$ to yield a quotient $q(x)$ and a remainder $r(x)$. The remainder is either 0, or if not, its degree is less than the degree of the divisor $b(x)$. For example, x^2 may be divided by $x - 2$ to give a quotient of $x + 2$ and a remainder of 4:

$$\underbrace{x^2}_{a(x)} = \underbrace{(x - 2)}_{b(x)} \underbrace{(x + 2)}_{q(x)} + \underbrace{4}_{r(x)}$$

This kind of polynomial division is familiar to every student of elementary algebra. It is customarily set up as follows:

```
                          x + 2      ◄────Quotient q (x)
Divisor ────►  x - 2 ⟌ x²            ◄────Dividend b (x)
  a (x)              x² - 2x
                    ─────────
                        2x
                        2x - 4
                    ─────────
                        4        ◄────Remainder r (x)
```

The process of polynomial division is formalized in the next theorem.

Theorem 3: Division algorithm for polynomials *If $a(x)$ and $b(x)$ are polynomials over a field F, and $b(x) \neq 0$, there exist polynomials $q(x)$ and $r(x)$ over*

F such that

$$a(x) = b(x)q(x) + r(x)$$

and $\qquad\qquad r(x) = 0 \qquad$ or $\qquad$ deg $r(x) <$ deg $b(x)$

Let $b(x)$ remain fixed, and let us show that every polynomial $a(x)$ satisfies the following condition:

(*) *There exist polynomials* $q(x)$ *and* $r(x)$ *over* F *such that* $a(x) = b(x)q(x) + r(x)$, *and* $r(x) = 0$ or deg $r(x) <$ deg $b(x)$.

We will assume there are polynomials $a(x)$ which do *not* fulfill Condition (*) and from this assumption we will derive a contradiction. Let $a(x)$ be a polynomial of *lowest degree* which fails to satisfy (*). Note that $a(x)$ cannot be zero, because we can express 0 as $0 = b(x) \cdot 0 + 0$, whereby $a(x)$ would satisfy (*). Furthermore, deg $a(x) \geq$ deg $b(x)$, for if deg $a(x) <$ deg $b(x)$ then we could write $a(x) = b(x) \cdot 0 + a(x)$, so again $a(x)$ would satisfy (*).

Let $a(x) = a_0 + \cdots + a_n x^n$ and $b(x) = b_0 + \cdots + b_m x^m$. Define a new polynomial

$$A(x) = a(x) - \frac{a_n}{b_m} x^{n-m} b(x) \qquad\qquad\qquad (**)$$

$$= a(x) - \left(b_0 \frac{a_n}{b_m} x^{n-m} + b_1 \frac{a_n}{b_m} x^{n-m+1} + \cdots + \underbrace{b_m \frac{a_n}{b_m} x^{n-m+m}}_{a_n x^n} \right)$$

This expression is the difference of two polynomials both of degree n and both having the same leading term $a_n x^n$. Because $a_n x^n$ cancels in the subtraction, $A(x)$ has degree *less than* n.

Remember that $a(x)$ is a polynomial of *least degree* which fails to satisfy (*), hence $A(x)$ *does* satisfy (*). This means there are polynomials $p(x)$ and $r(x)$ such that

$$A(x) = b(x)p(x) + r(x)$$

where $r(x) = 0$ or deg $r(x) <$ deg $b(x)$. But then

$$a(x) = A(x) + \frac{a_n}{b_m} x^{n-m} b(x) \qquad\qquad \text{by } (**)$$

$$= b(x)p(x) + r(x) + \frac{a_n}{b_m} x^{n-m} b(x)$$

$$= b(x)\left(p(x) + \frac{a_n}{b_m} x^{n-m} \right) + r(x)$$

If we let $p(x) + (a_n/b_m)x^{n-m}$ be renamed $q(x)$, then $a(x) = b(x)q(x) + r(x)$, so $a(x)$ fulfills Condition (*). This is a contradiction, as required.

EXERCISES

A. Elementary Computation in Domains of Polynomials

REMARK ON NOTATION: In some of the problems which follow, we consider polynomials with coefficients in $\mathbb{Z}_n$ for various n. To simplify notation, we denote the elements of $\mathbb{Z}_n$ by $1, 2, \ldots, n - 1$ rather than the more correct $\bar{1}, \bar{2}, \ldots, \overline{n-1}$.

1 Let $a(x) = 2x^2 + 3x + 1$ and $b(x) = x^3 + 5x^2 + x$. Compute $a(x) + b(x)$, $a(x) - b(x)$ and $a(x)b(x)$ in $\mathbb{Z}[x]$, $\mathbb{Z}_5[x]$, $\mathbb{Z}_6[x]$, and $\mathbb{Z}_7[x]$.

2 Find the quotient and remainder when $x^3 + x^2 + x + 1$ is divided by $x^2 + 3x + 2$ in $\mathbb{Z}[x]$ and in $\mathbb{Z}_5[x]$.

3 Find the quotient and remainder when $x^3 + 2$ is divided by $2x^2 + 3x + 4$ in $\mathbb{Z}[x]$, in $\mathbb{Z}_3[x]$, and in $\mathbb{Z}_5[x]$.

We call $b(x)$ a *factor* of $a(x)$ if $a(x) = b(x)q(x)$ for some $q(x)$, that is, if the remainder when $a(x)$ is divided by $b(x)$ is equal to zero.

4 Show that the following is true in $A[x]$ for any ring A: For any odd n,
  (a) $x + 1$ is a factor of $x^n + 1$
  (b) $x + 1$ is a factor of $x^n + x^{n-1} + \cdots + x + 1$

5 Prove the following: In $\mathbb{Z}_3[x]$, $x + 2$ is a factor of $x^m + 2$, for all m. In $\mathbb{Z}_n[x]$, $x + (n - 1)$ is a factor of $x^m + (n - 1)$, for all m and n.

6 Prove that there is no *integer* m such that $3x^2 + 4x + m$ is a factor of $6x^4 + 50$ in $\mathbb{Z}[x]$.

7 For what values of n is $x^2 + 1$ a factor of $x^5 + 5x + 6$ in $\mathbb{Z}_n[x]$?

B. Problems Involving Concepts and Definitions

1 Is $x^8 + 1 = x^3 + 1$ in $\mathbb{Z}_5[x]$? Explain your answer.

2 Is there any ring A such that in $A[x]$, some polynomial of degree 2 is equal to a polynomial of degree 4? Explain.

3 Write all the quadratic polynomials in $\mathbb{Z}_5[x]$. How many are there? How many cubic polynomials are there in $\mathbb{Z}_5[x]$? More generally, how many polynomials of degree n are there in $\mathbb{Z}_n[x]$?

4 Let A be an integral domain; prove the following:
If $(x + 1)^2 = x^2 + 1$ in $A[x]$, then A must have characteristic 2.
If $(x + 1)^4 = x^4 + 1$ in $A[x]$, then A must have characteristic 2.
If $(x + 1)^6 = x^6 + 2x^3 + 1$ in $A[x]$, then A must have characteristic 3.

5 Find an example of each of the following in $\mathbb{Z}_8[x]$: a divisor of zero, an invertible element, an idempotent element.

6 Explain why x cannot be invertible in any $A[x]$, hence no domain of polynomials can ever be a field.

7 There are rings such as P_3 in which every element $\neq 0$, 1 is a divisor of zero. Explain why this cannot happen in any ring of polynomials $A[x]$, even when A is *not* an integral domain.

8 Show that in every $A[x]$, there are elements $\neq 0$, 1 which are not idempotent, and elements $\neq 0$, 1 which are not nilpotent.

9 Prove that if $A[x]$ has an invertible element, so does A.

C. Rings $A[x]$ where A Is Not an Integral Domain

1 If A is not an integral domain, neither is $A[x]$. Prove this by showing that if A has divisors of zero, so does $A[x]$.

2 Give examples of divisors of zero, of degrees 0, 1, and 2, in $\mathbb{Z}_4[x]$.

3 In $\mathbb{Z}_{10}[x]$, $(2x + 2)(2x + 2) = (2x + 2)(5x^3 + 2x + 2)$, yet $(2x + 2)$ cannot be canceled in this equation. Explain why this is possible in $\mathbb{Z}_{10}[x]$, but not in $\mathbb{Z}_5[x]$.

4 Give examples in $\mathbb{Z}_4[x]$, in $\mathbb{Z}_6[x]$, and in $\mathbb{Z}_9[x]$ of polynomials $a(x)$ and $b(x)$ such that deg $a(x)b(x) <$ deg $a(x) +$ deg $b(x)$.

5 If A is an integral domain, we have seen that in $A[x]$,

$$\deg a(x)b(x) = \deg a(x) + \deg b(x)$$

Show that if A is *not* an integral domain, we can always find polynomials $a(x)$ and $b(x)$ such that deg $a(x)b(x) <$ deg $a(x) +$ deg $b(x)$.

6 Show that if A is an integral domain, the only invertible elements in $A[x]$ are the constant polynomials ± 1. Then show that in $\mathbb{Z}_4[x]$ there are invertible polynomials of all degrees.

7 Give all the ways of factoring x^2 in $\mathbb{Z}_9[x]$; in $\mathbb{Z}_5[x]$. Explain the difference in behavior.

8 Find all the square roots of $x^2 + x + 4$ in $\mathbb{Z}_5[x]$. Show that in $\mathbb{Z}_8[x]$, there are infinitely many square roots of 1.

D. Domains $A[x]$ where A Has Finite Characteristic

In each of the following, let A be an integral domain.

1 Prove that if A has characteristic p, then $A[x]$ has characteristic p.

2 Use part 1 to give an example of an infinite integral domain with finite characteristic.

3 Prove: If A has characteristic 3, then $x + 2$ is a factor of $x^m + 2$ for all m. More generally, if A has characteristic p, then $x + (p - 1)$ is a factor of $x^m + (p - 1)$ for all m.

4 Prove that if A has characteristic p, then in $A[x]$, $(x + c)^p = x^p + c^p$. (You may use essentially the same argument as in the proof of the binomial theorem.)

5 Explain why the following "proof" of part 4 is not valid: $(x + c)^p = x^p + c^p$ in $A[x]$ because $(a + c)^p = a^p + c^p$ for all $a, c \in A$. (Note the following example: in $\mathbb{Z}_2$, $a^2 + 1 = a^4 + 1$ for every a, yet $x^2 + 1 \neq x^4 + 1$ in $\mathbb{Z}_2[x]$.)

6 Use the same argument as in part 4 to prove that if A has characteristic p, then $(cx^i + dx^j)^p = c^p x^{ip} + d^p x^{jp}$. Use this to prove:

$$(a_0 + a_1 x + \cdots + a_n x^n)^p = a_0^p + a_1^p x^p + \cdots + a_n^p x^{np}$$

E. Subrings and Ideals in $A[x]$

1 Show that if B is a subring of A, then $B[x]$ is a subring of $A[x]$.

2 If B is an *ideal* of A, $B[x]$ is not necessarily an ideal of $A[x]$. Give an example to prove this contention.

3 Let S be the set of all the polynomials $a(x)$ in $A[x]$ for which every coefficient a_i for *odd* i is equal to zero. Show that S is a subring of $A[x]$. Why is the same not true when "odd" is replaced by "even"?

4 Let J consist of all the elements in $A[x]$ whose constant coefficient is equal to zero. Prove that J is an ideal of $A[x]$.

5 Let J consist of all the polynomials $a_0 + a_1 x + \cdots + a_n x^n$ in $A[x]$ such that $a_0 + a_1 + \cdots + a_n = 0$. Prove that J is an ideal of $A[x]$.

6 Prove that the ideals in both parts 4 and 5 are *prime* ideals.

F. Homomorphisms of Domains of Polynomials

Let A be an integral domain.

1 Let $h : A[x] \to A$ map every polynomial to its constant coefficient; that is,

$$h(a_0 + a_1 x + \cdots + a_n x^n) = a_0$$

Prove that h is a homomorphism from $A[x]$ *onto* A, and describe its kernel.

2 Explain why the kernel of h in part 1 consists of all the products $xa(x)$, for all $a(x) \in A[x]$. Why is this the same as the principal ideal $\langle x \rangle$ in $A[x]$?

3 Using parts 1 and 2, explain why $A[x]/\langle x \rangle \cong A$.

4 Let $g : A[x] \to A$ send every polynomial to the sum of its coefficients. Prove that g is a surjective homomorphism, and describe its kernel.

5 If $c \in A$, let $h : A[x] \to A[x]$ be defined by $h(a(x)) = a(cx)$, that is,

$$h(a_0 + a_1 x + \cdots + a_n x^n) = a_0 + a_1 cx + a_2 c^2 x^2 + \cdots + a_n c^n x^n$$

Prove that h is a homomorphism and describe its kernel.

6 If h is the homomorphism of part 5, prove that h is an automorphism (isomorphism from $A[x]$ to itself) iff c is invertible.

G. Homomorphisms of Polynomial Domains Induced by a Homomorphism of the Ring of Coefficients

Let A and B be rings and let $h: A \to B$ be a homomorphism with kernel K. Define $\bar{h}: A[x] \to B[x]$ by:

$$\bar{h}(a_0 + a_1 x + \cdots + a_n x^n) = h(a_0) + h(a_1)x + \cdots + h(a_n)x^n$$

(We say that $\bar{h}$ is *induced by h*.)

1 Prove that $\bar{h}$ is a homomorphism from $A[x]$ to $B[x]$.
2 Describe the kernel $\bar{K}$ of $\bar{h}$.
3 Prove that $\bar{h}$ is surjective iff h is surjective.
4 Prove that $\bar{h}$ is injective iff h is injective.
5 Prove that if $a(x)$ is a factor of $b(x)$, then $\bar{h}(a(x))$ is a factor of $\bar{h}(b(x))$.
6 If $h: \mathbb{Z} \to \mathbb{Z}_n$ is the natural homomorphism, let $\bar{h}: \mathbb{Z}[x] \to \mathbb{Z}_n[x]$ be the homomorphism induced by h. Prove that $\bar{h}(a(x)) = 0$ iff n divides every coefficient of $a(x)$.
7 Let $\bar{h}$ be as in part 6, and let n be a prime. Prove that if $a(x)b(x) \in \ker \bar{h}$, then either $a(x)$ or $b(x)$ is in $\ker \bar{h}$. (HINT: Use Exercise F2 of Chapter 19.)

H. Polynomials in Several Variables

$A[x_1, x_2]$ denotes the ring of all the polynomials in *two letters* x_1 and x_2 with coefficients in A. For example, $x^2 - 2xy + y^2 + x - 5$ is a quadratic polynomial in $\mathbb{Q}[x, y]$. More generally, $A[x_1, \ldots, x_n]$ is the ring of all the polynomials in n *letters* $x_1, \ldots, x_n$ with coefficients in A. Formally it is defined as follows: Let $A[x_1]$ be denoted by A_1; then $A_1[x_2]$ is $A[x_1, x_2]$. Continuing in this fashion, we may adjoin one new letter x_i at a time, to get $A[x_1, \ldots, x_n]$.

1 Prove that if A is an integral domain, then $A[x_1, \ldots, x_n]$ is an integral domain.
2 Give a reasonable definition of the *degree* of any polynomial $p(x, y)$ in $A[x, y]$ and then list all the polynomials of degree ≤ 3 in $\mathbb{Z}_3[x, y]$.

Let us denote an arbitrary polynomial $p(x, y)$ in $A[x, y]$ by $\sum a_{ij} x^i y^j$ where $\sum$ ranges over *some* pairs i, j of nonnegative integers.

3 Imitating the definitions of sum and product of polynomials in $A[x]$, give a definition of sum and product of polynomials in $A[x, y]$.
4 Prove that $\deg a(x, y)b(x, y) = \deg a(x, y) + \deg b(x, y)$ if A is an integral domain.

I. Fields of Polynomial Quotients

Let A be an integral domain. By the closing part of Chapter 20, every integral domain can be extended to a "field of quotients." Thus, $A[x]$ can be extended to a field of polynomial quotients, which is denoted by $A(x)$. Note that $A(x)$ consists of all the fractions $a(x)/b(x)$ for $a(x)$ and $b(x) \neq 0$ in $A[x]$, and these fractions are added, subtracted, multiplied, and divided in the customary way.

1 Show that $A(x)$ has the same characteristic as A.

2 Using part 1, explain why there is an infinite field of characteristic p, for every prime p.

3 If A and B are integral domains and $h: A \rightarrow B$ is an isomorphism, prove that h determines an isomorphism $\bar{h}: A(x) \rightarrow B(x)$.

J. Division Algorithm: Uniqueness of Quotient and Remainder

In the division algorithm, prove that $q(x)$ and $r(x)$ are uniquely determined. [HINT: Suppose $a(x) = b(x)q_1(x) + r_1(x) = b(x)q_2(x) + r_2(x)$, and subtract these two expressions, which are both equal to $a(x)$.]

FACTORING POLYNOMIALS

Just as every integer can be factored into primes, so every polynomial can be factored into "irreducible" polynomials which cannot be factored further. As a matter of fact, polynomials behave very much like integers when it comes to factoring them. This is especially true when the polynomials have all their coefficients in a *field*.

Throughout this chapter, we let F represent some field and we consider polynomials over F. It will be found that $F[x]$ has a considerable number of properties in common with $\mathbb{Z}$. To begin with, all the ideals of $F[x]$ are principal ideals, which was also the case for the ideals of $\mathbb{Z}$.

Note carefully that in $F[x]$, the principal ideal generated by a polynomial $a(x)$ consists of all the products $a(x)s(x)$ as $a(x)$ remains fixed and $s(x)$ ranges over all the members of $F[x]$.

Theorem 1 *Every ideal of $F[x]$ is principal.*

Let J be any ideal of $F[x]$. If J contains nothing but the zero polynomial, J is the principal generated by 0. If there are nonzero polynomials in J, let $b(x)$ be any polynomial of *lowest degree* in J. We will show that $J = \langle b(x) \rangle$, which is to say that every element of J is a polynomial multiple $b(x)q(x)$ of $b(x)$.

Indeed, if $a(x)$ is any element of J, we may use the division algorithm to write $a(x) = b(x)q(x) + r(x)$, where $r(x) = 0$ or deg $r(x) <$ deg $b(x)$. Now,

$r(x) = a(x) - b(x)q(x)$; but $a(x)$ was chosen in J, and $b(x) \in J$, hence $b(x)q(x) \in J$. It follows that $r(x)$ is in J.

If $r(x) \neq 0$, its degree is less than the degree of $b(x)$. But this is impossible because $b(x)$ is a polynomial of lowest degree in J. Therefore, of necessity, $r(x) = 0$.

Thus, finally, $a(x) = b(x)q(x)$; so every member of J is a multiple of $b(x)$, as claimed.

It follows that every ideal J of $F[x]$ is principal. In fact, as the proof above indicates, J is generated by any one of its members of lowest degree.

Throughout the discussion which follows, remember that we are considering polynomials in a fixed domain $F[x]$ where F is a *field*.

Let $a(x)$ and $b(x)$ be in $F[x]$. We say that $b(x)$ is a *multiple* of $a(x)$ if

$$b(x) = a(x)s(x)$$

for some polynomial $s(x)$ in $F[x]$. If $b(x)$ is a multiple of $a(x)$, we also say that $a(x)$ is a *factor* of $b(x)$, or that $a(x)$ *divides* $b(x)$. In symbols, we write

$$a(x) \,|\, b(x)$$

Every nonzero constant polynomial divides every polynomial. For if $c \neq 0$ is constant and $a(x) = a_0 + \cdots + a_n x^n$, then

$$a_0 + a_1 x + \cdots + a_n x^n = c\left(\frac{a_0}{c} + \frac{a_1}{c} x + \cdots + \frac{a_n}{c} x^n\right)$$

hence $c \,|\, a(x)$. A polynomial $a(x)$ is invertible iff it is a divisor of the unity polynomial 1. But if $a(x)b(x) = 1$, this means that $a(x)$ and $b(x)$ both have degree 0, that is, are constant polynomials: $a(x) = a$, $b(x) = b$, and $ab = 1$. Thus,

the invertible elements of $F[x]$ are all the nonzero constant polynomials.

A pair of nonzero polynomials $a(x)$ and $b(x)$ are called *associates* if they divide one another: $a(x) \,|\, b(x)$ and $b(x) \,|\, a(x)$. That is to say,

$$a(x) = b(x)c(x) \qquad \text{and} \qquad b(x) = a(x)d(x)$$

for some $c(x)$ and $d(x)$. If this happens to be the case, then

$$a(x) = b(x)c(x) = a(x)d(x)c(x)$$

hence $d(x)c(x) = 1$ because $F[x]$ is an integral domain. But then $c(x)$ and $d(x)$ are constant polynomials, and therefore $a(x)$ and $b(x)$ are constant multiples of each other. Thus, in $F[x]$,

$a(x)$ and $b(x)$ are associates iff they are constant multiples of each other.

If $a(x) = a_0 + \cdots + a_n x^n$, the associates of $a(x)$ are all its nonzero constant multiples. Among these multiples is the polynomial

$$\frac{a_0}{a_n} + \frac{a_1}{a_n} x + \cdots + x^n$$

which is equal to $(1/a_n)a(x)$, and which has 1 as its leading coefficient. Any polynomial whose leading coefficient is equal to 1 is called *monic*. Thus, *every nonzero polynomial $a(x)$ has a unique monic associate.* For example, the monic associate of $3 + 4x + 2x^3$ is $\frac{3}{2} + 2x + x^3$.

A polynomial $d(x)$ is called a *greatest common divisor* of $a(x)$ and $b(x)$ if $d(x)$ divides $a(x)$ and $b(x)$, and is a multiple of any other common divisor of $a(x)$ and $b(x)$; in other words,

(i) $d(x)\,|\,a(x)$ and $d(x)\,|\,b(x)$, and

(ii) For any $u(x)$ in $F[x]$, if $u(x)\,|\,a(x)$ and $u(x)\,|\,b(x)$, then $u(x)\,|\,d(x)$.

According to this definition, two different gcd's of $a(x)$ and $b(x)$ divide each other, that is, are associates. Of all the possible gcd's of $a(x)$ and $b(x)$, we select the monic one, call it *the* gcd of $a(x)$ and $b(x)$, and denote it by $\gcd[a(x), b(x)]$.

It is important to know that any pair of polynomials always *has* a greatest common divisor.

Theorem 2 *Any two nonzero polynomials $a(x)$ and $b(x)$ in $F[x]$ have a gcd $d(x)$. Furthermore, $d(x)$ can be expressed as a "linear combination"*

$$d(x) = r(x)a(x) + s(x)b(x)$$

where $r(x)$ and $s(x)$ are in $F[x]$.

The proof is analogous to the proof of the corresponding theorem for integers. If J is the set of all the linear combinations

$$u(x)a(x) + v(x)b(x)$$

as $u(x)$ and $v(x)$ range over $F[x]$, then J is an ideal of $F[x]$, say the ideal $\langle d(x) \rangle$ generated by $d(x)$. Now $a(x) = 1a(x) + 0b(x)$ and $b(x) = 0a(x) + 1b(x)$, so $a(x)$ and $b(x)$ are in J. But every element of J is a multiple of $d(x)$, so

$$d(x)\,|\,a(x) \qquad \text{and} \qquad d(x)\,|\,b(x)$$

If $k(x)$ is any common divisor of $a(x)$ and $b(x)$, this means there are polynomials $f(x)$ and $g(x)$ such that $a(x) = k(x)f(x)$ and $b(x) = k(x)g(x)$. Now, $d(x) \in J$, so $d(x)$ can be written as a linear combination

$$d(x) = r(x)a(x) + s(x)b(x)$$

$$= r(x)k(x)f(x) + s(x)k(x)g(x)$$

$$= k(x)[r(x)f(x) + s(x)g(x)]$$

hence $k(x) \mid d(x)$. This confirms that $d(x)$ is the gcd of $a(x)$ and $b(x)$.

Polynomials $a(x)$ and $b(x)$ in $F[x]$ are said to be *relatively prime* if their gcd is equal to 1. (This is equivalent to saying that their only common factors are constants in F.)

A polynomial $a(x)$ of positive degree is said to be *reducible over F* if there are polynomials $b(x)$ and $c(x)$ in $F[x]$, both of positive degree, such that

$$a(x) = b(x)c(x)$$

Because $b(x)$ and $c(x)$ both have positive degrees, and the sum of their degrees is deg $a(x)$, *each has degree less than* deg $a(x)$.

A polynomial $p(x)$ of positive degree in $F[x]$ is said to be *irreducible over F* if it cannot be expressed as the product of two polynomials of positive degree in $F[x]$. Thus, $p(x)$ is irreducible iff it is not reducible.

When we say that a polynomial $p(x)$ is irreducible, it is important that we specify *irreducible over the field F*. A polynomial may be irreducible over F, yet reducible over a larger field E. For example, $p(x) = x^2 + 1$ is irreducible over $\mathbb{R}$; but over $\mathbb{C}$ it has factors $(x + i)(x - i)$.

We next state the analogs for polynomials of Euclid's lemma and its corollaries. The proofs are almost identical to their counterparts in $\mathbb{Z}$; therefore they are left as exercises.

Euclid's lemma for polynomials *Let $p(x)$ be irreducible. If $p(x) \mid a(x)b(x)$, then $p(x) \mid a(x)$ or $p(x) \mid b(x)$.*

Corollary 1 *Let $p(x)$ be irreducible. If $p(x) \mid a_1(x)a_2(x) \cdots a_n(x)$, then $p(x) \mid a_i(x)$ for one of the factors $a_i(x)$ among $a_1(x), \ldots, a_n(x)$.*

Corollary 2 *Let $q_1(x), \ldots, q_r(x)$ and $p(x)$ be monic irreducible polynomials. If $p(x) \mid q_1(x) \cdots q_r(x)$, then $p(x)$ is equal to one of the factors $q_1(x), \ldots, q_r(x)$.*

Theorem 3: Factorization into irreducible polynomials *Every polynomial $a(x)$ of positive degree in $F[x]$ can be written as a product*

$$a(x) = kp_1(x)p_2(x) \cdots p_r(x)$$

where k is a constant in F and $p_1(x), \ldots, p_r(x)$ are monic irreducible polynomials of $F[x]$.

If this were not true, we could choose a polynomial $a(x)$ *of lowest degree* among those which cannot be factored into irreducibles. Then $a(x)$ is reducible, so $a(x) = b(x)c(x)$ where $b(x)$ and $c(x)$ have lower degree than $a(x)$. But this means that $b(x)$ and $c(x)$ *can* be factored into irreducibles, and therefore $a(x)$ can also.

Theorem 4: Unique factorization *If $a(x)$ can be written in two ways as a product of irreducibles, say*

$$a(x) = kp_1(x) \cdots p_r(x) = lq_1(x) \cdots q_s(x)$$

then $k = l$, $r = s$, and each $p_i(x)$ is equal to a $q_j(x)$.

The proof is the same, in all major respects, as the corresponding proof for $\mathbb{Z}$; it is left as an exercise.

In the next chapter we will be able to improve somewhat on the last two results in the special cases of $\mathbb{R}[x]$ and $\mathbb{C}[x]$. Also, we will learn more about factoring polynomials into irreducibles.

EXERCISES

A. Examples of Factoring into Irreducible Factors

1 Factor $x^4 - 4$ into irreducible factors over $\mathbb{Q}$, over $\mathbb{R}$, and over $\mathbb{C}$.

2 Factor $x^6 - 16$ into irreducible factors over $\mathbb{Q}$, over $\mathbb{R}$, and over $\mathbb{C}$.

3 Find all the irreducible polynomials of degree ≤ 4 in $\mathbb{Z}_2[x]$.

4 Show that $x^2 + 2$ is irreducible in $\mathbb{Z}_5[x]$. Then factor $x^4 - 4$ into irreducible factors in $\mathbb{Z}_5[x]$. (By Theorem 3, it is sufficient to search for monic factors.)

5 Factor $2x^3 + 4x + 1$ in $\mathbb{Z}_5[x]$. (Factor it as in Theorem 3.)

6 In $\mathbb{Z}_6[x]$, factor each of the following into two polynomials of degree 1: x, $x + 2$, $x + 3$. Why is this possible?

B. Short Questions Relating to Irreducible Polynomials

Let F be a field. *Explain why each of the following is true in $F[x]$.*

1 Every polynomial of degree 1 is irreducible.

2 If $a(x)$ and $b(x)$ are distinct monic polynomials, they cannot be associates.

3 Any two distinct irreducible polynomials are relatively prime.

4 If $a(x)$ is irreducible, any associate of $a(x)$ is irreducible.

5 If $a(x) \neq 0$, $a(x)$ cannot be an associate of 0.

6 In $\mathbb{Z}_p[x]$, every polynomial has exactly $p - 1$ associates.

7 $x^2 + 1$ is reducible in $\mathbb{Z}_p[x]$ iff $p = a + b$ where $ab = 1 \pmod{p}$.

C. Number of Irreducible Quadratics over a Finite Field

1 Without finding them, determine *how many* reducible monic quadratics there are in $\mathbb{Z}_5[x]$. [HINT: Every reducible monic quadratic can be uniquely factored as $(x + a)(x + b)$.]

2 How many reducible quadratics are there in $\mathbb{Z}_5[x]$? How many *irreducible* quadratics?

3 Generalize: How many irreducible quadratics are there over a finite field of n elements?

4 How many irreducible cubics are there over a field of n elements?

D. Ideals in Domains of Polynomials

Let F be a field, and let J designate any ideal of $F[x]$. *Prove the following*:

1 Any two generators of J are associates.

2 J has a *unique* monic generator $m(x)$. An arbitrary polynomial $a(x) \in F[x]$ is in J iff $m(x) \mid a(x)$.

3 J is a prime ideal iff it has an irreducible generator.

4 If $p(x)$ is irreducible, then $p(x)$ is a *maximal* ideal of $F[x]$. (See Chapter 18, Exercise H5.)

5 Let S be the set of all polynomials $a_0 + a_1 x + \cdots + a_n x^n$ in $F[x]$ which satisfy $a_0 + a_1 + \cdots + a_n = 0$. It has been shown (Chapter 24, Exercise E5) that S is an ideal of $F[x]$. Prove that $x - 1 \in S$, and explain why it follows that $S = \langle x - 1 \rangle$.

6 Conclude from parts 4 and 5 that $F[x]/\langle x - 1 \rangle \cong F$. (See Chapter 24, Exercise F4.)

7 Let $F[x, y]$ denote the domain of all the polynomials $\sum a_{ij} x^i y^j$ in *two* letters x and y, with coefficients in F. Let J be the ideal of $F[x, y]$ which contains all the polynomials whose constant coefficient is zero. Prove that J is not a principal ideal. Conclude that Theorem 1 is not true in $F[x, y]$.

E. Proof of the Unique Factorization Theorem

1 Prove Euclid's lemma for polynomials.

2 Prove the two corollaries to Euclid's lemma.

3 Prove the unique factorization theorem for polynomials.

F. A Method for Computing the gcd

Let $a(x)$ and $b(x)$ be polynomials of positive degree. By the division algorithm, we may divide $a(x)$ by $b(x)$:

$$a(x) = b(x)q_1(x) + r_1(x)$$

1 Prove that every common divisor of $a(x)$ and $b(x)$ is a common divisor of $b(x)$ and $r_1(x)$.

It follows from part 1 that the gcd of $a(x)$ and $b(x)$ is the same as the gcd of $b(x)$ and $r_1(x)$. This procedure can now be repeated on $b(x)$ and $r_1(x)$; divide $b(x)$ by $r_1(x)$:

$$b(x) = r_1(x)q_2(x) + r_2(x)$$

Next,
$$r_1(x) = r_2(x)q_3(x) + r_3(x)$$

$$\vdots$$

Finally,
$$r_{n-1}(x) = r_n(x)q_{n+1}(x) + 0$$

In other words, we continue to divide each remainder by the succeeding remainder. Since the remainders continually decrease in degree, there must ultimately be a zero remainder. But we have seen that

$$\gcd[a(x), b(x)] = \gcd[b(x), r_1(x)] = \cdots = \gcd[r_{n-1}(x), r_n(x)]$$

Since $r_n(x)$ is a divisor of $r_{n-1}(x)$, it must be the gcd of $r_n(x)$ and $r_{n-1}(x)$. Thus,

$$r_n(x) = \gcd[a(x), b(x)]$$

This method is called the *euclidean algorithm* for finding the gcd.

2 Find the gcd of $x^3 + 1$ and $x^4 + x^3 + 2x^2 + x - 1$. Express this gcd as a linear combination of the two polynomials.
3 Do the same for $x^{24} - 1$ and $x^{15} - 1$.
4 Find the gcd of $x^3 + x^2 + x + 1$ and $x^4 + x^3 + 2x^2 + 2x$ in $\mathbb{Z}_3[x]$.

G. An Automorphism of $F[x]$

Let $h: F[x] \to F[x]$ be defined by:

$$h(a_0 + a_1 x + \cdots + a_n x^n) = a_n + a_{n-1} x + \cdots + a_0 x^n$$

Prove the following:

1 h is a homomorphism.
2 h is injective and surjective, hence an automorphism of $F[x]$.
3 $a_0 + a_1 x + \cdots + a_n x^n$ is irreducible iff $a_n + a_{n-1} x + \cdots + a_0 x^n$ is irreducible.
4 Let $a_0 + a_1 x + \cdots + a_n x^n = (b_0 + \cdots + b_m x^m)(c_0 + \cdots + c_q x^q)$. Factor

$$a_n + a_{n-1} x + \cdots + a_0 x^n$$

5 Let $a(x) = a_0 + a_1 x + \cdots + a_n x^n$ and $\hat{a}(x) = a_n + a_{n-1} x + \cdots + a_0 x^n$. If $c \in F$, prove that $a(c) = 0$ iff $\hat{a}(1/c) = 0$.

TWENTY-SIX

SUBSTITUTION IN POLYNOMIALS

Up to now we have treated polynomials as formal expressions. If $a(x)$ is a polynomial over a field F, say

$$a(x) = a_0 + a_1 x + \cdots + a_n x^n$$

this means that the coefficients $a_0, a_1, \ldots, a_n$ are elements of the field F, while the letter x is a *placeholder* which plays no other role than to occupy a given position.

When we dealt with polynomials in elementary algebra, it was quite different. The letter x was called an *unknown* and was allowed to assume numerical values. This made $a(x)$ into a function having x as its independent variable. Such a function is called a *polynomial function.*

This chapter is devoted to the study of polynomial functions. We begin with a few careful definitions.

Let $a(x) = a_0 + a_1 x + \cdots + a_n x^n$ be a polynomial over F. If c is any element of F, then

$$a_0 + a_1 c + \cdots + a_n c^n$$

is also an element of F, obtained by *substituting* c for x in the polynomial $a(x)$. This element is denoted by $a(c)$. Thus,

$$a(c) = a_0 + a_1 c + \cdots + a_n c^n$$

Since we may substitute any element of F for x, we may regard $a(x)$ as a *function from F to F*. As such, it is called a *polynomial function on F.*

The difference between a polynomial and a polynomial function is mainly a difference of viewpoint. Given $a(x)$ with coefficients in F: if x is regarded merely as a placeholder, then $a(x)$ is a polynomial; if x is allowed to assume values in F, then $a(x)$ is a polynomial function. The difference is a small one, and we will not make an issue of it.

If $a(x)$ is a polynomial with coefficients in F, and c is an element of F such that

$$a(c) = 0$$

then we call c a *root* of $a(x)$. For example, 2 is a root of the polynomial $3x^2 + x - 14 \in \mathbb{R}[x]$, because $3 \cdot 2^2 + 2 - 14 = 0$.

There is an absolutely fundamental connection between *roots* of a polynomial and *factors* of that polynomial. This connection is explored in the following pages, beginning with the next theorem:

Let $a(x)$ be a polynomial over a field F.

Theorem 1 *c is a root of $a(x)$ iff $x - c$ is a factor of $a(x)$.*

If $x - c$ is a factor of $a(x)$, this means that $a(x) = (x - c)q(x)$ for some $q(x)$. Thus, $a(c) = (c - c)q(c) = 0$, so c is a root of $a(x)$. Conversely, if c is a root of $a(x)$, we may use the division algorithm to divide $a(x)$ by $x - c$: $a(x) = (x - c)q(x) + r(x)$. The remainder $r(x)$ is either 0 or a polynomial of lower degree than $x - c$; but *lower degree than $x - c$* means that $r(x)$ is a constant polynomial: $r(x) = r \geq 0$. Then

$$0 = a(c) = (c - c)q(c) + r = 0 + r = r$$

Thus, $r = 0$, and therefore $x - c$ is a factor of $a(x)$.

Theorem 1 tells us that if c is a root of $a(x)$, then $x - c$ is a *factor* of $a(x)$ (and vice versa). This is easily extended: if c_1 and c_2 are two roots of $a(x)$, then $x - c_1$ and $x - c_2$ are two factors of $a(x)$. Similarly, three roots give rise to three factors, four roots to four factors, and so on. This is stated concisely in the next theorem.

Theorem 2 *If $a(x)$ has distinct roots $c_1, \ldots, c_m$ in F, then $(x - c_1)(x - c_2) \cdots (x - c_m)$ is a factor of $a(x)$.*

To prove this, let us first make a simple observation: *if a polynomial $a(x)$ can be factored, any root of $a(x)$ must be a root of one of its factors.* Indeed, if $a(x) = s(x)t(x)$ and $a(c) = 0$, then $s(c) t(c) = 0$, and therefore either $s(c) = 0$ or $t(c) = 0$.

Let $c_1, \ldots, c_m$ be distinct roots of $a(x)$. By Theorem 1,

$$a(x) = (x - c_1)q_1(x)$$

By our observation in the preceding paragraph, c_2 must be a root of $x - c_1$ or of $q_1(x)$. It cannot be a root of $x - c_1$ because $c_2 - c_1 \neq 0$; so c_2 is a root of $q_1(x)$. Thus, $q_1(x) = (x - c_2)q_2(x)$, and therefore

$$a(x) = (x - c_1)(x - c_2)q_2(x)$$

Repeating this argument for each of the remaining roots gives us our result.

An immediate consequence is the following important fact:

Theorem 3 *If $a(x)$ has degree n, it has at most n roots.*

If $a(x)$ had $n + 1$ roots $c_1, \ldots, c_{n+1}$, then by Theorem 2, $(x - c_1) \cdots (x - c_{n+1})$ would be a factor of $a(x)$, and the degree of $a(x)$ would therefore be at least $n + 1$.

It was stated earlier in this chapter that the difference between polynomials and polynomial functions is mainly a difference of viewpoint. *Mainly*, but not entirely! Remember that two polynomials $a(x)$ and $b(x)$ are equal iff corresponding coefficients are equal, whereas two *functions* $a(x)$ and $b(x)$ are equal iff $a(x) = b(x)$ for every x in their domain. These two notions of equality do not always coincide!

For example, consider the following two polynomials in $\mathbb{Z}_5[x]$:

$$a(x) = x^5 + 1$$

$$b(x) = x - 4$$

You may check that $a(0) = b(0)$, $a(1) = b(1)$, $\ldots$, $a(4) = b(4)$, hence $a(x)$ and $b(x)$ are equal functions from $\mathbb{Z}_5$ to $\mathbb{Z}_5$. But as polynomials, $a(x)$ and $b(x)$ are quite distinct! (They do not even have the same degree.)

It is reassuring to know that this cannot happen when the field F is infinite. Suppose $a(x)$ and $b(x)$ are polynomials over a field F which has infinitely many elements. If $a(x)$ and $b(x)$ are equal as functions, this means that $a(c) = b(c)$ for every $c \in F$. Define the polynomial $d(x)$ to be the difference of $a(x)$ and $b(x) : d(x) = a(x) - b(x)$. Then $d(c) = 0$ for every $c \in F$. Now, if $d(x)$ were not the zero polynomial, it would be a polynomial (with some finite degree n) having *infinitely many roots*, and by Theorem 3 this is impossible! Thus, $d(x)$ is the zero polynomial (all its coefficients are equal to zero), and therefore $a(x)$ is the same polynomial as $b(x)$. (They have the same coefficients.)

This tells us that if F is a field with infinitely many elements (such as $\mathbb{Q}$, $\mathbb{R}$, or $\mathbb{C}$), there is no need to distinguish between polynomials and polynomial functions. The difference is, indeed, just a difference of viewpoint.

POLYNOMIALS OVER $\mathbb{Z}$ AND $\mathbb{Q}$

In scientific computation a great many functions can be approximated by polynomials, usually polynomials whose coefficients are integers or rational numbers. Such polynomials are therefore of great practical interest. It is easy to find the rational roots of such polynomials, and to determine if a polynomial over $\mathbb{Q}$ is irreducible over $\mathbb{Q}$. We will do these things next.

First, let us make an important observation:

Let $a(x)$ be a polynomial with rational coefficients, say

$$a(x) = \frac{k_0}{l_0} + \frac{k_1}{l_1} x + \cdots + \frac{k_n}{l_n} x^n$$

We may factor out the constant $1/l_0 l_1 \cdots l_n$ and get

$$a(x) = \frac{1}{l_0 \cdots l_n} \underbrace{(k_0 l_1 \cdots l_n + \cdots + k_n l_0 \cdots l_{n-1} x^n)}_{b(x)}$$

The polynomial $b(x)$ has *integer coefficients*; and since it differs from $a(x)$ only by a constant factor, it has the same roots as $a(x)$. Thus, *for every polynomial with rational coefficients, there is a polynomial with integer coefficients having the same roots.* Therefore, for the present we will confine our attention to polynomials with integer coefficients. The next theorem makes it easy to find all the rational roots of such polynomials:

Let s/t be a rational number in simplest form (that is, the integers s and t do not have a common factor greater than 1). Let $a(x) = a_0 + \cdots + a_n x^n$ be a polynomial with integer coefficients.

Theorem 4 *If s/t is a root of $a(x)$, then $s \mid a_0$ and $t \mid a_n$.*

If s/t is a root of $a(x)$, this means that

$$a_0 + a_1(s/t) + \cdots + a_n(s^n/t^n) = 0$$

Multiplying both sides of this equation by t^n we get

$$a_0 t^n + a_1 s t^{n-1} + \cdots + a_n s^n = 0 \tag{*}$$

We may now factor out s from all but the first term to get:

$$-a_0 t^n = s(a_1 t^{n-1} + \cdots + a_n s^{n-1})$$

Thus, $s \mid a_0 t^n$; and since s and t have no common factors, $s \mid a_0$. Similarly, in (∗), we may factor out t from all but the last term to get

$$t(a_0 t^{n-1} + \cdots + a_{n-1} s^{n-1}) = -a_n s^n$$

Thus, $t \mid a_n s^n$; and since s and t have no common factors, $t \mid a_n$.

As an example of the way Theorem 4 may be used, let us find the rational roots of $a(x) = 2x^4 + 7x^3 + 5x^2 + 7x + 3$. Any rational root must be a fraction s/t where s is a factor of 3 and t is a factor of 2. The possible roots are therefore ± 1, ± 2, $\pm\frac{1}{2}$ and $\pm\frac{3}{2}$. Testing each of these numbers by direct substitution into the equation $a(x) = 0$, we find that $-\frac{1}{2}$ and -3 are roots are therefore ± 1, ± 3, $\pm\frac{1}{2}$ and $\pm\frac{3}{2}$. Testing each of these numbers by $(x + \frac{1}{2})(x + 3)(x^2 + 1)$.

Before going the next step in our discussion, we note a simple but fairly surprising fact. Let $a(x) = b(x)c(x)$, where $a(x)$, $b(x)$, and $c(x)$ have integer coefficients.

Lemma *If a prime number p divides every coefficient of $a(x)$, it either divides every coefficient of $b(x)$ or every coefficient of $c(x)$.*

If this is not the case, let b_r be the first coefficient of $b(x)$ not divisible by p, and let c_t be the first coefficient of $c(x)$ not divisible by p. Now, $a(x) = b(x)c(x)$, so

$$a_{r+t} = b_0 c_{r+t} + \cdots + b_r c_t + \cdots + b_{r+t} c_0$$

Each term on the right, except $b_r c_t$, is a product $b_i c_j$ where either $i > r$ or $j > t$. By our choice of b_r and c_t, if $i > r$ then $p \mid b_i$, and if $j > t$ then $p \mid c_j$. Thus, p is a factor of every term on the right with the possible exception of $b_r c_t$, but p is also a factor of a_{r+t}. Thus, p *must* be a factor of $b_r c_t$, hence of either b_r or c_t, and this is impossible.

We saw (in the discussion immediately preceding Theorem 4) that any polynomial $a(x)$ with rational coefficients has a constant multiple $ka(x)$, *with integer coefficients*, which has the same roots as $a(x)$. We can go one better; let $a(x) \in \mathbb{Z}[x]$:

Theorem 5. *Suppose $a(x)$ can be factored as $a(x) = b(x)c(x)$, where $b(x)$ and $c(x)$ have rational coefficients. Then there are polynomials $B(x)$ and $C(x)$ with* *integer coefficients, which are constant multiples of b(x) and c(x), respectively, such that* $a(x) = B(x)C(x)$.

Let k and l be integers such that $kb(x)$ and $lc(x)$ have integer coefficients. Then $kla(x) = [kb(x)][lc(x)]$. By the lemma, each prime factor of kl may now be canceled with a factor of either $kb(x)$ or $lc(x)$.

Remember that a polynomial $a(x)$ of positive degree is said to be *reducible over F* if there are polynomials $b(x)$ and $c(x)$ in $F[x]$, both of positive degree, such that $a(x) = b(x)c(x)$. If there are *no* such polynomials, then $a(x)$ is *irreducible over F*.

If we use this terminology, Theorem 5 states that *any polynomial with integer coefficients which is reducible over* $\mathbb{Q}$ *is reducible already over* $\mathbb{Z}$.

In Chapter 25 we saw that every polynomial can be factored into irreducible polynomials. In order to factor a polynomial completely (that is, into irreducibles), we must be able to recognize an irreducible polynomial when we see one! This is not always an easy matter. But there is a method which works remarkably well for recognizing when a polynomial is irreducible over $\mathbb{Q}$:

Theorem 6: Eisenstein's irreducibility criterion *Let*

$$a(x) = a_0 + a_1 x + \cdots + a_n x^n$$

be a polynomial with integer coefficients. Suppose there is a prime number p which divides every coefficient of a(x) except the leading coefficient a_n; suppose p does **not** *divide a_n and p^2 does* **not** *divide a_0. Then a(x) is irreducible over* $\mathbb{Q}$.

If $a(x)$ can be factored over $\mathbb{Q}$ as $a(x) = b(x)c(x)$, then by Theorem 5 we may assume $b(x)$ and $c(x)$ have integer coefficients: say

$$b(x) = b_0 + \cdots + b_k x^k \qquad \text{and} \qquad c(x) = c_0 + \cdots + c_m x^m$$

Now, $a_0 = b_0 c_0$; p divides a_0 but p^2 does not, so only *one* of b_0, c_0 is divisible by p. Say $p \mid c_0$ and $p \nmid b_0$. Next, $a_n = b_k c_m$ and $p \nmid a_n$, so $p \nmid c_m$.

Let s be the smallest integer such that $p \nmid c_s$. We have

$$a_s = b_0 c_s + b_1 c_{s-1} + \cdots + b_s c_0$$

and by our choice of c_s, every term on the right except $b_0 c_s$ is divisible by p. But a_s also is divisible by p, and therefore $b_0 c_s$ must be divisible by p. This is impossible because $p \nmid b_0$ and $p \nmid c_s$. Thus, $a(x)$ cannot be factored.

For example, $x^3 + 2x^2 + 4x + 2$ is irreducible over $\mathbb{Q}$ because $p = 2$ satisfies the conditions of Eisenstein's criterion.

POLYNOMIALS OVER $\mathbb{R}$ AND $\mathbb{C}$

One of the most far-reaching theorems of classical mathematics concerns polynomials with complex coefficients. It is so important in the framework of traditional algebra that it is called the *fundamental theorem of algebra*. It states the following:

Every nonconstant polynomial with complex coefficients has a complex root.

(The proof of this theorem is based upon techniques of calculus and can be found in most books on complex analysis. It is omitted here.)

It follows immediately that *the irreducible polynomials in* $\mathbb{C}[x]$ *are exactly the polynomials of degree* 1. For if $a(x)$ is a polynomial of degree greater than 1 in $\mathbb{C}[x]$, then by the fundamental theorem of algebra it has a root c and therefore a factor $x - c$.

Now, every polynomial in $\mathbb{C}[x]$ can be factored into irreducibles. Since the irreducible polynomials are all of degree 1, it follows that if $a(x)$ is a polynomial of degree n over $\mathbb{C}$, it can be factored into

$$a(x) = k(x - c_1)(x - c_2) \cdots (x - c_n)$$

In particular, *if* $a(x)$ *has degree* n *it has* n (*not necessarily distinct*) *complex roots* $c_1, \ldots, c_n$.

Since every real number a is a complex number ($a = a + 0i$), what has just been stated applies equally to polynomials with real coefficients. Specifically, if $a(x)$ is a polynomial of degree n with real coefficients, it can be factored into $a(x) = k(x - c_1) \cdots (x - c_n)$, where $c_1, \ldots, c_n$ are complex numbers (some of which may be real).

For our closing comments, we need the following lemma:

Lemma. *Suppose* $a(x) \in \mathbb{R}[x]$. *If* $a + bi$ *is a root of* $a(x)$, *so is* $a - bi$.

Remember that $a - bi$ is called the conjugate of $a + bi$. If r is any complex number, we write $\bar{r}$ for its conjugate. It is easy to see that the function $f(r) = \bar{r}$ is a homomorphism from $\mathbb{C}$ to $\mathbb{C}$ (in fact, it is an isomorphism). For every *real* number $a, f(a) = a$. Thus, if $a(x)$ has real coefficients, then $f(a_0 + a_1 r + \cdots + a_n r^n) = a_0 + a_1 \bar{r} + \cdots + a_n \bar{r}^n$. Since $f(0) = 0$, it follows that if r is a root of $a(x)$, so is $\bar{r}$.

Now let $a(x)$ be any polynomial with real coefficients, and let $r = a + bi$ be a complex root of $a(x)$. Then $\bar{r}$ is also a root of $a(x)$, so

$$(x - r)(x - \bar{r}) = x^2 - 2ax + (a^2 + b^2)$$

and this is a quadratic polynomial *with real coefficients*! We have thus shown that *any polynomial with real coefficients can be factored into polynomials of degree* 1 *or* 2 *in* $\mathbb{R}$ *[x]*. In particular, the irreducible polynomials of $\mathbb{R}$ [x] are the linear polynomials and the irreducible quadratics (that is, the $ax^2 + bx + c$ where $b^2 - 4ac < 0$).

EXERCISES

A. Finding Roots of Polynomials over Finite Fields

In order to find a root of $a(x)$ in a finite field F, the simplest method (if F is small) is to test every element of F by substitution into the equation $a(x) = 0$.

1 Find all the roots of the following polynomials in $\mathbb{Z}_5[x]$, and factor the polynomials:

$$x^3 + x^2 + x + 1; \qquad 3x^4 + x^2 + 1; \qquad x^5 + 1; \qquad x^4 + 1; \qquad x^4 + 4$$

2 Use Fermat's theorem to find all the roots of the following polynomials in $\mathbb{Z}_7[x]$:

$$x^{100} - 1; \qquad 3x^{98} + x^{19} + 3; \qquad 2x^{74} - x^{55} + 2x + 6$$

3 Using Fermat's theorem, find polynomials of degree ≤ 6 which determine the same functions as the following polynomials in $\mathbb{Z}_7[x]$:

$$3x^{75} - 5x^{54} + 2x^{13} - x^2; \quad 4x^{108} + 6x^{101} - 2x^{81}; \quad 3x^{103} - x^{73} + 3x^{55} - x^{25}$$

4 Explain why every polynomial in $\mathbb{Z}_p[x]$ has the same roots as a polynomial of degree $< p$.

B. Finding Roots of Polynomials over $\mathbb{Q}$

1 Find all the rational roots of the following polynomials, and factor them into irreducible polynomials in $\mathbb{Q}[x]$:

$$9x^3 + 18x^2 - 4x - 8; \qquad 4x^3 - 3x^2 - 8x + 6;$$

$$2x^4 + 3x^3 - 8x - 12; \qquad 6x^4 - 7x^3 + 8x^2 - 7x + 2$$

2 Factor each of the preceding polynomials in $\mathbb{R}[x]$ and in $\mathbb{C}[x]$.

3 Find associates with integer coefficients for each of the following polynomials:

$$x^3 + \frac{3}{2}x^2 - \frac{4}{9}x - \frac{2}{3}; \quad \frac{1}{2}x^3 - \frac{1}{4}x^2 - \frac{1}{2}x + \frac{1}{4}; \quad \sqrt{3}x^3 + \frac{1}{\sqrt{3}}x^2 - \sqrt{3}x - \frac{1}{\sqrt{3}}$$

4 Find all the rational roots of the polynomials in part 3 and factor them over $\mathbb{R}$.

5 Does $2x^4 + 3x^2 - 2$ have any rational roots? Can it be factored into two polynomials of lower degree in $\mathbb{Q}[x]$? Explain.

C. Short Questions Relating to Roots

Let F be a field. *Prove that each of the following is true in $F[x]$.*

1 The remainder of $p(x)$, when divided by $x - c$, is $p(c)$.

2 $(x - c) \mid (p(x) - p(c))$

3 Every polynomial has the same roots as any of its associates.

4 If $a(x)$ and $b(x)$ have the same roots in F, are they necessarily associates? Explain.

5 If $a(x) \mid b(x)$, then $a(c) \mid b(c)$ for any $c \in F$.

6 If $a(x)$ is a monic polynomial of degree n, and $a(x)$ has n roots $c_1, \ldots, c_n \in F$, then $a(x) = (x - c_1) \cdots (x - c_n)$.

7 Suppose $a(x)$ and $b(x)$ have degree $< n$. If $a(c) = b(c)$ for n values of c, then $a(x) = b(x)$.

8 There are infinitely many irreducible polynomials in $\mathbb{Z}_5[x]$.

9 How many roots does $x^2 - x$ have in $\mathbb{Z}_{10}$? In $\mathbb{Z}_{11}$? Explain the difference.

D. Irreducible Polynomials in $\mathbb{Q}[x]$ by Eisenstein's Criterion (and Variations on the Theme)

1 Show that each of the following polynomials is irreducible over $\mathbb{Q}$:

$$3x^4 - 8x^3 + 6x^2 - 4x + 6; \qquad \frac{2}{3}x^5 + \frac{1}{2}x^4 - 2x^2 + \frac{1}{2};$$

$$\frac{1}{5}x^4 - \frac{1}{3}x^3 - \frac{2}{3}x + 1; \qquad \frac{1}{2}x^4 + \frac{4}{3}x^3 - \frac{2}{3}x^2 + 1$$

2 It often happens that a polynomial $a(y)$, as it stands, does not satisfy the conditions of Eisenstein's criterion, but with a simple change of variable $y = x + c$, it does. It is important to note that if $a(x)$ can be factored into $p(x)q(x)$, then certainly $a(x + c)$ can be factored into $p(x + c)q(x + c)$. Thus, the irreducibility of $a(x + c)$ implies the irreducibility of $a(x)$.

(*a*) Use the change of variable $y = x + 1$ to show that $x^4 + 4x + 1$ is irreducible in $\mathbb{Q}[x]$. [In other words, test $(x + 1)^4 + 4(x + 1) + 1$ by Eisenstein's criterion.]

(*b*) Find an appropriate change of variable to prove that the following are irreducible in $\mathbb{Q}[x]$:

$$x^4 + 2x^2 - 1; \qquad x^3 - 3x + 1; \qquad x^4 + 1; \qquad x^4 - 10x^2 + 1$$

3 Prove that for any prime p, $x^{p-1} + x^{p-2} + \cdots + x + 1$ is irreducible in $\mathbb{Q}[x]$. [HINT: By elementary algebra,

$$(x - 1)(x^{p-1} + x^{p-2} + \cdots + x + 1) = x^p - 1$$

hence $\qquad\qquad x^{p-1} + x^{p-2} + \cdots + x + 1 = \dfrac{x^p - 1}{x - 1}$

Use the change of variable $y = x + 1$, and expand by the binomial theorem.

4 By Exercise G4 of Chapter 25, the function

$$h(a_0 + a_1 x + \cdots + a_n x^n) = a_n + a_{n-1} x + \cdots + a_0 x^n$$

is an isomorphism of $\mathbb{Z}[x]$ onto $\mathbb{Z}[x]$. In particular, h matches irreducible polynomials with irreducible polynomials. Use this fact to state a dual version of Eisenstein's irreducibility criterion.

5 Use part 4 to show that each of the following polynomials is irreducible in $\mathbb{Q}[x]$:

$$6x^4 + 4x^3 - 6x^2 - 8x + 5; \qquad x^4 - \frac{1}{2}x^2 + \frac{3}{2}x - \frac{4}{3}; \qquad x^3 + \frac{1}{2}x^2 - \frac{3}{2}x + \frac{6}{5}$$

E. Irreducibility of Polynomials of Degree ≤ 4

1 Let F be any field. Explain why, if $a(x)$ is a quadratic or cubic polynomial in $F[x]$, $a(x)$ is irreducible in $F[x]$ iff $a(x)$ has no roots in F.

2 Prove that the following polynomials are irreducible in $\mathbb{Q}[x]$:

$$\frac{1}{2}x^3 + 2x - \frac{3}{2}; \quad 3x^2 - 2x - 4; \quad x^3 + x^2 + \frac{3}{2}x + \frac{1}{2}; \quad x^3 + \frac{1}{2}; \quad x^2 - \frac{5}{2}x + \frac{3}{2}$$

3 Suppose a monic polynomial $a(x)$ of degree 4 in $F[x]$ has no roots in F. Then $a(x)$ is reducible iff it is a product of two quadratics $x^2 + ax + b$ and $x^2 + cx + d$, that is, iff

$$a(x) = x^4 + (a + c)x^3 + (ac + b + d)x^2 + (bc + ad)x + bd$$

If the coefficients of $a(x)$ cannot be so expressed (in terms of *any* $a, b, c, d \in F$) then $a(x)$ must be irreducible.

Example $a(x) = x^4 + 2x^3 + x + 1;$ then $bd = 1,$ so $b = d = \pm 1;$ thus, $bc + ad = \pm(a + c)$, *but* $a + c = 2$ *and* $bc + ad = 1$, *which is impossible.*

Prove that the following polynomials are irreducible in $\mathbb{Q}[x]$ (use Theorem 5, searching only for integer values of a, b, c, d):

$$x^4 - 5x^2 + 1; \qquad 3x^4 - x^2 - 2; \qquad x^4 + x^3 + 3x + 1$$

4 Prove that the following polynomials are irreducible in $\mathbb{Z}_5[x]$:

$$2x^3 + x^2 + 4x + 1; \qquad x^4 + 2; \qquad x^4 + 4x^2 + 2; \qquad x^4 + 1$$

F. Mapping onto $\mathbb{Z}_n$ to Determine Irreducibility over $\mathbb{Q}$

If $h: \mathbb{Z} \to \mathbb{Z}_n$ is the natural homomorphism, let $\bar{h}: \mathbb{Z}[x] \to \mathbb{Z}_n[x]$ be defined by

$$\bar{h}(a_0 + a_1 x + \cdots + a_n x^n) = h(a_0) + h(a_1)x + \cdots + h(a_n)x^n$$

In Chapter 24, Exercise G, it is proved that $\bar{h}$ is a homomorphism. *Assume this fact to prove the following :*

1 If $\bar{h}(a(x))$ is irreducible in $\mathbb{Z}_n[x]$, then $a(x)$ is irreducible in $\mathbb{Z}[x]$.
2 Prove that $x^4 + 10x^3 + 7$ is irreducible in $\mathbb{Q}[x]$ by using the natural homomorphism from $\mathbb{Z}$ to $\mathbb{Z}_5$.
3 Prove that the following are irreducible in $\mathbb{Q}[x]$ (find the right value of n and use the natural homomorphism from $\mathbb{Z}$ to $\mathbb{Z}_n$):

$$x^4 - 10x^2 + 1; \qquad x^4 + 7x^3 + 14x^2 + 3; \qquad x^5 + 1$$

G. Roots and Factors in $A[x]$ when A Is an Integral Domain

It is a useful fact that Theorems 1, 2, and 3 are still true in $A[x]$ when A is not a field, but merely an integral domain. The proof of Theorem 1 must be altered a bit to avoid using the division algorithm. We proceed as follows:
If $a(x) = a_0 + a_1 x + \cdots + a_n x^n$ and c is a root of $a(x)$, consider

$$a(x) - a(c) = a_1(x - c) + a_2(x^2 - c^2) + \cdots + a_n(x^n - c^n)$$

1 Prove that for $k = 1, \ldots, n$:

$$a_k(x^k - c^k) = a_k(x - c)(x^{k-1} + x^{k-2}c + \cdots + c^{k-1})$$

2 Conclude from part 1 that $a(x) - a(c) = (x - c)q(x)$ for some $q(x)$.

3 Complete the proof of Theorem 1, explaining why this particular proof is valid when A is an integral domain, not necessarily a field.

4 Check that Theorems 2 and 3 are true in $A[x]$ when A is an integral domain.

H. Polynomial Functions over a Finite Field

1 Find three polynomials in $\mathbb{Z}_5[x]$ which determine the same function as

$$x^2 - x + 1$$

2 Prove that $x^p - x$ has p roots in $\mathbb{Z}_p[x]$, for any prime p. Draw the conclusion that in $\mathbb{Z}_p[x]$, $x^p - x$ can be factored as:

$$x^p - x = x(x - 1)(x - 2) \cdots [x - (p - 1)]$$

3 Prove that if $a(x)$ and $b(x)$ determine the same function in $\mathbb{Z}_p[x]$, then

$$(x^p - x) \mid (a(x) - b(x))$$

In the next four problems, let F be any finite field. Prove the following :

4 Let $a(x)$ and $b(x)$ be in $F[x]$. If $a(x)$ and $b(x)$ determine the same function, and if the number of elements in F exceeds the degree of $a(x)$ as well as the degree of $b(x)$, then $a(x) = b(x)$.

5 The set of all $a(x)$ which determine the zero function is an ideal of $F[x]$. What is its generator?

6 Let $\mathscr{F}(F)$ be the ring of all functions from F to F, defined in the same way as $\mathscr{F}(\mathbb{R})$. Let $h: F[x] \rightarrow \mathscr{F}(F)$ send every polynomial $a(x)$ to the polynomial function which it determines. Show that h is a homomorphism from $F[x]$ onto $\mathscr{F}(F)$.

7 Let $F = \{c_1, \cdots, c_n\}$ and $p(x) = (x - c_1) \cdots (x - c_n)$. Prove that

$$F[x]/\langle p(x)\rangle \cong \mathscr{F}(F)$$

I. Polynomial Interpolation

One of the most important applications of polynomials is to problems where we are given several values of x (say $x = a_0, a_1, \ldots, a_n$) and corresponding values of y (say $y = b_0, b_1, \ldots, b_n$), and we need to find a function $y = f(x)$ such that $f(a_0) = b_0$, $f(a_1) = b_1, \ldots, f(a_n) = b_n$. The simplest and most useful kind of function for this purpose is a polynomial function of the lowest possible degree.

We now consider a commonly used technique for constructing a polynomial $p(x)$ of degree n which assumes given values $b_0, b_1, \ldots, b_n$ at given points $a_0, a_1, \ldots, a_n$. That is,

$$p(a_0) = b_0, \, p(a_1) = b_1, \ldots, p(a_n) = b_n$$

First, for each $i = 0, 1, \ldots, n$, let

$$q_i(x) = (x - a_0) \ldots (x - a_{i-1})(x - a_{i+1}) \cdots (x - a_n)$$

1 Show that $q_i(a_j) = 0$ for $j \neq i$, and $q_i(a_i) \neq 0$.

Let $q_i(a_i) = c_i$, and define $p(x)$ as follows:

$$p(x) = \sum_{i=0}^{n} \frac{b_i}{c_i} q_i(x) = \frac{b_0}{c_0} q_0(x) + \cdots + \frac{b_n}{c_n} q_n(x)$$

(This is called the *Lagrange interpolation formula*.)

2 Explain why $p(a_0) = b_0, p(a_1) = b_1, \ldots, p(a_n) = b_n$.

3 Prove that there is one and *only* one polynomial $p(x)$ of degree $\leq n$ such that $p(a_0) = b_0, \ldots, p(a_n) = b_n$.

4 Use the Lagrange interpolation formula to prove that if F is a finite field, every function from F to F is equal to a polynomial function. (In fact, the degree of this polynomial is less than the number of elements in F.)

5 If $t(x)$ is any polynomial in $F[x]$, and $a_0, \ldots, a_n \in F$, the unique polynomial $p(x)$ of degree $\leq n$ such that $p(a_0) = t(a_0), \ldots, p(a_n) = t(a_n)$ is called the *Lagrange interpolator* for $t(x)$ and $a_0, \ldots, a_n$. Prove that the remainder, when $t(x)$ is divided by $(x - a_0)(x - a_1) \cdots (x - a_n)$, is the Lagrange interpolator.

TWENTY-SEVEN
EXTENSIONS OF FIELDS

In the first 26 chapters of this book we introduced the cast and set the scene on a vast and complex stage. Now it is time for the action to begin. We will be surprised to discover that none of our effort has been wasted; for every notion which was defined with such meticulous care, every subtlety, every fine distinction, will have its use and play its prescribed role in the story which is about to unfold.

We will see modern algebra reaching out and merging with other disciplines of mathematics; we will see its machinery put to use for solving a wide range of problems which, on the surface, have nothing whatever to do with modern algebra. Some of these problems—ancient problems of geometry, riddles about numbers, questions concerning the solutions of equations—reach back to the very beginnings of mathematics. Great masters of the art of mathematics puzzled over them in every age and left them unsolved, for the machinery to solve them was not there. Now, with a light touch modern algebra uncovers the answers.

Modern algebra was not built in an ivory tower but was created part and parcel with the rest of mathematics—tied to it, drawing from it, and offering it solutions. Clearly it did not develop as methodically as it has been presented here. It would be pointless, in a first course in abstract algebra, to replicate all the currents and crosscurrents, all the hits and misses and false starts. Instead, we are provided with a finished product in which the agonies and efforts that went into creating it cannot be discerned.

There is a disadvantage to this: without knowing the origin of a given concept, without knowing the specific problems which gave it birth, the student often wonders what it means and why it was ever invented.

We hope, beginning now, to shed light on that kind of question, to justify what we have already done, and to demonstrate that the concepts introduced in earlier chapters are correctly designed for their intended purposes.

Most of classical mathematics is set in a framework consisting of fields, especially $\mathbb{Q}$, $\mathbb{R}$, and $\mathbb{C}$. The theory of equations deals with polynomials over $\mathbb{R}$ and $\mathbb{C}$, calculus is concerned with functions over $\mathbb{R}$, and plane geometry is set in $\mathbb{R} \times \mathbb{R}$. It is not surprising, therefore, that modern efforts to generalize and unify these subjects should also center around the study of fields. It turns out that a great variety of problems, ranging from geometry to practical computation, can be translated into the language of fields and formulated entirely in terms of the theory of fields. The study of fields will therefore be our central concern in the remaining chapters, though we will see other themes merging and flowing into it like the tributaries of a great river.

If F is a field, then a *subfield* of F is any nonempty subset of F which is closed with respect to addition and subtraction, multiplication and division. (It would be equivalent to say: closed with respect to addition and negatives, multiplication and multiplicative inverses.) As we already know, if K is a subfield of F, then K is a field in its own right.

If K is a subfield of F, we say also that F is an *extension field* of K. When it is clear in context that both F and K are fields, we say simply that F is an *extension* of K.

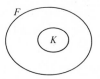

Given a field F, we may *look inward* from F at all the subfields of F. On the other hand, we may *look outward* from F at all the extensions of F. Just as there are relationships between F and its subfields, there are also interesting relationships between F and its extensions. One of these relationships, as we shall see later, is highly reminiscent of Lagrange's theorem—an inside-out version of it.

Why should we be interested in looking at the extensions of fields? There are several reasons, but one is very special. If F is an arbitrary field, there are, in general, polynomials over F which have no roots in F. For example, $x^2 + 1$ has no roots in $\mathbb{R}$. This situation is unfortunate but, it turns out, not hopeless. For, as we shall soon see, every polynomial over any field F, *has roots*. If these roots are not already *in F*, they are in a suitable extension of F. For example, $x^2 + 1 = 0$ has solutions in $\mathbb{C}$.

In the matter of factoring polynomials and extracting their roots, $\mathbb{C}$ is utopia! In $\mathbb{C}$ every polynomial $a(x)$ of degree n has exactly n roots $c_1, \ldots, c_n$ and can therefore be factored as $a(x) = k(x - c_1)(x - c_2) \cdots (x - c_n)$. This ideal situation is not enjoyed by all fields—far from it! In an arbitrary field F, a polynomial of degree n may have any number of roots, from no roots to n roots, and there may be irreducible polynomials of any degree whatever. This is a messy situation, which does not hold the promise of an elegant theory of solutions to polynomial equations. However, it turns out that F *always has a suitable extension E* such that any polynomial $a(x)$ of degree n over F has *exactly n solutions in E*. Therefore, $a(x)$ can be factored in $E[x]$ as

$$a(x) = k(x - c_1)(x - c_2) \cdots (x - c_n)$$

Thus, paradise is regained by the expedient of enlarging the field F. This is one of the strongest reasons for our interest in field extensions. They will give us a trim and elegant theory of solutions to polynomial equations.

Now, let us get to work! Let E be a field, F a subfield of E, and c any

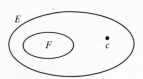

element of E. We define the *substitution function* σ_c as follows:

For every polynomial $a(x)$ in $F[x]$,

$$\sigma_c(a(x)) = a(c)$$

Thus, σ_c is the function "substitute c for x." It is a function from $F[x]$ into E. In fact, σ_c *is a homomorphism*. This is true because

$$\underbrace{\sigma_c(a(x) + b(x))}_{a(c) + b(c)} = \underbrace{\sigma_c(a(x))}_{a(c)} + \underbrace{\sigma_c(b(x))}_{b(c)}$$

and

$$\underbrace{\sigma_c(a(x)b(x))}_{a(c)b(c)} = \underbrace{\sigma_c(a(x))}_{a(c)} \ \underbrace{\sigma_c(b(x))}_{b(c)}$$

The kernel of the homomorphism σ_c is the set of all the polynomials $a(x)$ such that $a(c) = \sigma_c(a(x)) = 0$. That is, *the kernel of σ_c consists of all the polynomials $a(x)$ in $F[x]$ such that c is a root of $a(x)$.*

Let J_c denote the kernel of σ_c; since the kernel of any homomorphism is an ideal, J_c is an ideal of $F[x]$.

An element c in E is called *algebraic over F* if it is the root of some nonzero polynomial $a(x)$ in $F[x]$. Otherwise, c is called *transcendental over F*. Obviously c is algebraic over F iff J_c contains nonzero polynomials, and transcendental over F iff $J_c = \{0\}$.

We will confine our attention now to the case where c is algebraic. The transcendental case will be examined in Exercise G at the end of this chapter.

Thus, let c be algebraic over F, and let J_c be the kernal of σ_c (where σ_c is the function "substitute c for x"). Remember that in $F[x]$ every ideal is a principal ideal, hence $J_c = \langle p(x) \rangle =$ the set of all the multiples of $p(x)$, for some polynomial $p(x)$. Since every polynomial in J_c is a multiple of $p(x)$, $p(x)$ is a polynomial *of lowest degree among all the nonzero polynomials in J_c*. It is easy to see that $p(x)$ *is irreducible*; otherwise we could factor it into polynomials of lower degree, say $p(x) = f(x)g(x)$. But then $0 = p(c) = f(c)g(c)$, so $f(c) = 0$ or $g(c) = 0$, and therefore either $f(x)$ or $g(x)$ is in J_c. This is impossible, because we have just seen that $p(x)$ has the *lowest degree* among all the polynomials in J_c, whereas $f(x)$ and $g(x)$ both have lower degree than $p(x)$.

Since every *constant* multiple of $p(x)$ is in J_c, we may take $p(x)$ to be monic, that is, to have leading coefficient 1. Then $p(x)$ is the *unique* monic polynomial of lowest degree in J_c. (Also, it is the only monic irreducible polynomial in J_c.) This polynomial $p(x)$ is called the *minimum polynomial* of c over F, and will be of considerable importance in our discussions in a later chapter.

Let us look at an example: $\mathbb{R}$ is an extension field of $\mathbb{Q}$, and $\mathbb{R}$ contains the irrational number $\sqrt{2}$. The function $\sigma_{\sqrt{2}}$ is the function "substitute $\sqrt{2}$ for x"; for example $\sigma_{\sqrt{2}}(x^4 - 3x^2 + 1) = \sqrt{2}^4 - 3\sqrt{2}^2 + 1 = -1$. By our discussion above, $\sigma_{\sqrt{2}} : \mathbb{Q}[x] \to \mathbb{R}$ is a homomorphism and its kernel consists of all the polynomials in $\mathbb{Q}[x]$ which have $\sqrt{2}$ as one of their roots. The monic polynomial of least degree in $\mathbb{Q}[x]$ having $\sqrt{2}$ as a root is $p(x) = x^2 - 2$; hence $x^2 - 2$ is the minimum polynomial of $\sqrt{2}$ over $\mathbb{Q}$.

Now, let us turn our attention to the *range* of σ_c. Since σ_c is a homomorphism, its range is obviously closed with respect to addition, multiplication, and negatives, but it is not obviously closed with respect to multiplicative inverses. Not obviously, but in fact it *is* closed for multiplicative inverses, which is far from self-evident, and quite a remarkable fact. In order to prove this, let $f(c)$ be any nonzero element in the range of σ_c. Since $f(c) \neq 0$, $f(x)$ is not in the kernel of σ_c. Thus, $f(x)$ is not a multiple of $p(x)$, and since $p(x)$ is irreducible, it follows that $f(x)$ and $p(x)$ are *relatively prime*. Therefore there are polynomials $s(x)$ and $t(x)$ such that $s(x)f(x) + t(x)p(x) = 1$. But then

$$s(c)f(c) + \underbrace{t(c)p(c)}_{=0} = 1$$

and therefore $s(c)$ is the multiplicative inverse of $f(c)$.

We have just shown that *the range of σ_c is a subfield of E*. Now, the range of σ_c is the set of all the elements $a(c)$, for all $a(x)$ in $F[x]$:

$$\text{Range } \sigma_c = \{a(c) : a(x) \in F[x]\}$$

We have just seen that range σ_c is a field. In fact, it is the *smallest field containing F and c*: indeed, any other field containing F and c would inevitably contain every element of the form

$$a_0 + a_1 c + \cdots + a_n c^n \qquad (a_0, \ldots, a_n \in F)$$

in other words, would contain every element in the range of σ_c.

By the *smallest field containing F and c* we mean the field which contains F and c and is contained in any other field containing F and c. It is called the field *generated by F and c*, and is denoted by the important symbol

$$F(c)$$

Now, here is what we have, in a nutshell: σ_c is a homomorphism with domain $F[x]$, range $F(c)$, and kernel $J_c = \langle p(x) \rangle$. Thus, by the fundamental homomorphism theorem,

$$\boxed{F(c) \cong F[x]/\langle p(x) \rangle}$$

Finally, here is an interesting sidelight: if c and d are both roots of $p(x)$, then, by what we have just proved, $F(c)$ and $F(d)$ are both isomorphic to $F[x]/\langle p(x) \rangle$, and therefore isomorphic to each other:

(*) If c and d are roots of the same irreducible polynomial $p(x)$ in $F[x]$, then $F(c) \cong F(d)$.

In particular, this shows that, given F and c, $F(c)$ is unique up to isomorphism.

It is time now to recall our main objective: if $a(x)$ is a polynomial in $F[x]$ which has no roots in F, we wish to enlarge F to a field E which contains a root of $a(x)$. How can we manage this?

An observation is in order: finding extensions of F is not as easy as finding subfields of F. A subfield of F is a subset of an *existing* set: *it is there*! But an extension of F is *not yet there*. We must somehow build it around F.

Let $p(x)$ be an irreducible polynomial in $F[x]$. We have just seen that if F can be enlarged to a field E containing a root c of $p(x)$, then $F(c)$ is already what we are looking for: it is an extension of F containing a root of $p(x)$. Furthermore, $F(c)$ is isomorphic to $F[x]/\langle p(x)\rangle$. Thus, *the field extension we are searching for is precisely $F[x]/\langle p(x)\rangle$*. Our result is summarized in the next theorem.

Basic theorem of field extensions *Let F be a field and $a(x)$ a nonconstant polynomial in $F[x]$. There exists an extension field E of F and an element c in E such that c is a root of $a(x)$.*

To begin with, $a(x)$ can be factored into irreducible polynomials in $F[x]$. If $p(x)$ is any nonconstant irreducible factor of $a(x)$, it is clearly sufficient to find an extension of F containing a root of $p(x)$, since it will also be a root of $a(x)$.

In Exercise D4 of Chapter 25, the reader was asked to supply the simple proof that, if $p(x)$ is irreducible in $F[x]$, then $\langle p(x)\rangle$ is a maximal ideal of $F[x]$. Furthermore, by the argument at the end of Chapter 19, if $\langle p(x)\rangle$ is a maximal ideal of $F[x]$, then the quotient ring $F[x]/\langle p(x)\rangle$ is a field.

It remains only to prove that $F[x]/\langle p(x)\rangle$ is the desired field extension of F. When we write $J = \langle p(x)\rangle$, let us remember that every element of $F[x]/J$ is a coset of J. We will prove that $F[x]/J$ is an extension of F by identifying each element a in F with its coset $J + a$.

To be precise, define $h: F \rightarrow F[x]/J$ by $h(a) = J + a$. Note that h is the function which matches every a in F with its coset $J + a$ in $F[x]/J$. We will now show that h is an isomorphism.

By the familiar rules of coset addition and multiplication, h is a homomorphism. Now, every homomorphism between fields is injective. (This is true because the kernel of a homomorphism is an ideal, and a field has no nontrivial ideals.) Thus, h is an isomorphism between its domain and its range.

What is the range of h? It consists of all the cosets $J + a$ where $a \in F$, that is, all the cosets of constant polynomials. (If a is in F, then a is a constant polynomial.) Thus, *F is isomorphic to the subfield of $F[x]/J$ containing all the cosets of constant polynomials*. This subfield is therefore an isomorphic copy of F, which may be identified with F, so $F[x]/J$ is an extension of F.

Finally, if $p(x) = a_0 + a_1 x + \cdots + a_n x^n$, let us show that the coset $J + x$ is a root of $p(x)$ in $F[x]/J$. Of course, in $F[x]/J$, the coefficients are not actually $a_0, a_1, \ldots, a_n$, but their cosets $J + a_0$, $J + a_1$, ..., $J + a_n$. Writing

$$J + a_0 = \bar{a}_0, \ldots, J + a_n = \bar{a}_n \quad \text{and} \quad J + x = \bar{x}$$

we must prove that

$$\bar{a}_0 + \bar{a}_1 \bar{x} + \cdots + \bar{a}_n \bar{x}^n = J$$

Well,

$$\bar{a}_0 + \bar{a}_1 \bar{x} + \cdots + \bar{a}_n \bar{x}^n = (J + a_0) + (J + a_1)(J + x) + \cdots + (J + a_n)(J + x)^n$$

$$= (J + a_0) + (J + a_1 x) + \cdots + (J + a_n x^n)$$

$$= J + p(x)$$

$$= J \quad [\text{because } p(x) \in J]$$

This completes the proof of the basic theorem of field extensions. Observe that we may use this theorem several times in succession to get the following:

Let $a(x)$ be a polynomial of degree n in $F[x]$. There is an extension field E of F which contains all n roots of $a(x)$.

EXERCISES

A. Recognizing Algebraic Elements

Example *To show that $\sqrt{1 + \sqrt{2}}$ is algebraic over* $\mathbb{Q}$, *one must find a polynomial* $p(x) \in \mathbb{Q}[x]$ *such that* $\sqrt{1 + \sqrt{2}}$ *is a root of $p(x)$.*

Let $a = \sqrt{1 + \sqrt{2}}$; then $a^2 = 1 + \sqrt{2}$, $a^2 - 1 = \sqrt{2}$, and finally, $(a^2 - 1)^2 = 2$. Thus, a satisfies $p(x) = x^4 - 2x^2 - 1 = 0$.

1 Prove that each of the following numbers is algebraic over $\mathbb{Q}$:

(a) i (b) $\sqrt{2}$ (c) $2 + 3i$ (d) $\sqrt{1 + \sqrt[3]{2}}$

(e) $\sqrt{i - \sqrt{2}}$ (f) $\sqrt{2} + \sqrt{3}$ (g) $\sqrt{2} + \sqrt[3]{4}$

2 Prove that each of the following numbers is algebraic over the given field:

(a) $\sqrt{\pi}$ over $\mathbb{Q}(\pi)$ (b) $\sqrt{\pi}$ over $\mathbb{Q}(\pi^2)$ (c) $\pi^2 - 1$ over $\mathbb{Q}(\pi^3)$

NOTE: Recognizing a transcendental element is much more difficult, since it requires proving that the element cannot be a root of *any* polynomial over the given field. In recent times it has been proved, using sophisticated mathematical machinery, that π and e are transcendental over $\mathbb{Q}$.

B. Finding the Minimum Polynomial

1 Find the minimum polynomial of each of the following numbers over $\mathbb{Q}$. (Where appropriate, use the methods of Chapter 26, Exercises D, E, and F to ensure that your polynomial is irreducible.)

(a) $1 + 2i$ (b) $1 + \sqrt{2}$ (c) $1 + \sqrt{2}i$

(d) $\sqrt{2 + \sqrt[3]{3}}$ (e) $\sqrt{3} + \sqrt{5}$ (f) $\sqrt{i + \sqrt{2}}$

2 Show that the minimum polynomial of $\sqrt{2} + i$ is:

(a) $x^2 - 2\sqrt{2}x + 3$ over $\mathbb{R}$ (b) $x^4 - 2x^2 + 9$ over $\mathbb{Q}$ (c) $x^2 - 2ix - 3$ over $\mathbb{Q}(i)$

3 Find the minimum polynomial of the following numbers over the indicated fields:

$$\sqrt{3} + i \quad \text{over } \mathbb{R}; \text{ over } \mathbb{Q}: \text{ over } \mathbb{Q}(i); \text{ over } \mathbb{Q}(\sqrt{3})$$

$$\sqrt{i + \sqrt{2}} \quad \text{over } \mathbb{R}; \text{ over } \mathbb{Q}(i); \text{ over } \mathbb{Q}(\sqrt{2}); \text{ over } \mathbb{Q}$$

4 For each of the following polynomials $p(x)$, find a number a such that $p(x)$ is the minimum polynomial of a over $\mathbb{Q}$:

(a) $x^2 + 2x - 7$ (b) $x^4 + 2x^2 - 1$ (c) $x^4 - 10x^2 + 1$

5 Find a monic irreducible polynomial $p(x)$ such that $\mathbb{Q}[x]/\langle p(x) \rangle$ is isomorphic to:

(a) $\mathbb{Q}(\sqrt{2})$ (b) $\mathbb{Q}(1 + \sqrt{2})$ (c) $\mathbb{Q}(\sqrt{1 + \sqrt{2}})$

C. The Structure of Fields $F[x]/\langle p(x) \rangle$

Let $p(x)$ be an irreducible polynomial of degree n over F. Let c denote a root of $p(x)$ in some extension of F (as in the basic theorem on field extensions). *Prove*:

1 Every element in $F(c)$ can be written as $r(c)$, for some $r(x)$ of degree $<n$ in $F[x]$. [HINT: Given any element $t(c) \in F(c)$, use the division algorithm to divide $t(x)$ by $p(x)$.]

2 If $s(c) = t(c)$ in $F(c)$, where $s(x)$ and $t(x)$ have degree $<n$, then $s(x) = t(x)$.

3 Conclude from parts 1 and 2 that every element in $F(c)$ can be written *uniquely* as $r(c)$, with deg $r(x) < n$.

4 Using part 3, explain why there are exactly four elements in $\mathbb{Z}_2[x]/\langle x^2 + x + 1 \rangle$. List these four elements, and give their addition and multiplication tables. [HINT: Identify $\mathbb{Z}_2[x]/\langle x^2 + x + 1 \rangle$ with $\mathbb{Z}_2(c)$, where c is a root of $x^2 + x + 1$. Write the elements of $\mathbb{Z}_2(c)$ as in part 3. When computing the multiplication table, use the fact that $c^2 + c + 1 = 0$.]

5 Describe $\mathbb{Z}_2[x]/\langle x^3 + x + 1 \rangle$, as in part 4.

6 Describe $\mathbb{Z}_3[x]/\langle x^3 + x^2 + 2 \rangle$, as in part 4.

D. Short Questions Relating to Field Extensions

Let F be any field. *Prove each of the following:*

1 If c is algebraic over F, so are $c + 1$ and kc (where $k \in F$).

2 If cd is algebraic over F, then c is algebraic over $F(d)$. If $c + d$ is algebraic over F, then c is algebraic over $F(d)$.

3 If the minimum polynomial of a over F is of degree 1, then $a \in F$, and conversely.

4 Suppose $F \subseteq K$ and $a \in K$. If $p(x)$ is a monic irreducible polynomial in $F[x]$, and $p(a) = 0$, then $p(x)$ is the minimum polynomial of a over F.

5 Name a field ($\neq \mathbb{R}$ or $\mathbb{C}$) which contains a root of $x^5 + 2x^3 + 4x^2 + 6$.

6 $\mathbb{Q}(1 + i) \cong \mathbb{Q}(1 - i)$. However, $\mathbb{Q}(\sqrt{2}) \not\cong \mathbb{Q}(\sqrt{3})$.

7 If $p(x)$ has degree 2, then $\mathbb{Q}[x]/\langle p(x) \rangle$ contains *both* roots of $p(x)$.

E. Simple Extensions

Recall the definition of $F(a)$. It is a field such that (i) $F \subseteq F(a)$; (ii) $a \in F(a)$; (iii) any field containing F and a contains $F(a)$. *Use this definition to prove each of the following, where $F \subseteq K, c \in F$, and $a \in K$:*

1 $F(a) = F(a + c)$ and $F(a) = F(ca)$.

2 $F(a^2) \subseteq F(a)$ and $F(a + b) \subseteq F(a, b)$. [$F(a, b)$ is the field containing F, a, and b, and contained in any other field containing F, a, and b.] Why are the reverse inclusions not necessarily true?

3 $a + c$ is a root of $p(x)$ iff a is a root of $p(x + c)$; ca is a root of $p(x)$ iff a is a root of $p(cx)$.

4 Let a be a root of $p(x + c)$. Then $F[x]/\langle p(x + c) \rangle \cong F(a)$ and

$$F[x]/\langle p(x) \rangle \cong F(a + c)$$

Conclude that $F[x]/\langle p(x + c) \rangle \cong F[x]/\langle p(x) \rangle$.

5 Let a be a root of $p(cx)$. Then $F[x]/\langle p(cx) \rangle \cong F(a)$ and $F[x]/\langle p(x) \rangle \cong F(ca)$. Conclude that $F[x]/\langle p(cx) \rangle \cong F[x]/\langle p(x) \rangle$.

6 Use parts 4 and 5 to prove the following:

(a) $\mathbb{Z}_{11}[x]/\langle x^2 + 1 \rangle \cong \mathbb{Z}_{11}[x]/\langle x^2 + x + 4 \rangle$.

(b) If a is a root of $x^2 - 2$ and b is a root of $x^2 - 4x + 2$, then $\mathbb{Q}(a) \cong \mathbb{Q}(b)$.

(c) If a is a root of $x^2 - 2$ and b is a root of $x^2 - \frac{1}{2}$, then $\mathbb{Q}(a) \cong \mathbb{Q}(b)$.

† F. Quadratic Extensions

If the minimum polynomial of a over F has degree 2, we call $F(a)$ a quadratic extension of F. Prove the following, where F is any field whose characteristic is $\neq 2$.

1 Any quadratic extension of F is of the form $F(\sqrt{a})$, for some $a \in F$. (HINT: Complete the square, and use Exercise E4.)

Let F be a finite field, and F^* the multiplicative group of nonzero elements of F. Obviously $H = \{x^2 : x \in F^*\}$ is a subgroup of F^*; since every square x^2 in F^* is the square of only two different elements, namely $\pm x$, exactly half the elements of F^* are in H. Thus, H has exactly two cosets: H itself, containing all the squares, and aH (where $a \notin H$), containing all the nonsquares. If a and b are nonsquares, then by Chapter 15, Theorem 5a,

$$ab^{-1} = \frac{a}{b} \in H$$

Thus: if a and b are nonsquares, a/b is a square. *Use these remarks in the following*:

2 Let F be a finite field. If a, $b \in F$, let $p(x) = x^2 - a$ and $q(x) = x^2 - b$ be irreducible in $F[x]$, and let $\sqrt{a}$ and $\sqrt{b}$ denote roots of $p(x)$ and $q(x)$ in an extension of F. Explain why a/b is a square, say $a/b = c^2$ for some $c \in F$. Prove that $\sqrt{b}$ is a root of $p(cx)$.

3 Use part 2 to prove that $F[x]/\langle p(cx)\rangle \cong F(\sqrt{b})$; then use Exercise E5 to conclude that $F(\sqrt{a}) \cong F(\sqrt{b})$.

4 Use part 3 to prove: Any two quadratic extensions of a finite field are isomorphic.

5 If a and b are nonsquares in $\mathbb{R}$, a/b is a square (why?). Use the same argument as in part 4 to prove that any two simple extensions of $\mathbb{R}$ are isomorphic (hence isomorphic to $\mathbb{C}$).

G. Questions Relating to Transcendental Elements

Let F be a field, and let c be transcendental over F. *Prove the following*:

1 $\{a(c) : a(x) \in F[x]\}$ is an integral domain isomorphic to $F[x]$.

2 $F(c)$ is the field of quotients of $\{a(c) : a(x) \in F[x]\}$, and is isomorphic to $F(x)$, the field of quotients of $F[x]$.

3 If c is transcendental over F, so are $c + 1$, kc (where $k \in F$), and c^2.

4 If c is transcendental over F, every element in $F(c)$ but not in F is transcendental over F.

† H. Common Factors of Two Polynomials: Over F and over Extensions of F

Let F be a field, and let $a(x), b(x) \in F[x]$. *Prove the following*:

1 If $a(x)$ and $b(x)$ have a common root c in some extension of F, they have a common factor in $F[x]$. [Use the fact that $a(x), b(x) \in \ker \sigma_c$.]

2 If $a(x)$ and $b(x)$ are relatively prime in $F[x]$, they are relatively prime in $K[x]$, for any extension K of F.

3 Let $K \subseteq L$ be fields containing the coefficients of $a(x)$ and $b(x)$. Then $a(x)$ and $b(x)$ are relatively prime in $K[x]$ iff they are relatively prime in $L[x]$.

† I. Derivatives and Their Properties

Let $a(x) = a_0 + a_1 x + \cdots + a_n x^n \in F[x]$. The *derivative* of $a(x)$ is the following polynomial $a'(x) \in F[x]$:

$$a'(x) = a_1 + 2a_2 x + \cdots + na_n x^{n-1}$$

(This is the same as the derivative of a polynomial in calculus.) We now prove the analogs of the formal rules of differentiation, familiar from calculus.

Let $a(x), b(x) \in F[x]$, and let $k \in F$. *Prove the following*:

1 $[a(x) + b(x)]' = a'(x) + b'(x)$

2 $[a(x)b(x)]' = a'(x)b(x) + a(x)b'(x)$

3 $[ka(x)]' = ka'(x)$

4 If F has characteristic 0 and $a'(x) = 0$, then $a(x)$ is a constant polynomial. Why is this conclusion not necessarily true if F has characteristic $p \neq 0$?

5 Find the derivative of the following polynomials in $\mathbb{Z}_5[x]$:

$$x^6 + 2x^3 + x + 1 \qquad x^5 + 3x^2 + 1 \qquad x^{15} + 3x^{10} + 4x^5 + 1$$

6 If F has characteristic $p \neq 0$, and $a'(x) = 0$, prove that the only nonzero terms of $a(x)$ are of the form $a_{mp} x^{mp}$ for some m. [That is, $a(x)$ is a polynomial in powers of x^p.]

† J. Multiple Roots

Suppose $a(x) \in F[x]$, and K is an extension of F. An element $c \in K$ is called a *multiple root* of $a(x)$ if $(x - c)^m \mid a(x)$ for some $m > 1$. It is often important to know if all the roots of a polynomial are different, or not. We now consider a method for determining whether an arbitrary polynomial $a(x) \in F[x]$ has multiple roots in any extension of F.

Let K be any field containing all the roots of $a(x)$. Suppose $a(x)$ has a multiple root c.

1 Prove that $a(x) = (x - c)^2 q(x)$ for some $q(x) \in K[x]$.

2 Compute $a'(x)$, using part 1.

3 Show that $x - c$ is a common factor of $a(x)$ and $a'(x)$. Use Exercise H1 to conclude that $a(x)$ and $a'(x)$ have a common factor of degree > 1 in $F[x]$.

Thus, if $a(x)$ has a multiple root, then $a(x)$ and $a'(x)$ have a common factor in $F[x]$. To prove the converse, suppose $a(x)$ has *no* multiple roots. Then $a(x)$ can be factored as $a(x) = (x - c_1) \cdots (x - c_n)$ where $c_1, \ldots, c_n$ are all different.

4 Explain why $a'(x)$ is a sum of terms of the form

$$(x - c_1) \cdots (x - c_{i-1})(x - c_{i+1}) \cdots (x - c_n).$$

5 Using part 4, explain why none of the roots $c_1, \ldots, c_n$ of $a(x)$ are roots of $a'(x)$.

6 Conclude that $a(x)$ and $a'(x)$ have no common factor of degree > 1 in $F[x]$.

This important result is stated as follows: *A polynomial $a(x)$ in $F[x]$ has a multiple root iff $a(x)$ and $a'(x)$ have a common factor of degree > 1 in $F[x]$.*

7 Show that each of the following polynomials has *no* multiple roots in any extension of its field of coefficients:

$$x^3 - 7x^2 + 8 \in \mathbb{Q}[x] \qquad x^2 + x + 1 \in \mathbb{Z}_5[x] \qquad x^{100} - 1 \in \mathbb{Z}_7[x]$$

The preceding example is most interesting: it shows that there are 100 *different* hundredth roots of 1 over $\mathbb{Z}_7$. (The roots ± 1 are in $\mathbb{Z}_7$, while the remaining 98 roots are in extensions of $\mathbb{Z}_7$.) Corresponding results hold for most other fields.

TWENTY-EIGHT

VECTOR SPACES

Many physical quantities, such as length, area, weight, and temperature, are completely described by a single real number. On the other hand, many other quantities arising in scientific measurement and everyday reckoning are best described by a combination of *several* numbers. For example, a point in space is specified by giving its three coordinates with respect to an *xyz* coordinate system.

Here is an example of a different kind: A store handles 100 items; its monthly inventory is a sequence of 100 numbers $(a_1, a_2, ..., a_{100})$ specifying the quantities of each of the 100 items currently in stock. Such a sequence of numbers is usually called a *vector*. When the store is restocked, a vector is *added* to the current inventory vector. At the end of a good month of sales, a vector is *subtracted*.

As this example shows, it is natural to add vectors by adding corresponding components, and subtract vectors by subtracting corresponding components. If the store manager in the preceding example decided to double inventory, each component of the inventory vector would be multiplied by 2. This shows that a natural way of multiplying a vector by a real number k is to multiply each component by k. This kind of multiplication is commonly called *scalar multiplication*.

Historically, as the use of vectors became widespread and they came to be an indispensable tool of science, vector algebra grew to be one of the major branches of mathematics. Today it forms the basis for much of advanced calculus, the theory and practice of differential equations, statistics, and vast areas of applied mathematics. Scientific computation is enormously simplified by vector methods; for example, 3, or 300, or 3000 individual readings of scientific instruments can be expressed as a single vector.

In any branch of mathematics it is elegant and desirable (but not always possible) to find a *simple* list of axioms from which all the required theorems may be proved. In the specific case of vector algebra, we wish to select as axioms only those particular properties of vectors which are absolutely necessary for proving further properties of vectors. And we must select a sufficiently complete list of axioms so that, by using them and them alone, we can prove all the properties of vectors needed in mathematics.

A delightfully simple list of axioms is available for vector algebra. The remarkable fact about this axiom system is that, although we conceive of vectors as finite sequences $(a_1, a_2, \ldots, a_n)$ of numbers, nothing in the axioms actually requires them to be such sequences! Instead, vectors are treated simply as elements in a set, satisfying certain equations. Here is our basic definition:

A *vector space* over a field F is a set V, with two operations $+$ and $\cdot$ called *vector addition* and *scalar multiplication*, such that

1. V with vector addition is an abelian group.
2. For any $k \in F$ and $\mathbf{a} \in V$, the scalar product $k\mathbf{a}$ is an element of V, subject to the following conditions: for all $k, l \in F$ and $\mathbf{a}, \mathbf{b} \in V$,
   (a) $k(\mathbf{a} + \mathbf{b}) = k\mathbf{a} + k\mathbf{b}$,
   (b) $(k + l)\mathbf{a} = k\mathbf{a} + l\mathbf{a}$,
   (c) $k(l\mathbf{a}) = (kl)\mathbf{a}$,
   (d) $1\mathbf{a} = \mathbf{a}$.

The elements of V are called *vectors* and the elements of the field F are called *scalars*.

In the following exposition the field F will not be specifically referred to unless the context requires it. For notational clarity, vectors will be written in **bold type** and scalars in *italics*.

The traditional example of a vector space is the set $\mathbb{R}^n$ of all n-tuples of real numbers, $(a_1, a_2, \ldots, a_n)$, with the operations

$$(a_1, a_2, \ldots, a_n) + (b_1, b_2, \ldots, b_n) = (a_1 + b_1, a_2 + b_2, \ldots, a_n + b_n)$$

and
$$k(a_1, a_2, \ldots, a_n) = (ka_1, ka_2, \ldots, ka_n)$$

For example, $\mathbb{R}^2$ is the set of all two-dimensional vectors (a, b), while $\mathbb{R}^3$ is the set of all vectors (a, b, c) in euclidean space.

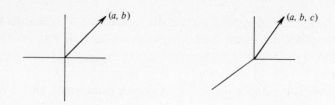

However, these are not the only vector spaces! Our definition of vector space is so very simple that many other things, quite different in appearance from the traditional vector spaces, satisfy the conditions of our definition and are therefore, legitimately, vector spaces.

For example, $\mathscr{F}(\mathbb{R})$, you may recall, is the set of all functions from $\mathbb{R}$ to $\mathbb{R}$. We define the sum $f + g$ of two functions by the rule

$$[f + g](x) = f(x) + g(x)$$

and we define the product af, of a real number a and a function f, by

$$[af](x) = af(x)$$

It is very easy to verify that $\mathscr{F}(\mathbb{R})$, with these operations, satisfies all the conditions needed in order to be a vector space over the field $\mathbb{R}$.

As another example, let $\mathscr{Pl}$ denote the set of all polynomials with real coefficients. Polynomials are added as usual, and scalar multiplication is defined by

$$k(a_0 + a_1 x + \cdots + a_n x^n) = (ka_0) + (ka_1)x + \cdots + (ka_n)x^n$$

Again, it is not hard to see that $\mathscr{Pl}$ is a vector space over $\mathbb{R}$.

Let V be a vector space. Since V with addition alone is an abelian group, there is a zero element in V called the *zero vector*, written as $\mathbf{0}$. Every vector $\mathbf{a}$ in V has a negative, written as $-\mathbf{a}$. Finally, since V with vector addition is an abelian group, it satisfies the following conditions which are true in all abelian groups:

$$\mathbf{a} + \mathbf{b} = \mathbf{a} + \mathbf{c} \quad \text{implies} \quad \mathbf{b} = \mathbf{c} \tag{*}$$

$$\mathbf{a} + \mathbf{b} = \mathbf{0} \quad \text{implies} \quad \mathbf{a} = -\mathbf{b} \quad \text{and} \quad \mathbf{b} = -\mathbf{a} \tag{**}$$

$$-(\mathbf{a} + \mathbf{b}) = (-\mathbf{a}) + (-\mathbf{b}) \quad \text{and} \quad -(-\mathbf{a}) = \mathbf{a} \tag{***}$$

There are simple, obvious rules for multiplication by zero and by negative scalars. They are contained in the next theorem.

Theorem 1 *If V is a vector space, then:*
(a) $0\mathbf{a} = \mathbf{0}$, *for every* $\mathbf{a} \in V$.

(b) $k\mathbf{0} = \mathbf{0}$, *for every scalar k.*

(c) *If $k\mathbf{a} = \mathbf{0}$, then $k = 0$ or $\mathbf{a} = \mathbf{0}$.*

(d) $(-1)\mathbf{a} = -\mathbf{a}$ *for every $\mathbf{a} \in V$.*

To prove Rule (a), we observe that

$$0\mathbf{a} = (0 + 0)\mathbf{a} = 0\mathbf{a} + 0\mathbf{a}$$

hence $\mathbf{0} + 0\mathbf{a} = 0\mathbf{a} + 0\mathbf{a}$. It follows by (*) that $\mathbf{0} = 0\mathbf{a}$.

Rule (b) is proved similarly. As for Rule (c), if $k = 0$, we are done. If $k \neq 0$, we may multiply $k\mathbf{a} = \mathbf{0}$ by $1/k$ to get $a = \mathbf{0}$. Finally, for Rule (d), we have:

$$\mathbf{a} + (-1)\mathbf{a} = 1\mathbf{a} + (-1)\mathbf{a} = (1 + (-1))\mathbf{a} = 0\mathbf{a} = \mathbf{0}$$

so by (**), $(-1)\mathbf{a} = -\mathbf{a}$.

Let V be a vector space, and $U \subseteq V$. We say that U is *closed with respect to scalar multiplication* if $k\mathbf{a} \in U$ for every scalar k and every $\mathbf{a} \in U$. We call U a *subspace* of V if U is *closed with respect to addition and scalar multiplication*. It is easy to see that if V is a vector space over the field F, and U is a subspace of V, then U is a vector space over the same field F.

If $\mathbf{a}_1, \mathbf{a}_2, \ldots, \mathbf{a}_n$ are in V and $k_1, k_2, \ldots, k_n$ are scalars, then the vector

$$k_1\mathbf{a}_1 + k_2\mathbf{a}_2 + \cdots + k_n\mathbf{a}_n$$

is called a *linear combination* of $\mathbf{a}_1, \mathbf{a}_2, \ldots, \mathbf{a}_n$. The set of all the linear combinations of $\mathbf{a}_1, \mathbf{a}_2, \ldots, \mathbf{a}_n$ is a subspace of V. (This fact is exceedingly easy to verify.)

If U is the subspace consisting of all the linear combinations of $\mathbf{a}_1, \mathbf{a}_2, \ldots, \mathbf{a}_n$, we call U the subspace *spanned by* $\mathbf{a}_1, \mathbf{a}_2, \ldots, \mathbf{a}_n$. An equivalent way of saying the same thing is as follows: a space (or subspace) U is spanned by $\mathbf{a}_1, \mathbf{a}_2, \ldots, \mathbf{a}_n$ iff every vector in U is a linear combination of $\mathbf{a}_1, \mathbf{a}_2, \ldots, \mathbf{a}_n$.

If U is spanned by $\mathbf{a}_1, \mathbf{a}_2, \ldots, \mathbf{a}_n$, we also say that $\mathbf{a}_1, \mathbf{a}_2, \ldots, \mathbf{a}_n$ *span* U.

Let $S = \{\mathbf{a}_1, \mathbf{a}_2, \ldots, \mathbf{a}_n\}$ be a set of distinct vectors in a vector space V. Then S is said to be *linearly dependent* if there are scalars $k_1, \ldots, k_n$, *not all zero*, such that

$$k_1\mathbf{a}_1 + k_2\mathbf{a}_2 + \cdots + k_n\mathbf{a}_n = \mathbf{0} \tag{1}$$

Obviously this is the same as saying that at least one of the vectors in S is a linear combination of the remaining ones. [Solve for any vector $\mathbf{a}_i$ in (1) having a nonzero coefficient.]

If $S = \{\mathbf{a}_1, \mathbf{a}_2, \ldots, \mathbf{a}_n\}$ is not linearly dependent, then it is *linearly inde-*

pendent. That is, S is linearly independent iff

$$k_1 \mathbf{a}_1 + k_2 \mathbf{a}_2 + \cdots + k_n \mathbf{a}_n = 0 \qquad \text{implies} \qquad k_1 = k_2 = \cdots = k_n = 0$$

This is the same as saying that *no* vector in S is equal to a linear combination of the other vectors in S.

It is obvious from these definitions that any set of vectors containing the zero vector is linearly dependent. Furthermore, the set $\{\mathbf{a}\}$, containing a single nonzero vector $\mathbf{a}$, is linearly independent.

The next two lemmas, although very easy and at first glance rather trite, are used to prove the most fundamental theorems of this subject.

Lemma 1 *If $\{\mathbf{a}_1, \mathbf{a}_2, \ldots, \mathbf{a}_n\}$ is linearly dependent, then some $\mathbf{a}_i$ is a linear combination of the preceding ones, $\mathbf{a}_1, \mathbf{a}_2, \ldots, \mathbf{a}_{i-1}$.*

Indeed, if $\{\mathbf{a}_1, \mathbf{a}_2, \ldots, \mathbf{a}_n\}$ is linearly dependent, then $k_1 \mathbf{a}_1 + \cdots + k_n \mathbf{a}_n = 0$ for coefficients $k_1, k_2, \ldots, k_n$ which are not all zero. If k_i is the last nonzero coefficient among them, then $k_1 \mathbf{a}_1 + \cdots + k_i \mathbf{a}_i = 0$, and this equation can be used to solve for $\mathbf{a}_i$ in terms of $\mathbf{a}_1, \ldots, \mathbf{a}_{i-1}$.

Let $\{\mathbf{a}_1, \mathbf{a}_2, \ldots, \cancel{\mathbf{a}}_i, \ldots, \mathbf{a}_n\}$ denote the set $\{\mathbf{a}_1, \mathbf{a}_2, \ldots, \mathbf{a}_n\}$ after removal of $\mathbf{a}_i$.

Lemma 2 *If $\{\mathbf{a}_1, \mathbf{a}_2, \ldots, \mathbf{a}_n\}$ spans V, and $\mathbf{a}_i$ is a linear combination of preceding vectors, then $\{\mathbf{a}_1, \ldots, \cancel{\mathbf{a}}_i, \ldots, \mathbf{a}_n\}$ still spans V.*

Our assumption is that $\mathbf{a}_i = k_1 \mathbf{a}_1 + \cdots + k_{i-1} \mathbf{a}_{i-1}$ for some scalars $k_1, \ldots, k_{i-1}$. Since every vector $\mathbf{b} \in V$ is a linear combination

$$\mathbf{b} = l_1 \mathbf{a}_1 + \cdots + l_i \mathbf{a}_i + \cdots + l_n \mathbf{a}_n.$$

it can also be written as a linear combination

$$\mathbf{b} = l_1 \mathbf{a}_1 + \cdots + l_i(k_1 \mathbf{a}_1 + \cdots + k_{i-1} \mathbf{a}_{i-1}) + \cdots + l_n \mathbf{a}_n$$

in which $\mathbf{a}_i$ does not figure.

A set of vectors $\{\mathbf{a}_1, \ldots, \mathbf{a}_n\}$ in V is called a *basis* of V if it is *linearly independent and spans V.*

For example, the vectors $\varepsilon_1 = (1, 0, 0)$, $\varepsilon_2 = (0, 1, 0)$, and $\varepsilon_3 = (0, 0, 1)$ form a basis of $\mathbb{R}^3$. They are linearly independent because, obviously, no vector in $\{\varepsilon_1, \varepsilon_2, \varepsilon_3\}$ is equal to a linear combination of preceding ones. [Any linear combination of ε_1 and ε_2 is of the form $a\varepsilon_1 + b\varepsilon_2 = (a, b, 0)$, whereas ε_3 is not of this form; similarly, any linear combination of ε_1 alone is of the form $a\varepsilon_1 = (a, 0, 0)$, and ε_2 is not of that form.] The vectors ε_1, ε_2, ε_3 span $\mathbb{R}^3$ because any vector (a, b, c) in $\mathbb{R}$ can be written as $(a, b, c) = a\varepsilon_1 + b\varepsilon_2 + c\varepsilon_3$.

Actually, $\{\varepsilon_1, \varepsilon_2, \varepsilon_3\}$ is not the only basis of $\mathbb{R}^3$. Another basis of $\mathbb{R}^3$ consists of the vectors $(1, 2, 3)$, $(1, 0, 2)$, and $(3, 2, 1)$; in fact, there are infinitely many different bases of $\mathbb{R}^3$. Nevertheless, all bases of $\mathbb{R}^3$ have one thing in common: they contain exactly three vectors! This is a consequence of our next theorem:

Theorem 2 *Any two bases of a vector space V have the same number of elements.*

Suppose, on the contrary, that V has a basis $A = \{\mathbf{a}_1, \ldots, \mathbf{a}_n\}$ and a basis $B = \{\mathbf{b}_1, \ldots, \mathbf{b}_m\}$ where $m \neq n$. To be specific, suppose $n < m$. From this assumption we will derive a contradiction.

Put the vector $\mathbf{b}_1$ in the set A, so A now contains $\{\mathbf{b}_1, \mathbf{a}_1, \mathbf{a}_2, \ldots, \mathbf{a}_n\}$. This set is linearly dependent because $\mathbf{b}_1$ is a linear combination of $\mathbf{a}_1, \ldots, \mathbf{a}_n$. But then, by Lemma 1, some $\mathbf{a}_i$ is a linear combination of preceding vectors. By Lemma 2 we may expel this $\mathbf{a}_i$, and the remaining set $\{\mathbf{b}_1, \mathbf{a}_1, \ldots, \cancel{\mathbf{a}_i}, \ldots, \mathbf{a}_n\}$ still spans V.

Repeat this argument a second time by putting $\mathbf{b}_2$ in A, so A now contains $\{\mathbf{b}_2, \mathbf{b}_1, \mathbf{a}_1, \mathbf{a}_2, \ldots, \cancel{\mathbf{a}_i}, \ldots, \mathbf{a}_n\}$. This set is linearly dependent because $\{\mathbf{b}_1, \mathbf{a}_1, \ldots, \cancel{\mathbf{a}_i}, \ldots, \mathbf{a}_n\}$ spans V and therefore $\mathbf{b}_2$ is a linear combination of $\mathbf{b}_1, \mathbf{a}_1, \ldots, \cancel{\mathbf{a}_i}, \ldots, \mathbf{a}_n$. By Lemma 1, some $\mathbf{a}_j$ is a linear combination of preceding vectors in A, so by Lemma 2 we may remove $\mathbf{a}_j$, and $\{\mathbf{b}_2, \mathbf{b}_1, \mathbf{a}_1, \mathbf{a}_2, \ldots, \cancel{\mathbf{a}_i}, \ldots, \cancel{\mathbf{a}_j}, \ldots, \mathbf{a}_n\}$ still spans V.

This argument is repeated n times. Each time, a vector from B is put into A and a vector $\mathbf{a}_k$ is removed. At the end of the nth repetition, A contains only $\mathbf{b}_1, \ldots, \mathbf{b}_n$, and $\{\mathbf{b}_1, \ldots, \mathbf{b}_n\}$ still spans V. But this is impossible because it implies that $\mathbf{b}_{n+1}$ is a linear combination of $\mathbf{b}_1, \ldots, \mathbf{b}_n$, whereas in fact, $B = \{\mathbf{b}_1, \ldots, \mathbf{b}_n, \ldots, \mathbf{b}_m\}$ is linearly independent!

This contradiction proves that any two bases of V must contain *the same number of elements!*

If V has a basis $\{\mathbf{a}_1, \ldots, \mathbf{a}_n\}$, we call V a *finite-dimensional vector space* and say that V is of *dimension n*. In that case, by Theorem 2 every basis of V has exactly n elements.

In the sequel we consider only finite-dimensional vector spaces. The next two lemmas are quite interesting. The first one states that if $\{\mathbf{a}_1, \ldots, \mathbf{a}_m\}$ spans V, there is a way of removing vectors from this set, one by one, until we are left with an *independent* set which still spans V.

Lemma 3 *If the set* $\{\mathbf{a}_1, \ldots, \mathbf{a}_m\}$ *spans* V, *it contains a basis of* V.

If $\{\mathbf{a}_1, \ldots, \mathbf{a}_m\}$ is an independent set, it is a basis, and we are done. If not, some $\mathbf{a}_i$ is a linear combination of preceding ones, so $\{\mathbf{a}_1, \ldots, \mathbf{a}_i, \ldots, \mathbf{a}_m\}$ still spans V. Repeating this process, we discard vectors one by one from $\{\mathbf{a}_1, \ldots, \mathbf{a}_m\}$ and, each time, the remaining vectors still span V. We keep doing this until the remaining set is independent. (In the worst case, this will happen when only one vector is left.)

The next lemma asserts that if $\{\mathbf{a}_1, \ldots, \mathbf{a}_s\}$ is an independent set of vectors in V, there is a way of adding vectors to this set so as to get a basis of V.

Lemma 4 *If the set* $\{\mathbf{a}_1, \ldots, \mathbf{a}_s\}$ *is linearly independent, it can be extended to a basis of* V.

If $\{\mathbf{b}_1, \ldots, \mathbf{b}_n\}$ is any basis of V, then $\{\mathbf{a}_1, \ldots, \mathbf{a}_s, \mathbf{b}_1, \ldots, \mathbf{b}_n\}$ spans V. By the proof of Lemma 3, we may discard vectors from this set until we get a basis of V. Note that we never discard any $\mathbf{a}_i$, because, by hypothesis, $\mathbf{a}_i$ is not a linear combination of preceding vectors.

The next theorem is an immediate consequence of Lemmas 3 and 4.

Theorem 3 *Let* V *have dimension* n. *If* $\{\mathbf{a}_1, \ldots, \mathbf{a}_n\}$ *is an independent set, it is already a basis of* V. *If* $\{\mathbf{b}_1, \ldots, \mathbf{b}_n\}$ *spans* V, *it is already a basis of* V.

If $\{\mathbf{a}_1, \ldots, \mathbf{a}_n\}$ is a basis of V, then every vector $\mathbf{c}$ in V has a *unique* expression $\mathbf{c} = k_1 \mathbf{a}_1 + \cdots + k_n \mathbf{a}_n$ as a linear combination of $\mathbf{a}_1, \ldots, \mathbf{a}_n$. Indeed, if

$$\mathbf{c} = k_1 \mathbf{a}_1 + \cdots + k_n \mathbf{a}_n = l_1 \mathbf{a}_1 + \cdots + l_n \mathbf{a}_n$$

then

$$(k_1 - l_1)\mathbf{a}_1 + \cdots + (k_n - l_n)\mathbf{a}_n = \mathbf{0}$$

hence

$$k_1 - l_1 = \cdots = k_n - l_n = 0$$

so $k_1 = l_1, \ldots k_n = l_n$. If $\mathbf{c} = k_1 \mathbf{a}_1 + \cdots + k_n \mathbf{a}_n$, the coefficients $k_1, \ldots, k_n$ are called the *coordinates* of $\mathbf{c}$ with respect to the basis $\{\mathbf{a}_1, \ldots, \mathbf{a}_n\}$. It is then convenient to represent $\mathbf{c}$ as the n-tuple

$$\mathbf{c} = (k_1, \ldots, k_n)$$

If U and V are vector spaces over a field F, a function $h: U \rightarrow V$ is a *homomorphism* if it satisfies the following two conditions:

$$h(\mathbf{a} + \mathbf{b}) = h(\mathbf{a}) + h(\mathbf{b})$$

and
$$h(k\mathbf{a}) = kh(\mathbf{a})$$

A homomorphism of vector spaces is also called a *linear transformation*.

If $h: U \rightarrow V$ is a linear transformation, its kernel [that is, the set of all $\mathbf{a} \in U$ such that $h(\mathbf{a}) = \mathbf{0}$] is a subspace of U, called the *null space* of h. Homomorphisms of vector spaces behave very much like homomorphisms of groups and rings. Their properties are presented in the exercises.

EXERCISES

A. Examples of Vector Spaces

1 Prove that $\mathbb{R}^n$, as defined on page 283, satisfies all the conditions for being a vector space over $\mathbb{R}$.

2 Prove that $\mathscr{F}(\mathbb{R})$, as defined on page 284 is a vector space over $\mathbb{R}$.

3 Prove that $\mathscr{P}\ell$, as defined on page 284, is a vector space over $\mathbb{R}$.

4 Prove that $\mathscr{M}_2(\mathbb{R})$, the set of all 2×2 matrices of real numbers, with matrix addition and the scalar multiplication

$$k\begin{pmatrix} a & b \\ c & d \end{pmatrix} = \begin{pmatrix} ka & kb \\ kc & kd \end{pmatrix}$$

is a vector space over $\mathbb{R}$.

B. Examples of Subspaces

1 Prove that $\{(a, b, c) : 2a - 3b + c = 0\}$ is a subspace of $\mathbb{R}^3$.

2 Prove that the set of all $(x, y, z) \in \mathbb{R}^3$ which satisfy the pair of equations $ax + by + c = 0, dx + ey + f = 0$ is a subspace of $\mathbb{R}^3$.

3 Prove that $\{f : f(1) = 0\}$ is a subspace of $\mathscr{F}(\mathbb{R})$.

4 Prove that $\{f : f$ is a constant on the interval $[0, 1]\}$ is a subspace of $\mathscr{F}(\mathbb{R})$.

5 Prove that the set of all even functions [that is, functions f such that $f(x) = f(-x)$] is a subspace of $\mathscr{F}(\mathbb{R})$. Is the same true for the set of all the odd functions [that is, functions f such that $f(-x) = -f(x)$]?

6 Prove that the set of all polynomials of degree $\leq n$ is a subspace of $\mathscr{P}\ell$.

C. Examples of Linear Independence and Bases

1 Prove that $\{(0, 0, 0, 1), (0, 0, 1, 1), (0, 1, 1, 1), (1, 1, 1, 1)\}$ is a basis of $\mathbb{R}^4$.

2 If $\mathbf{a} = (1, 2, 3, 4)$ and $\mathbf{b} = (4, 3, 2, 1)$, explain why $\{\mathbf{a}, \mathbf{b}\}$ may be extended to a basis of $\mathbb{R}^4$. Then find a basis of $\mathbb{R}^4$ which includes $\mathbf{a}$ and $\mathbf{b}$.

3 Let A be the set of eight vectors (x, y, z) where $x, y, z = 1, 2$. Prove that A spans $\mathbb{R}^3$, and find a subset of A which is a basis of $\mathbb{R}^3$.

4 If $\mathscr{P}\ell_n$ is the subspace of $\mathscr{P}\ell$ consisting of all polynomials of degree $\leq n$, prove that $\{1, x, x^2, \ldots, x^n\}$ is a basis of $\mathscr{P}\ell_n$. Then find another basis of $\mathscr{P}\ell_n$.

5 Find a basis for each of the following subspaces of $\mathbb{R}^3$:

(a) $S_1 = \{(x,y,z) : 3x - 2y + z = 0\}$ (b) $S_2 = \{(x,y,z) : x + y - z = 0$ and $2x - y + z = 0\}$

6 Find a basis for the subspace of $\mathbb{R}^3$ spanned by the set of vectors (x, y, z) such that $x^2 + y^2 + z^2 = 1$.

7 Let U be the subspace of $\mathscr{F}(\mathbb{R})$ spanned by $\{\cos^2 x, \sin^2 x, \cos 2x\}$. Find the dimension of U, and then find a basis of U.

8 Find a basis for the subspace of $\mathscr{P}\ell$ spanned by

$$\{x^3 + x^2 + x + 1, x^2 + 1, x^3 - x^2 + x - 1, x^2 - 1\}$$

D. Properties of Subspaces and Bases

Let V be a finite-dimensional vector space Let dim V designate the dimension of V. *Prove each of the following*:

1 If U is a subspace of V, then dim $U \leq$ dim V.

2 If U is a subspace of V, and dim $U =$ dim V, then $U = V$.

3 Any set of vectors containing $\mathbf{0}$ is linearly dependent.

4 The set $\{\mathbf{a}\}$, containing only one nonzero vector $\mathbf{a}$, is linearly independent.

5 Any subset of an independent set is independent. Any set of vectors containing a dependent set is dependent.

6 If $\{\mathbf{a}, \mathbf{b}, \mathbf{c}\}$ is linearly independent, so is $\{\mathbf{a} + \mathbf{b}, \mathbf{b} + \mathbf{c}, \mathbf{a} + \mathbf{c}\}$.

7 If $\{\mathbf{a}_1, \ldots, \mathbf{a}_n\}$ is a basis of V, so is $\{k_1\mathbf{a}_1, \ldots, k_n\mathbf{a}_n\}$ for any nonzero scalars $k_1, \ldots, k_n$.

8 The space spanned by $\{\mathbf{a}_1, \ldots, \mathbf{a}_n\}$ is the same as the space spanned by $\{\mathbf{b}_1, \ldots, \mathbf{b}_m\}$ iff each $\mathbf{a}_i$ is a linear combination of $\mathbf{b}_1, \ldots, \mathbf{b}_m$, and each $\mathbf{b}_j$ is a linear combination of $\mathbf{a}_1, \ldots, \mathbf{a}_n$.

E. Properties of Linear Transformations

Let U and V be finite-dimensional vector spaces over a field F, and let $h : U \to V$ be a linear transformation. *Prove each of the following*:

1 The kernel of h is a subspace of U. (It is called the *null space* of h.)

2 The range of h is a subspace of V. (It is called the *range space* of h.)

3 h is injective iff the null space of h is equal to $\{0\}$.

Let $\mathcal{N}$ be the null space of h, and $\mathcal{R}$ the range space of h. Let $\{a_1, \ldots, a_r\}$ be a basis of $\mathcal{N}$. Extend it to a basis $\{a_1, \ldots, a_r, \ldots, a_n\}$ of U. Prove:

4 Every vector $\mathbf{b} \in \mathcal{R}$ is a linear combination of $h(a_{r+1}), \ldots, h(a_n)$.

5 $\{h(a_{r+1}), \ldots, h(a_n)\}$ is linearly independent.

6 The dimension of $\mathcal{R}$ is $n - r$.

7 Conclude as follows: for any linear transformation h, dim (domain h) = dim (null space of h) + dim (range space of h).

8 Let U and V have the same dimension n. Use part 7 to prove that h is injective iff h is surjective.

F. Isomorphism of Vector Spaces

Let U and V be vector spaces over the field F, with dim $U = n$ and dim $V = m$. Let $h : U \to V$ be a homomorphism. *Prove the following*:

1 Let h be injective. If $\{a_1, \ldots, a_r\}$ is a linearly independent subset of U, then $\{h(a_1), \ldots, h(a_r)\}$ is a linearly independent subset of V.

2 h is injective iff dim U = dim $h(U)$.

3 Suppose dim U = dim V; h is an isomorphism (that is, a bijective homomorphism) iff h is injective iff h is surjective.

4 Any n-dimensional vector space V over F is isomorphic to the space F^n of all n-tuples of elements of F.

† G. Sums of Vector Spaces

Let T and U be subspaces of V. The *sum* of T and U, denoted by $T + U$, is the set of all vectors $\mathbf{a} + \mathbf{b}$, where $\mathbf{a} \in T$ and $\mathbf{b} \in U$. *Prove the following*:

1 $T + U$ and $T \cap U$ are subspaces of V.

V is said to be the *direct sum* of T and U if $V = T + U$ and $T \cap U = \{0\}$. In that case, we write $V = T \oplus U$. *Prove the following*:

2 $V = T \oplus U$ iff every vector $\mathbf{c} \in V$ can be written, in a unique manner, as a sum $\mathbf{c} = \mathbf{a} + \mathbf{b}$ where $\mathbf{a} \in T$ and $\mathbf{b} \in U$.

3 Let T be a k-dimensional subspace of an n-dimensional space V. Prove that an $(n - k)$-dimensional subspace U exists such that $V = T \oplus U$.

4 If T and U are arbitrary subspaces of V, prove that

$$\dim (T + U) = \dim T + \dim U - \dim (T \cap U)$$

TWENTY-NINE
DEGREES OF FIELD EXTENSIONS

In this chapter we will see how the machinery of vector spaces can be applied to the study of field extensions.

Let F and K be fields. If K is an extension of F, we may regard K as being a vector space over F. We may treat the elements in K as "vectors"

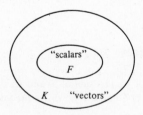

and the elements in F as "scalars." That is, when we add elements in K, we think of it as vector addition; when we add and multiply elements in F, we think of this as addition and multiplication of scalars; and finally, when we multiply an element of F by an element of K, we think of it as scalar multiplication.

We will be especially interested in the case where the resulting vector space is of finite dimension. If K, as a vector space over F, is of finite dimension, we call K a *finite extension* of F. If the dimension of the vector space K is n, we say that K is an *extension of degree n* over F. This is symbolized by writing

$$[K : F] = n$$

which should be read, "the degree of K over F is equal to n."

Let us recall that $F(c)$ denotes the smallest field which contains F and c. This means that $F(c)$ contains F and c, and that any other field K containing F and c must contain $F(c)$. We saw in Chapter 27 that if c is algebraic over F, then $F(c)$ consists of all the elements of the form $a(c)$, for all $a(x)$ in $F[x]$. Since $F(c)$ is an extension of F, we may regard it as a vector space over F. Is $F(c)$ a finite extension of F?

Well, let c be algebraic over F, and let $p(x)$ be the minimum polynomial of c over F. [That is, $p(x)$ is the monic polynomial of lowest degree having c as a root.] Let the degree of the polynomial $p(x)$ be equal to n. It turns out, then, that the n elements

$$1, c, c^2, \ldots, c^{n-1}$$

are linearly independent and span $F(c)$. We will prove this fact in a moment, but meanwhile let us record what it means. It means that the set of n "vectors" $\{1, c, c^2, \ldots, c^{n-1}\}$ is a basis of $F(c)$, hence $F(c)$ is a vector space of dimension n over the field F. This may be summed up concisely as follows:

Theorem 1 *The degree of $F(c)$ over F is equal to the degree of the minimum polynomial of c over F.*

It remains only to show that the n elements $1, c, \ldots, c^{n-1}$ span $F(c)$ and are linearly independent. Well, if $a(c)$ is any element of $F(c)$, use the division algorithm to divide $a(x)$ by $p(x)$:

$$a(x) = p(x)q(x) + r(x) \qquad \text{where deg } r(x) \leq n - 1$$

Therefore, $\quad a(c) = \underbrace{p(c)q(c)}_{=0} + r(c) = 0 + r(c) = r(c)$

This shows that every element of $F(c)$ is of the form $r(c)$ where $r(x)$ has degree $n - 1$ or less. Thus, every element of $F(c)$ can be written in the form

$$a_0 + a_1 c + \ldots + a_{n-1} c^{n-1}$$

which is a linear combination of $1, c, c^2, \ldots, c^{n-1}$.

Finally, to prove that $1, c, c^2, \ldots, c^{n-1}$ are linearly independent, suppose that $a_0 + a_1 c + \cdots + a_{n-1} c^{n-1} = 0$. If the coefficients $a_0, a_1, \ldots, a_{n-1}$ were not all zero, c would be the root of a nonzero polynomial of degree $n - 1$ or less, which is impossible because the minimum polynomial of c over F has degree n. Thus, $a_0 = a_1 = \cdots = a_{n-1} = 0$.

For example, let us look at $\mathbb{Q}(\sqrt{2})$: the number $\sqrt{2}$ is not a root of any monic polynomial of degree 1 over $\mathbb{Q}$. For such a polynomial would have to be $x - \sqrt{2}$, and the latter is not in $\mathbb{Q}[x]$ because $\sqrt{2}$ is irrational. However, $\sqrt{2}$ is a root of $x^2 - 2$, which is therefore the minimum polynomial of $\sqrt{2}$ over $\mathbb{Q}$, and which has degree 2. Thus,

$$[\mathbb{Q}(\sqrt{2}) : \mathbb{Q}] = 2$$

In particular, every element in $\mathbb{Q}(\sqrt{2})$ is therefore a linear combination of 1 and $\sqrt{2}$, that is, a number of the form $a + b\sqrt{2}$ where $a, b \in \mathbb{Q}$.

As another example, i is a root of the irreducible polynomial $x^2 + 1$ in $\mathbb{R}[x]$. Therefore $x^2 + 1$ is the minimum polynomial of i over $\mathbb{R}$; $x^2 + 1$ has degree 2, so $[\mathbb{R}(i) : \mathbb{R}] = 2$. Thus, $\mathbb{R}(i)$ consists of all the linear combinations of 1 and i with real coefficients, that is, all the $a + bi$ where $a, b \in \mathbb{R}$. Clearly then, $\mathbb{R}(i) = \mathbb{C}$, so the degree of $\mathbb{C}$ over $\mathbb{R}$ is equal to 2.

In the sequel we will often encounter the following situation: E is a finite extension of K, where K is a finite extension of F. If we know the

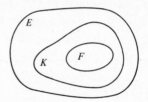

degree of E over K and the degree of K over F, can we determine the degree of E over F? This is a question of major importance! Fortunately, it has an easy answer, based on the following lemma:

Lemma *Let $a_1, a_2, \ldots, a_m$ be a basis of the vector space K over F, and let b_1, $b_2, \ldots, b_n$ be a basis of the vector space E over K. Then the set of mn products $\{a_i b_j\}$ is a basis of the vector space E over the field F.*

To prove that the set $\{a_i b_j\}$ spans E, note that each element c in E can be written as a linear combination $c = k_1 b_1 + \cdots + k_n b_n$ with coefficients k_i in K. But each k_i, because it is in K, is a linear combination

$$k_i = l_{i1} a_1 + \cdots + l_{im} a_m$$

with coefficients l_{ij} in F. Substituting,

$$c = (l_{11} a_1 + \cdots + l_{1m} a_m) b_1 + \cdots + (l_{n1} a_1 + \cdots + l_{nm} a_m) b_n$$

$$= \sum l_{ij} a_i b_j$$

and this is a linear combination of the products $a_i b_j$ with coefficient l_{ij} in F.

To prove that $\{a_i b_j\}$ is linearly independent, suppose $\sum l_{ij} a_i b_j = 0$. This can be written as

$$(l_{11}a_1 + \cdots + l_{1m}a_m)b_1 + \cdots + (l_{n1}a_1 + \cdots + l_{nm}a_m)b_n = 0$$

and since $b_1, \ldots, b_n$ are independent, $l_{i1}a_1 + \cdots + l_{im}a_m = 0$ for each i. But $a_1, \ldots, a_m$ are also independent, so every $l_{ij} = 0$.

With this result we can now conclude the following:

Theorem 2 *Suppose $F \subseteq K \subseteq E$ where E is a finite extension of K and K is a finite extension of F. Then E is a finite extension of F, and*

$$[E : F] = [E : K][K : F]$$

This theorem is a powerful tool in our study of fields. It plays a role in field theory analogous to the role of Lagrange's theorem in group theory. See what it says about any two extensions, K and E, of a fixed "base field" F: If K is a subfield of E, then the degree of K (over F) divides the degree of E (over F).

If c is algebraic over F, we say that $F(c)$ is obtained by *adjoining* c to F. If c and d are algebraic over F, we may first adjoin c to F, thereby obtaining $F(c)$, and then adjoin d to $F(c)$. The resulting field is denoted $F(c, d)$, and is the smallest field containing F, c, and d. [Indeed, any field containing F, c, and d must contain $F(c)$, hence also $F(c, d)$.] It does not matter whether we first adjoin c and then d, or vice versa.

If $c_1, \ldots, c_n$ are algebraic over F, we let $F(c_1, \ldots, c_n)$ be the smallest field containing F and $c_1, \ldots, c_n$. We call it the field obtained by *adjoining* $c_1, \ldots, c_n$ to F. We may form $F(c_1, \ldots, c_n)$ step by step, adjoining one c_i at a time, and the order of adjoining the c_i is irrelevant.

An extension $F(c)$ formed by adjoining a single element to F is called a *simple extension* of F. An extension $F(c_1, \ldots, c_n)$, formed by adjoining a finite number of elements $c_1, \ldots, c_n$, is called an *iterated extension*. It is called "iterated" because it can be formed step by step, one *simple* extension at a time:

$$F \subseteq F(c_1) \subseteq F(c_1, c_2) \subseteq F(c_1, c_2, c_3) \subseteq \cdots \subseteq F(c_1, \ldots, c_n) \qquad (*)$$

If $c_1, \ldots, c_n$ are algebraic over F, then by Theorem 1, each extension in $(*)$ is a finite extension. By Theorem 2, $F(c_1, c_2)$ is a finite extension of F; applying Theorem 2 again, $F(c_1, c_2, c_3)$ is a finite extension of F; and so on. So finally, *if $c_1, \ldots, c_n$ are algebraic over F, then $F(c_1, \ldots, c_n)$ is a finite extension of F.*

Actually, the converse is true too: *every finite extension is an iterated extension.* This is obvious: for if K is a finite extension of F, say an extension of degree n, then K has a basis $\{a_1, \ldots, a_n\}$ over F. This means that every element in K is a linear combination of $a_1, \ldots, a_n$ with coefficients in F; but any field containing F and $a_1, \ldots, a_n$ obviously contains all the linear combinations of $a_1, \ldots, a_n$, hence K is the smallest field containing F and $a_1, \ldots, a_n$. That is, $K = F(a_1, \ldots, a_n)$.

In fact, if K is a finite extension of F and $K = F(a_1, \ldots, a_n)$, then $a_1, \ldots, a_n$ have to be *algebraic* over F. This is a consequence of a simple but important little theorem:

Theorem 3 *If K is a finite extension of F, every element of K is algebraic over F.*

Indeed, suppose K is of degree n over F, and let c be any element of K. Then the set $\{1, c, c^2, \ldots, c^n\}$ is linearly dependent, because it has $n + 1$ elements in a vector space K of dimension n. Consequently there are scalars $a_0, \ldots, a_n \in F$, not all zero, such that $a_0 + a_1c + \cdots + a_nc^n = 0$. Therefore c is a root of the polynomial $a(x) = a_0 + a_1x + \cdots + a_nx^n$ in $F[x]$.

Let us sum up: *Every iterated extension $F(c_1, \ldots, c_n)$, where $c_1, \ldots, c_n$ are algebraic over F, is a finite extension of F. Conversely, every finite extension of F is an iterated extension $F(c_1, \ldots, c_n)$, where $c_1, \ldots, c_n$ are algebraic over F.*

Here is an example of the concepts presented in this chapter. We have already seen that $\mathbb{Q}(\sqrt{2})$ is of degree 2 over $\mathbb{Q}$, and therefore $\mathbb{Q}(\sqrt{2})$ consists of all the numbers $a + b\sqrt{2}$ where $a, b \in \mathbb{Q}$. Observe that $\sqrt{3}$ cannot be in $\mathbb{Q}(\sqrt{2})$; for if it were, we would have $\sqrt{3} = a + b\sqrt{2}$ for rational a and b; squaring both sides and solving for $\sqrt{2}$ would give us $\sqrt{2} =$ a rational number, which is impossible.

Since $\sqrt{3}$ is not in $\mathbb{Q}(\sqrt{2})$, $\sqrt{3}$ cannot be a root of a polynomial of degree 1 over $\mathbb{Q}(\sqrt{2})$ (such a polynomial would have to be $x - \sqrt{3}$). But $\sqrt{3}$ is a root of $x^2 - 3$, which is therefore the minimum polynomial of $\sqrt{3}$ over $\mathbb{Q}(\sqrt{2})$. Thus, $\mathbb{Q}(\sqrt{2}, \sqrt{3})$ is of degree 2 over $\mathbb{Q}(\sqrt{2})$, and therefore by Theorem 2, $\mathbb{Q}(\sqrt{2}, \sqrt{3})$ is of degree 4 over $\mathbb{Q}$.

By the comments preceding Theorem 1, $\{1, \sqrt{2}\}$ is a basis of $\mathbb{Q}(\sqrt{2})$ over $\mathbb{Q}$, and $\{1, \sqrt{3}\}$ is a basis of $\mathbb{Q}(\sqrt{2}, \sqrt{3})$ over $\mathbb{Q}(\sqrt{2})$. Thus, by the lemma of this chapter, $\{1, \sqrt{2}, \sqrt{3}, \sqrt{6}\}$ is a basis of $\mathbb{Q}(\sqrt{2}, \sqrt{3})$ over $\mathbb{Q}$. This means that $\mathbb{Q}(\sqrt{2}, \sqrt{3})$ consists of all the numbers $a + b\sqrt{2} + c\sqrt{3} + d\sqrt{6}$, for all a, b, c, and d in $\mathbb{Q}$.

For later reference. The technical observation which follows will be needed later.

By the comments immediately preceding Theorem 1, every element of $F(c_1)$ is a linear combination of powers of c_1, with coefficients in F. That is, every element of $F(c_1)$ is of the form

$$\sum_i k_i c_1^i \tag{**}$$

where the k_i are in F. For the same reason, every element of $F(c_1, c_2)$ is of the form

$$\sum_j l_j c_2^j$$

where the coefficients l_j are in $F(c_1)$. Thus, each coefficient l_j is equal to a sum of the form (**). But then, clearing brackets, it follows that every element of $F(c_1, c_2)$ is of the form

$$\sum_{i,j} k_{ij} c_1^i c_2^j$$

where the coefficients k_{ij} are in F.

If we continue this process, it is easy to see that every element of $F(c_1, c_2, \ldots, c_n)$ is a sum of terms of the form

$$k c_1^{i_1} c_2^{i_2} \cdots c_n^{i_n}$$

where the coefficient k of each term is in F.

EXERCISES

A. Examples of Finite Extensions

1 Find a basis for $\mathbb{Q}(i\sqrt{2})$ over $\mathbb{Q}$, and describe the elements of $\mathbb{Q}(i\sqrt{2})$. (See the two examples immediately following Theorem 1.)

2 Show that every element of $\mathbb{R}(2 + 3i)$ can be written as $a + bi$, where $a, b \in \mathbb{R}$. Conclude that $\mathbb{R}(2 + 3i) = \mathbb{C}$.

3 If $a = \sqrt{1 + \sqrt[3]{2}}$, show that $\{1, 2^{1/3}, 2^{2/3}, a, 2^{1/3}a, 2^{2/3}a\}$ is a basis of $\mathbb{Q}(a)$ over $\mathbb{Q}$. Describe the elements of $\mathbb{Q}(a)$.

4 Find a basis of $\mathbb{Q}(\sqrt{2} + \sqrt[3]{4})$ over $\mathbb{Q}$, and describe the elements of $\mathbb{Q}(\sqrt{2} + \sqrt[3]{4})$.

5 Find a basis of $\mathbb{Q}(\sqrt{5}, \sqrt{7})$ over $\mathbb{Q}$, and describe the elements of $\mathbb{Q}(\sqrt{5}, \sqrt{7})$. (See the example at the end of this chapter.)

6 Find a basis of $\mathbb{Q}(\sqrt{2}, \sqrt{3}, \sqrt{5})$ over $\mathbb{Q}$, and describe the elements of $\mathbb{Q}(\sqrt{2}, \sqrt{3}, \sqrt{5})$.

7 Name a finite extension of $\mathbb{Q}$ over which π is algebraic of degree 3.

B. Further Examples of Finite Extensions

Let F be a field of characteristic $\neq 2$. Let $a \neq b$ be in F. *Prove the following*:

1 Any field F containing $\sqrt{a} + \sqrt{b}$ also contains $\sqrt{a}$ and $\sqrt{b}$. [HINT: Compute $(\sqrt{a} + \sqrt{b})^2$ and show that $\sqrt{ab} \in F$. Then compute $\sqrt{ab}(\sqrt{a} + \sqrt{b})$, which is also in F.] Conclude that $F(\sqrt{a} + \sqrt{b}) = F(\sqrt{a}, \sqrt{b})$.

2 If $b \neq x^2 a$ for any $x \in F$, then $\sqrt{b} \notin F(\sqrt{a})$. Conclude that $F(\sqrt{a}, \sqrt{b})$ is of degree 4 over F.

3 Show that $x = \sqrt{a} + \sqrt{b}$ satisfies $x^4 - 2(a + b)x^2 + (a - b)^2 = 0$. Show that $x = \sqrt{a + b + 2\sqrt{ab}}$ also satisfies this equation. Conclude that

$$F(\sqrt{a + b + 2\sqrt{ab}}) = F(\sqrt{a}, \sqrt{b})$$

4 Using parts 1 to 3, find an uncomplicated basis for $\mathbb{Q}(d)$ over $\mathbb{Q}$, where d is a root of $x^4 - 14x^2 + 9$. Then find a basis for $\mathbb{Q}(\sqrt{7 + 2\sqrt{10}})$ over $\mathbb{Q}$.

C. Finite Extensions of Finite Fields

By the proof of the basic theorem of field extensions, if $p(x)$ is an irreducible polynomial of degree n in $F[x]$, then $F[x]/\langle p(x)\rangle \cong F(c)$ where c is a root of $p(x)$. By Theorem 1 in this chapter, $F(c)$ is of degree n over F. *Use the paragraph preceding Theorem 1 to prove the following*:

1 Every element of $F(c)$ can be written *uniquely* as $a_0 + a_1 c + \cdots + a_{n-1}c^{n-1}$, for some $a_0, \ldots, a_{n-1} \in F$.

2 Construct a field of four elements. (It is to be an extension of $\mathbb{Z}_2$.) Describe its elements, and supply its addition and multiplication tables.

3 Construct a field of eight elements. (It is to be an extension of $\mathbb{Z}_2$.)

4 If F has q elements, and a is algebraic over F of degree n, then $F(a)$ has q^n elements.

5 For every prime number p, there is an irreducible quadratic in $\mathbb{Z}_p[x]$. Conclude that for every prime p, there is a field with p^2 elements.

D. Degrees of Extensions (Applications of Theorem 2)

Let F be a field, and K a finite extension of F. *Prove the following*:

1 $[K : F] = 1$ iff $K = F$.

2 If $[K : F]$ is a prime number, there is no field properly between F and K (that is, there is no field L such that $F \subsetneq L \subsetneq K$).

3 If $[K : F]$ is a prime, then $K = F(a)$ for every $a \in K - F$.

4 Suppose $a, b \in K$ are algebraic over F with degrees m and n, where m and n are relatively prime. Then:

    (i) $F(a, b)$ is of degree mn over F.

    (ii) $F(a) \cap F(b) = F$.

5 If the degree of $F(a)$ over F is a prime, then $F(a) = F(a^n)$ for any n (on the condition that $a^n \notin F$).

6 If an irreducible polynomial $p(x) \in F[x]$ has a root in K, then deg $p(x) | [K : F]$.

E. Short Questions Relating to Degrees of Extensions

Let F be a field. *Prove the following*:

1 The degree of a over F is the same as the degree of $1/a$ over F. It is also the same as the degrees of $a + c$ and ac over F, for any $c \in F$.

2 a is of degree 1 over F iff $a \in F$.

3 If a real number c is a root of an irreducible polynomial of degree > 1 in $\mathbb{Q}[x]$, then c is irrational.

4 Use part 3 and Eisenstein's irreducibility criterion to prove that $\sqrt{m/n}$ (where m, $n \in \mathbb{Z}$) is irrational if there is a prime number which divides m but not n, and whose square does not divide m.

5 Show that part 4 remains true for $\sqrt[q]{m/n}$, where $q > 1$.

6 If a and b are algebraic over F, then $F(a, b)$ is a finite extension of F.

† F. Further Properties of Degrees of Extensions

Let F be a field, and K a finite extension of F. *Prove each of the following*:

1 Any element algebraic over K is algebraic over F, and conversely.

2 If b is algebraic over K, then $[F(b) : F] | [K(b) : F]$.

3 If b is algebraic over K, then $[K(b) : K] | [F(b) : F]$. (HINT: The minimum polynomial of b over F may factor in $K[x]$, and b will then be a root of one of its irreducible factors.)

4 If b is algebraic over K, then $[K(b) : F(b)] | [K : F]$. [HINT: Note that $F \subseteq K \subseteq K(b)$ and $F \subseteq F(b) \subseteq K(b)$. Relate the degrees of the four extensions involved here, using part 3.]

5 Let $p(x)$ be irreducible in $F[x]$. If $[K : F]$ and deg $p(x)$ are relatively prime, then $p(x)$ is irreducible in $K[x]$. [HINT: Suppose $p(x)$ is reducible in $K[x]$. Let $a(x)$ be an irreducible factor of $p(x)$ in $K[x]$, and let c be a root of $p(x)$ in some extension of K. Relate $[K : F]$ to $[F(c) : F]$, $[K(c) : K]$, and $[K(c) : F]$.]

† G. Fields of Algebraic Elements: Algebraic Numbers

Let $F \subseteq K$ and $a, b \in K$. We have seen on page 295 that if a and b are algebraic over F, then $F(a, b)$ is a finite extension of F. *Use this to prove the following :*

1 If a and b are algebraic over F, then $a + b$, $a - b$, ab, and a/b are algebraic over F.

2 The set $\{x \in K : x \text{ is algebraic over } F\}$ is a subfield of K, containing F.

Any complex number which is algebraic over $\mathbb{Q}$ is called an *algebraic number*. By part 2, the set of all the algebraic numbers is a field, which we shall designate by $\mathbb{A}$.

Let $a(x) = a_0 + a_1 x + \cdots + a_n x^n$ be in $\mathbb{A}[x]$, and let c be any root of $a(x)$. We will prove that $c \in \mathbb{A}$. To begin with, all the coefficients of $a(x)$ are in $\mathbb{Q}(a_0, a_1, \ldots, a_n)$. *Now prove the following :*

3 $\mathbb{Q}(a_0, a_1, \ldots, a_n)$ is a finite extension of $\mathbb{Q}$.

Let $\mathbb{Q}(a_0, \ldots, a_n) = \mathbb{Q}_1$. Since $a(x) \in \mathbb{Q}_1[x]$, c is algebraic over $\mathbb{Q}_1$. *Prove :*

4 $\mathbb{Q}_1(c)$ is a finite extension of $\mathbb{Q}_1$, hence a finite extension of $\mathbb{Q}$. (Why?)

5 $c \in \mathbb{A}$.

6 Conclude: The roots of any polynomial whose coefficients are algebraic numbers are themselves algebraic numbers.

A field F is called *algebraically closed* if the roots of every polynomial in $F[x]$ are in F. We have thus proved that A is algebraically closed.

THIRTY
RULER AND COMPASS

The ancient Greek geometers considered the circle and straight line to be the most basic of all geometric figures, other figures being merely variants and combinations of these basic ones. To understand this view we must remember that construction played a very important role in Greek geometry: when a figure was defined, a method was also given for constructing it. Certainly the circle and the straight line are the easiest figures to construct, for they require only the most rudimentary of all geometric instruments: the ruler and the compass. Furthermore, the ruler, in this case, is a simple, unmarked straightedge.

Rudimentary as these instruments may be, they can be used to carry out a surprising variety of geometric constructions. Lines can be divided into any number of equal segments, and any angle can be bisected. From any polygon it is possible to construct a square having the same area, or twice or three times the area. With amazing ingenuity, Greek geometers devised ways to cleverly use the ruler and compass, unaided by any other instrument, to perform all kinds of intricate and beautiful constructions. They were so successful that it was hard to believe they were unable to perform three little tasks which, at first sight, appear to be very simple: *doubling the cube, trisecting any angle,* and *squaring the circle.* The first task demands that a cube be constructed having twice the volume of a given cube. The second asks that any angle be divided into three equal parts. The third requires the construction of a square whose area is equal to that of a given circle. Remember, only a ruler and compass are to be used!

Mathematicians, in Greek antiquity and throughout the Renaissance, devoted a great deal of attention to these problems, and came up with many brilliant ideas. But they never found ways of performing the above three constructions. This is not surprising, for these constructions are impossible! Of course, the Greeks had no way of knowing that fact, for the mathematical machinery needed to prove that these constructions are impossible—in fact, the very notion that one could prove a construction to be impossible—was still two millennia away.

The final resolution of these problems, by proving that the required constructions are impossible, came from a most unlikely source: it was a by-product of the arcane study of field extensions, in the upper reaches of modern algebra.

To understand how all this works, we will see how the process of ruler-and-compass constructions can be placed in the framework of field theory. Clearly, we will be making use of analytic geometry.

If $\mathscr{A}$ is any set of points in the plane, consider operations of the following two kinds:

1. *Ruler operation*: Through any two points in $\mathscr{A}$, draw a straight line.
2. *Compass operation*: Given three points A, B, and C in $\mathscr{A}$, draw a circle with center C and radius equal in length to the segment AB.

The points of intersection of any two of these figures (line-line, line-circle, or circle-circle) are said to be *constructible in one step* from $\mathscr{A}$. A point P is called *constructible* from $\mathscr{A}$ if there are points $P_1, P_2, \ldots, P_n = P$ such that P_1 is constructible in one step from $\mathscr{A}$, P_2 is constructible in one step from $\mathscr{A} \cup \{P_1\}$, and so on, so that P_i is constructible in one step from $\mathscr{A} \cup \{P_1, \ldots, P_{i-1}\}$.

As a simple example, let us see that the midpoint of a line segment AB is constructible from the two points A and B in the above sense. Well, given A and B, first draw the line AB. Then, draw the circle with center A and

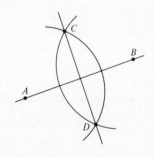

radius $\overline{AB}$ and the circle with center B and radius $\overline{AB}$; let C and D be the points of intersection of these circles. C and D are constructible in one step from $\{A, B\}$. Finally, draw the line through C and D; the intersection of this line with AB is the required midpoint. It is constructible from $\{A, B\}$.

As this example shows, the notion of constructible points is the correct formalization of the intuitive idea of ruler-and-compass constructions.

We call a point in the plane *constructible* if it is constructible from $\mathbb{Q} \times \mathbb{Q}$, that is, from the set of all points in the plane with rational coefficients.

How does field theory fit into this scheme? Obviously by associating with every point its coordinates. More exactly, with every constructible point P we associate a certain field extension of $\mathbb{Q}$, obtained as follows:

Suppose P has coordinates (a, b) and is constructed from $\mathbb{Q} \times \mathbb{Q}$ in one step. We associate with P the field $\mathbb{Q}(a, b)$, obtained by adjoining to $\mathbb{Q}$ the coordinates of P. More generally, suppose P is constructible from $\mathbb{Q} \times \mathbb{Q}$ in n steps: there are then n points $P_1, P_2, \ldots, P_n = P$ such that each P_i is constructible in one step from $\mathbb{Q} \times \mathbb{Q} \cup \{P_1, \ldots, P_{i-1}\}$. Let the coordinates of $P_1, \ldots, P_n$ be $(a_1, b_1), \ldots, (a_n, b_n)$, respectively. With the points $P_1, \ldots, P_n$ we associate fields $K_1, \ldots, K_n$ where $K_1 = \mathbb{Q}(a_1, b_1)$, and for each $i > 1$,

$$K_i = K_{i-1}(a_i, b_i)$$

Thus, $K_1 = \mathbb{Q}(a_1, b_1)$, $K_2 = K_1(a_2, b_2)$, and so on: beginning with $\mathbb{Q}$, we adjoin first the coordinates of P_1, then the coordinates of P_2, and so on successively, yielding the sequence of extensions

$$\mathbb{Q} \subseteq K_1 \subseteq K_2 \subseteq \cdots \subseteq K_n = K$$

We call K the *field extension associated with the point P.*

Everything we will have to say in the sequel follows easily from the next lemma.

Lemma *If $K_1, \ldots, K_n$ are as defined previously, then $[K_i : K_{i-1}] = 1, 2,$ or 4.*

Remember that K_{i-1} already contains the coordinates of $P_1, \ldots, P_{i-1}$, and K_i is obtained by adjoining to K_{i-1} the coordinates x_i, y_i of P_i. But P_i is constructible in one step from $\mathbb{Q} \times \mathbb{Q} \cup \{P_1, \ldots, P_{i-1}\}$, so we must consider three cases, corresponding to the three kinds of intersection which may produce P_i, namely: line intersects line, line intersects circle, and circle intersects circle.

Line intersects line: Suppose one line passes through the points (a_1, a_2) and (b_1, b_2), and the other line passes through (c_1, c_2) and (d_1, d_2). We may write equations for these lines in terms of the constants $a_1, a_2, b_1, b_2, c_1, c_2$ and d_1, d_2 (all of which are in K_{i-1}), and then solve these equations simultaneously to give the coordinates x, y of the point of intersection. Clearly, these values of x and y are expressed in terms of $a_1, a_2, b_1, b_2, c_1, c_2, d_1$, d_2, hence are still in K_{i-1}. Thus, $K_i = K_{i-1}$.

Line intersects circle: Consider the line AB and the circle with center C and radius equal to the distance $k = \overline{DE}$. Let A, B, C have coordinates (a_1, a_2), (b_1, b_2), and (c_1, c_2), respectively. By hypothesis, K_{i-1} contains the numbers $a_1, a_2, b_1, b_2, c_1, c_2$, as well as $k^2 = $ the square of the distance $\overline{DE}$. (To understand the last assertion, remember that K_{i-1} contains the coordinates of D and E; see the figure and use the pythagorean theorem.)

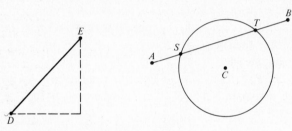

Now, the line AB has equation

$$\frac{y - b_2}{x - b_1} = \frac{b_2 - a_2}{b_1 - a_1} \tag{1}$$

and the circle has equation

$$(x - c_1)^2 + (y - c_2)^2 = k^2 \tag{2}$$

Solving for x in (1) and substituting into (2) gives:

$$(x - c_1)^2 + \frac{b_2 - a_2}{b_1 - a_1}(x - b_1) - b_2^2 - c_2^2 = k^2$$

This is obviously a quadratic equation, and its roots are the x coordinates of S and T. Thus, the x coordinates of both points of intersection are roots of a quadratic polynomial with coefficients in K_{i-1}. The same is true of the y coordinates. Thus, if $K_i = K_{i-1}(x_i, y_i)$ where (x_i, y_i) is one of the points of intersection, then

$$[K_{i-1}(x_i, y_i) : K_{i-1}] = [K_{i-1}(x_i, y_i) : K_{i-1}(x_i)][K_{i-1}(x_i) : K_{i-1}]$$
$$= 2 \times 2 = 4$$

(This assumes that $x_i, y_i \notin K_{i-1}$. If either x_i or y_i or both are already in K_{i-1}, then $[K_{i-1}(x_i, y_i) : K_{i-1}] = 1$ or 2.)

Circle intersects circle: Suppose the two circles have equations

$$x^2 + y^2 + ax + by + c = 0 \qquad (3)$$

and
$$x^2 + y^2 + dx + ey + f = 0 \qquad (4)$$

Then both points of intersection satisfy

$$(a - d)x + (b - e)y + (c - f) = 0 \qquad (5)$$

obtained simply by subtracting (4) from (3). Thus, x and y may be found by solving (4) and (5) simultaneously, which is exactly the preceding case.

We are now in a position to prove the main result of this chapter:

Theorem 1: Basic theorem on constructible points *If the point with coordinates (a, b) is constructible, then the degree of $\mathbb{Q}(a)$ over $\mathbb{Q}$ is a power of 2, and likewise for the degree of $\mathbb{Q}(b)$ over $\mathbb{Q}$.*

Let P be a constructible point; by definition, there are points $P_1, \ldots, P_n$ with coordinates $(a_1, b_1), \ldots, (a_n, b_n)$ such that each P_i is constructible in one step from $\mathbb{Q} \times \mathbb{Q} \cup \{P_1, \ldots, P_{i-1}\}$, and $P_n = P$. Let the fields associated with $P_1, \ldots, P_n$ be $K_1, \ldots, K_n$. Then

$$[K_n : \mathbb{Q}] = [K_n : K_{n-1}][K_{n-1} : K_{n-2}] \cdots [K_1 : \mathbb{Q}]$$

and by the preceding lemma this is a power of 2, say 2^m. But

$$[K_n : \mathbb{Q}] = [K_n : \mathbb{Q}(a)][\mathbb{Q}(a) : \mathbb{Q}]$$

hence $[\mathbb{Q}(a) : \mathbb{Q}]$ is a factor of 2^m, hence also a power of 2.

We will now use this theorem to prove that ruler-and-compass constructions cannot possibly exist for the three classical problems described in the opening to this chapter.

Theorem 2 *"Doubling the cube" is impossible by ruler and compass.*

Let us place the cube on a coordinate system so that one edge of the cube coincides with the unit interval on the x axis. That is, its endpoints are

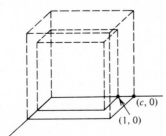

(0, 0) and (1, 0). If we were able to double the cube by ruler and compass, this means we could construct a point $(c, 0)$ such that $c^3 = 2$. However, by Theorem 1, $[\mathbb{Q}(c) : \mathbb{Q}]$ would have to be a power of 2, whereas in fact it is obviously 3. This contradiction proves that it is impossible to double the cube using only a ruler and compass.

Theorem 3 *"Trisecting the angle" by ruler and compass is impossible. That is, there exist angles which cannot be trisected using a ruler and compass.*

We will show specifically that an angle of 60° cannot be trisected. If we *could* trisect an angle of 60°, we would be able to construct a point $(c, 0)$ (see figure), where $c = \cos 20°$, hence certainly we could construct $(b, 0)$ where $b = 2 \cos 20°$.

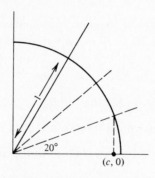

$$(c, 0)$$

But from elementary trigonometry

$$\cos 3\theta = 4 \cos^3 \theta - 3 \cos \theta$$

hence

$$\underbrace{\cos 60°}_{1/2} = 4 \cos^3 20° - 3 \cos 20°$$

Thus, $b = 2 \cos 20°$ satisfies $b^3 - 3b - 1 = 0$. The polynomial

$$p(x) = x^3 - 3x - 1$$

is irreducible over $\mathbb{Q}$ because $p(x + 1) = x^3 + 3x^2 - 3$ is irreducible by Eisenstein's criterion. It follows that $\mathbb{Q}(b)$ has degree 3 over $\mathbb{Q}$, contradicting the requirement (in Theorem 1) that this degree has to be a power of 2.

Theorem 4 *"Squaring the circle" by ruler and compass is impossible.*

If we were able to square the circle by ruler and compass, it would be possible to construct the point $(0, \sqrt{\pi})$, hence by Theorem 1, $[\mathbb{Q}(\sqrt{\pi}) : \mathbb{Q}]$ would be a power of 2. But it is well known that π is transcendental over $\mathbb{Q}$. By Theorem 4 of Chapter 29, the square of an algebraic element is algebraic, hence $\sqrt{\pi}$ is transcendental. It follows that $\mathbb{Q}(\sqrt{\pi})$ is not even a finite extension of $\mathbb{Q}$, much less an extension of some degree 2^m as required.

EXERCISES

† A. Constructible Numbers

If O and I are any two points in the plane, consider a coordinate system such that

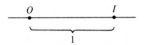

the interval OI coincides with the unit interval on the x axis. Let $\mathbb{D}$ be the set of real numbers such that $a \in \mathbb{D}$ iff the point $(a, 0)$ is constructible from $\{O, I\}$.
Prove the following:

1 If $a, b \in \mathbb{D}$, then $a + b \in \mathbb{D}$ and $a - b \in \mathbb{D}$.
2 If $a, b \in \mathbb{D}$, then $ab \in \mathbb{D}$. (HINT: Use similar triangles. See the accompanying figure.)

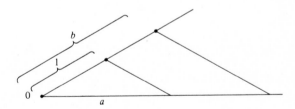

3 If $a, b \in \mathbb{D}$, then $a/b \in \mathbb{D}$. (Use the same figure as in part 2.)
4 If $a > 0$ and $a \in \mathbb{D}$, then $\sqrt{a} \in \mathbb{D}$. (HINT: Use the pythagorean theorem.)

It follows from parts 1 to 4 that $\mathbb{D}$ is a field, closed with respect to taking square roots. $\mathbb{D}$ is called the *field of constructible numbers*.

5 $\mathbb{Q} \subseteq \mathbb{D}$.
6 If a is a root of any quadratic polynomial with coefficients in $\mathbb{D}$, then $a \in \mathbb{D}$. (HINT: Complete the square and use part 4.)

† B. Constructible Points and Constructible Numbers

Prove each of the following :

1 Let $\mathscr{A}$ be any set of points in the plane; (a, b) is constructible from $\mathscr{A}$ iff $(a, 0)$ and $(0, b)$ are constructible from $\mathscr{A}$.

2 If a point P is constructible from $\{O, I\}$ [that is, from $(0, 0)$ and $(1, 0)$], then P is constructible from $\mathbb{Q} \times \mathbb{Q}$.

3 Every point in $\mathbb{Q} \times \mathbb{Q}$ is constructible from $\{O, I\}$. (Use A5 and the definition of $\mathbb{D}$.)

4 If a point P is constructible from $\mathbb{Q} \times \mathbb{Q}$, it is constructible from $\{O, I\}$.

By combining parts 2 and 4, we get the following important fact: Any point P is constructible from $\mathbb{Q} \times \mathbb{Q}$ iff P is constructible from $\{O, I\}$. Thus, we may define a point to be *constructible* iff it is constructible from $\{O, I\}$.

5 A point P is constructible iff both its coordinates are constructible numbers.

C. Constructible Angles

An angle α is called *constructible* iff there exist constructible points A, B, and C such that $\angle ABC = \alpha$. *Prove the following*:

1 The angle α is constructible iff $\sin \alpha$ and $\cos \alpha$ are constructible numbers.

2 $\cos \alpha \in \mathbb{D}$ iff $\sin \alpha \in \mathbb{D}$.

3 If $\cos \alpha, \cos \beta \in \mathbb{D}$, then $\cos (\alpha + \beta), \cos (\alpha - \beta) \in \mathbb{D}$.

4 $\cos (2\alpha) \in \mathbb{D}$ iff $\cos \alpha \in \mathbb{D}$.

5 If α and β are constructible angles, so are $\alpha + \beta, \alpha - \beta, \frac{1}{2}\alpha$, and $n\alpha$ for any positive integer n.

6 The following angles are constructible: $30°, 75°, 22\frac{1}{2}°$.

7 The following angles are *not* constructible: $20°, 40°, 140°$. (HINT: Use Theorem 3.)

D. Constructible Polygons

A polygon is called *constructible* iff its vertices are constructible points. *Prove the following*:

1 The regular n-gon is constructible iff the angle $2\pi/n$ is constructible.

2 The regular hexagon is constructible.

3 The regular polygon of nine sides is *not* constructible.

† E. A Constructible Polygon

We will show that $2\pi/5$ is a constructible angle, and it will follow that the regular pentagon is constructible.

1 If $r = \cos k + i \sin k$ is a complex number, prove that $1/r = \cos k - i \sin k$. Conclude that $r + 1/r = 2 \cos k$.

By de Moivre's theorem,

$$\omega = \cos \frac{2\pi}{5} + i \sin \frac{2\pi}{5}$$

is a complex fifth root of unity. Since

$$x^5 - 1 = (x - 1)(x^4 + x^3 + x^2 + x + 1)$$

ω is a root of $p(x) = x^4 + x^3 + x^2 + x + 1$.

2 Prove that $\omega^2 + \omega + 1 + \omega^{-1} + \omega^{-2} = 0$.
3 Prove that

$$4 \cos^2 \frac{2\pi}{5} + 2 \cos \frac{2\pi}{5} - 1 = 0$$

(HINT: Use parts 1 and 2.) Conclude that $\cos (2\pi/5)$ is a root of the quadratic $4x^2 - 2x - 1$.
4 Use part 3 and A6 to prove that $\cos (2\pi/5)$ is a constructible number.
5 Prove that $2\pi/5$ is a constructible angle.
6 Prove that the regular pentagon is constructible.

† F. A Nonconstructible Polygon

By de Moivre's theorem,

$$\omega = \cos \frac{2\pi}{7} + i \sin \frac{2\pi}{7}$$

is a complex seventh root of unity. Since

$$x^7 - 1 = (x - 1)(x^6 + x^5 + x^4 + x^3 + x^2 + x + 1)$$

ω is a root of $x^6 + x^5 + x^4 + x^3 + x^2 + x + 1$.

1 Prove that $\omega^3 + \omega^2 + \omega + 1 + \omega^{-1} + \omega^{-2} + \omega^{-3} = 0$.
2 Prove that

$$8 \cos^3 \frac{2\pi}{7} + 4 \cos^2 \frac{2\pi}{7} - 4 \cos \frac{2\pi}{7} - 1 = 0$$

(Use 1 and E1.) Conclude that $\cos (2\pi/7)$ is a root of $8x^3 + 4x^2 - 4x - 1$.

3 Prove that $8x^3 + 4x^2 - 4x - 1$ has no rational roots. Conclude that it is irreducible over $\mathbb{Q}$.

4 Conclude from part 3 that $\cos (2\pi/7)$ is not a constructible number.

5 Prove that $2\pi/7$ is not a constructible angle.

6 Prove that the regular polygon of seven sides is not constructible.

G. Further Properties of Constructible Numbers and Figures

Prove each of the following:

1 If the number a is a root of an irreducible polynomial $p(x) \in \mathbb{Q}[x]$ whose degree is not a power of 2, then a is not a constructible number.

2 Any constructible number can be obtained from rational numbers by repeated addition, subtraction, multiplication, division, and taking square roots.

3 Use part 2 and Exercise A to prove: $\mathbb{D}$ is the smallest field extension of $\mathbb{Q}$ closed with respect to square roots (that is, any field extension of $\mathbb{Q}$ closed with respect to square roots contains $\mathbb{D}$).

4 All the roots of the polynomial $x^4 + 2x^2 - 2$ are constructible numbers.

A line is called constructible if it passes through two constructible points. A circle is called constructible if its center and radius are constructible.

5 The line $ax + by + c = 0$ is constructible iff $a, b, c \in \mathbb{D}$.

6 The circle $x^2 + y^2 + ax + by + c = 0$ is constructible iff $a, b, c \in \mathbb{D}$.

GALOIS THEORY: PREAMBLE

Field extensions were used in Chapter 30 to settle some of the most puzzling questions of classical geometry. Now they will be used to solve a problem equally ancient and important: they will give us a definitive and elegant theory of solutions of polynomial equations.

We will be concerned not so much with *finding* solutions (which is a problem of computation) as with the nature and properties of these solutions. As we shall discover, these properties turn out to depend less on the polynomials themselves than on the fields which contain their solutions. This fact should be kept in mind if we want to clearly understand the discussions in this chapter and Chapter 32. We will be speaking of field extensions, but polynomials will always be lurking in the background. Every extension will be generated by roots of a polynomial, and every theorem about these extensions will actually be saying something about the polynomials.

Let us quickly review what we already know of field extensions, filling in a gap or two as we go along. Let F be a field; an element a (in an extension of F) is *algebraic over* F if a is a root of some polynomial with its coefficients in F. The *minimum polynomial* of a over F is the polynomial of lowest degree in $F[x]$ having a as a root; every other polynomial in $F[x]$ having a as a root is a multiple of the minimum polynomial.

The basic theorem of field extensions tells us that any polynomial of degree n in $F[x]$ has exactly n roots in a suitable extension of F. However, this does not necessarily mean n *distinct* roots. For example, in $\mathbb{R}[x]$ the polynomial $(x - 2)^5$ has five roots *all equal* to 2. Such roots are called *multiple roots*. It is perfectly obvious that we can come up with polynomials such as $(x - 2)^5$ having multiple roots; but are there any *irreducible* polynomials with multiple roots? Certainly the answer is not obvious. Here it is:

Theorem 1 *If F has characteristic* 0, *irreducible polynomials over F can never have multiple roots.*

To prove this, we must define the *derivative* of the polynomial $a(x) = a_0 + a_1 x + \cdots + a_n x^n$. It is $a'(x) = a_1 + 2a_2 x + \cdots + na_n x^{n-1}$. As in elementary calculus, it is easily checked that for any two polynomials $f(x)$ and $g(x)$,

$$(f + g)' = f' + g' \qquad \text{and} \qquad (fg)' = fg' + f'g$$

Now suppose $a(x)$ is irreducible in $F[x]$ and has a multiple root c: then in a suitable extension we can factor $a(x)$ as $a(x) = (x - c)^2 q(x)$, and therefore $a'(x) = 2(x - c)q(x) + (x - c)^2 q'(x)$. So $x - c$ is a factor of $a'(x)$, and therefore c is a root of $a'(x)$. Let $p(x)$ be the minimum polynomial of c over F; since both $a(x)$ and $a'(x)$ have c as a root, they are both multiples of $p(x)$.

But $a(x)$ is irreducible: its only nonconstant divisor is itself; so $p(x)$ must be $a(x)$. However, $a(x)$ cannot divide $a'(x)$ unless $a'(x) = 0$ because $a'(x)$ is of *lower degree* than $a(x)$. So $a'(x) = 0$ and therefore its coefficient na_n is 0. Here is where characteristic 0 comes in: if $na_n = 0$ then $a_n = 0$, and this is impossible because a_n is the leading coefficient of $a(x)$.

In the remaining three chapters we will confine our attention to *fields of characteristic* 0. Thus, by Theorem 1, any irreducible polynomial of degree n has n *distinct* roots.

Let us move on with our review. Let E be an extension of F. We call E a *finite extension* of F if E, as a vector space with scalars in F, has finite dimension. Specifically, if E has dimension n, we say that the degree of E over F is equal to n, and we symbolize this by writing $[E : F] = n$. If c is algebraic over F, the degreee of $F(c)$ over F turns out to be equal to the degree of $p(x)$, the minimum polynomial of c over F.

$F(c)$, obtained by adjoining an algebraic element c to F, is called a *simple* extension of F. $F(c_1, \ldots, c_n)$, obtained by adjoining n algebraic elements in succession to F, is called an *iterated* extension of F. Any iterated extension of F is finite, and, conversely, any finite extension of F is an iterated extension $F(c_1, \ldots, c_n)$. In fact, even more is true; let F be of characteristic 0.

Theorem 2 *Every finite extension of F is a **simple** extension F(c).*

We already know that every finite extension is an iterated extension. We will now show that any extension $F(a, b)$ is equal to $F(c)$ for some c. Using this result several times in succession yields our theorem. (At each stage, we reduce by 1 the number of elements that must be adjoined to F in order to get the desired extension.)

Well, given $F(a, b)$, let $A(x)$ be the minimum polynomial of a over F, and let $B(x)$ be the minimum polynomial of b over F. Let K denote any extension of F which contains all the roots $a_1, \ldots, a_n$ of $A(x)$ as well as all the roots $b_1, \ldots, b_m$ of $B(x)$. Let a_1 be a and let b_1 be b.

Let t be any nonzero element of F such that

$$t \neq \frac{a_i - a}{b - b_j} \qquad \text{for every } i \neq 1 \text{ and } j \neq 1$$

Cross multiplying and setting $c = a + tb$, it follows that $c \neq a_i + tb_j$, that is,

$$c - tb_j \neq a_i \qquad \text{for all } i \neq 1 \text{ and } j \neq 1$$

Define $h(x)$ by letting $h(x) = A(c - tx)$; then

$$h(b) = A(\underbrace{c - tb}_{= a}) = 0$$

while for every $j \neq 1$,

$$h(b_j) = A(\underbrace{c - tb_j}_{\neq \text{ any } a_i}) \neq 0$$

Thus, b is the *only common root* of $h(x)$ and $B(x)$.

We will prove that $b \in F(c)$, hence also $a = c - tb \in F(c)$, and therefore $F(a, b) \subseteq F(c)$. But $c \in F(a, b)$, so $F(c) \subseteq F(a, b)$. Thus $F(a, b) = F(c)$.

So, it remains only to prove that $b \in F(c)$. Let $p(x)$ be the minimum polynomial of b over $F(c)$. If the degree of $p(x)$ is 1, then $p(x)$ is $x - b$, so $b \in F(c)$, and we are done. Let us suppose $\deg p(x) \geq 2$ and get a contradiction: observe that $h(x)$ and $B(x)$ must both be multiples of $p(x)$ because both have b as a root, and $p(x)$ is the minimum polynomial of b. But if $h(x)$ and $B(x)$ have a common factor of degree ≥ 2, they must have two or more roots in common, contrary to the fact that b is their only common root. Our proof is complete.

For example, we may apply this theorem directly to $\mathbb{Q}(\sqrt{2}, \sqrt{3})$. Taking $t = 1$, we get $c = \sqrt{2} + \sqrt{3}$, hence $\mathbb{Q}(\sqrt{2}, \sqrt{3}) = \mathbb{Q}(\sqrt{2} + \sqrt{3})$.

If $a(x)$ is a polynomial of degree n in $F[x]$, let its roots be $c_1, \ldots, c_n$. Then $F(c_1, \ldots, c_n)$ is clearly *the smallest extension of F containing all the roots of $a(x)$*. $F(c_1, \ldots, c_n)$ is called the *root field of $a(x)$ over F*. We will have a great deal to say about root fields in this and subsequent chapters.

Isomorphisms were important when we were dealing with groups, and they are important also for fields. You will remember that if F_1 and F_2 are fields, an *isomorphism* from F_1 to F_2 is a bijective function $h : F_1 \rightarrow F_2$ satisfying

$$h(a + b) = h(a) + h(b) \qquad \text{and} \qquad h(ab) = h(a)h(b)$$

From these equations it follows that $h(0) = 0$, $h(1) = 1$, $h(-a) = -h(a)$, and $h(a^{-1}) = (h(a))^{-1}$.

Suppose F_1 and F_2 are fields, and $h : F_1 \rightarrow F_2$ is an isomorphism. Let K_1 and K_2 be extensions of F_1 and F_2, and let $\bar{h} : K_1 \rightarrow K_2$ also be an isomorphism. We call $\bar{h}$ an *extension* of h if $\bar{h}(x) = h(x)$ for every x in F_1, that is, if h and $\bar{h}$ are the same on F_1. ($\bar{h}$ is an extension of h in the plain sense that it is formed by "adding on" to h.)

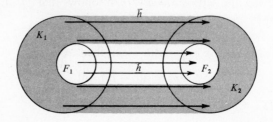

As an example, given any isomorphism $h : F_1 \rightarrow F_2$, we can extend h to an isomorphism $\bar{h} : F_1[x] \rightarrow F_2[x]$. (Note that $F[x]$ is an extension of F when we think of the elements of F as constant polynomials; of course, $F[x]$ is not a field, simply an integral domain, but in the present example this fact is unimportant.) Now we ask: What is an obvious and natural way of extending h? The answer, quite clearly, is to let $\bar{h}$ send the polynomial with coefficients a_0, a_1, ..., a_n to the polynomial with coefficients $h(a_0)$, $h(a_1)$, ..., $h(a_n)$:

$$\bar{h}(a_0 + a_1 x + \cdots + a_n x^n) = h(a_0) + h(a_1)x + \cdots + h(a_n)x^n$$

It is child's play to verify formally that $\bar{h}$ is an isomorphism from $F_1[x]$ to $F_2[x]$. In the sequel, the polynomial $\bar{h}(a(x))$, obtained in this fashion, will be denoted simply by $ha(x)$. Because $\bar{h}$ is an isomorphism, $a(x)$ is irreducible iff $ha(x)$ is irreducible.

A very similar isomorphism extension is given in the next theorem.

Theorem 3 *Let $h : F_1 \rightarrow F_2$ be an isomorphism, and let $p(x)$ be irreducible in $F_1[x]$. Suppose a is a root of $p(x)$, and b a root of $hp(x)$. Then h can be extended to an isomorphism*

$$\bar{h} : F_1(a) \to F_2(b)$$

Furthermore, $\bar{h}(a) = b$.

Remember that every element of $F_1(a)$ is of the form

$$c_0 + c_1 a + \cdots + c_n a^n$$

where $c_0, \ldots, c_n$ are in F_1, and every element of $F_2(b)$ is of the form $d_0 + d_1 b + \cdots + d_n b^n$ where $d_0, \ldots, d_n$ are in F_2. Imitating what we did successfully in the preceding example, we let h send the expression with coefficients $c_0, \ldots, c_n$ to the expression with coefficients $h(c_0), \ldots, h(c_n)$:

$$h(c_0 + c_1 a + \cdots + c_n a^n) = h(c_0) + h(c_1)b + \cdots + h(c_n)b^n$$

Again, it is routine to verify that h is an isomorphism. Details are laid out in Exercise H at the end of this chapter.

Most often we use Theorem 3 in the special case where F_1 and F_2 are the same field—let us call it F—and h is the identity function $\varepsilon : F \to F$. [Remember that the identity function is $\varepsilon(x) = x$.] When we apply Theorem 3 to the identity function $\varepsilon : F \to F$, we get:

Theorem 4 *Suppose a and b are roots of the same irreducible polynomial $p(x)$ in $F[x]$. Then there is an isomorphism $g : F(a) \to F(b)$ such that $g(x) = x$ for every x in F, and $g(a) = b$.*

From Theorem 3 we can also deduce another fact about extending isomorphisms:

Theorem 5 *Suppose K is a finite extension of F. Any isomorphism $h : F \to h(F)$ can be extended to an isomorphism $\bar{h} : K \to \bar{h}(K)$.*

By Theorem 2, K is a simple extension of F, say $K = F(a)$. Writing F' for $h(F)$, we can use Theorem 3 to extend $h : F \to F'$ to

$$\bar{h} : \underbrace{F(a)}_{K} \to \underbrace{F'(b)}_{\bar{h}(K)}$$

Let K be an extension of F. If h is any isomorphism with domain K, and $h(x) = x$ for every x in F, we say that h *fixes* F. Let c be an element of K; if h fixes F, and c is a root of some polynomial $a(x) = a_0 + \cdots + a_n x^n$ in $F[x]$, $h(c)$ *also is a root of $a(x)$*. It is easy to see why: the coefficients of $a(x)$ are in F and are therefore not changed by h. So if $a(c) = 0$, then

$$a(h(c)) = a_0 + a_1 h(c) + \cdots + a_n h(c)^n$$
$$= h(a_0 + a_1 c + \cdots + a_n c^n) = h(0) = 0$$

What we have just shown may be expressed as follows:

(*) Let $a(x)$ be any polynomial in $F[x]$. Any isomorphism which fixes F sends roots of $a(x)$ to roots of $a(x)$.

If K happens to be the root field of $a(x)$ over F, the situation becomes even more interesting. Say $K = F(c_1, c_2, \ldots, c_n)$, where $c_1, c_2 \ldots, c_n$ are the roots of $a(x)$. If $h : K \to h(K)$ is any isomorphism which fixes F, then by (*), h permutes $c_1, c_2, \ldots, c_n$. Now, by the brief discussion headed "For later reference" on page 297, every element of $F(c_1, \ldots, c_n)$ is a sum of terms of the form

$$kc_1^{i_1} c_2^{i_2} \cdots c_n^{i_n}$$

where the coefficient k is in F. Because h fixes F, $h(k) = k$. Furthermore, c_1, $c_2, \ldots, c_n$ are the roots of $a(x)$, so by (*), the product $c_1^{i_1} c_2^{i_2} \cdots c_n^{i_n}$ is transformed by h into another product of the same form. Thus, h sends every element of $F(c_1, c_2, \ldots, c_n)$ to another element of $F(c_1, c_2, \ldots, c_n)$.

The above comments are summarized in the next theorem:

Theorem 6 Let K be the root field of some polynomial over F. If h is any isomorphism with domain K which fixes F, then $h(K) \subseteq K$.

For later reference. The following results, which are of a somewhat technical nature, will be needed later. The first presents a surprisingly strong property of root fields.

Theorem 7 Let K be the root field of some polynomial over F. For **every** irreducible polynomial $p(x)$ in $F[x]$, if $p(x)$ has one root in K, then $p(x)$ must have **all** of its roots in K.

Indeed, suppose $p(x)$ has a root a in K, and let b be any other root of $p(x)$. By Theorems 4 and 5 [apply Theorem 5 to $F(a)$ instead of F], there is an isomorphism $i : K \to i(K)$ fixing F, with $i(a) = b$. By Theorem 6, $i(K) \subseteq K$. Thus, $b = i(a) \in K$.

Theorem 8 Suppose $I \subseteq E \subseteq K$, where E is a finite extension of I and K is a finite extension of E. If K is the root field of some polynomial over E, then K is also the root field of some polynomial over I.

Suppose K is a root field of some polynomial over E, and let $K = I(a)$. If $p(x)$ is the minimum polynomial of a over I, its coefficients are certainly in E and it has a root a in K, so by Theorem 7, all its roots are in K. Therefore, K is the root field of $p(x)$ over I.

EXERCISES

A. Examples of Root Fields over $\mathbb{Q}$

Illustration *Find the root field of* $a(x) = (x^2 - 3)(x^3 - 1)$ *over* $\mathbb{Q}$.

ANSWER The complex roots of $a(x)$ are $\pm\sqrt{3}$, 1, $\frac{1}{2}(-1 \pm \sqrt{3}i)$, so the root field is $\mathbb{Q}(\pm\sqrt{3}, 1, \frac{1}{2}(-1 \pm \sqrt{3}i))$. The same field can be written more simply as $\mathbb{Q}(\sqrt{3}, i)$.

1 Show that $\mathbb{Q}(\sqrt{3}, i)$ is the root field of $(x^2 - 2x - 2)(x^2 + 1)$ over $\mathbb{Q}$.

Comparing part 1 with the illustration, we note that different polynomials may have the same root field. This is true even if the polynomials are irreducible:

2 Prove that $x^2 - 3$ and $x^2 - 2x - 2$ are both irreducible over $\mathbb{Q}$. Then find their root fields over $\mathbb{Q}$ and show they are the same.

3 Find the root field of $x^4 - 2$, first over $\mathbb{Q}$, then over $\mathbb{R}$.

4 Explain: $\mathbb{Q}(i, \sqrt{2})$ is the root field of $x^4 - 2x^2 + 9$ over $\mathbb{Q}$, and is the root field of $x^2 - 2\sqrt{2}x + 3$ over $\mathbb{Q}(\sqrt{2})$.

5 Find irreducible polynomials $a(x)$ over $\mathbb{Q}$, and $b(x)$ over $\mathbb{Q}(i)$, such that $\mathbb{Q}(i, \sqrt{3})$ is the root field of $a(x)$ over $\mathbb{Q}$, and is the root field of $b(x)$ over $\mathbb{Q}(i)$. Then do the same for $\mathbb{Q}(\sqrt{-2}, \sqrt{-3})$.

6 Which of the following extensions are root fields over $\mathbb{Q}$? Justify your answer: $\mathbb{Q}(i)$; $\mathbb{Q}(\sqrt{2})$; $\mathbb{Q}(\sqrt[3]{2})$, where $\sqrt[3]{2}$ is the *real* cube root of 2; $\mathbb{Q}(2 + \sqrt{5})$; $\mathbb{Q}(i + \sqrt{3})$; $\mathbb{Q}(i, \sqrt{2}, \sqrt{3})$.

B. Examples of Root Fields over $\mathbb{Z}_p$

Illustration *Find the root field of* $x^2 + 1$ *over* $\mathbb{Z}_3$.

ANSWER By the basic theorem of field extensions,

$$\mathbb{Z}_3[x]/\langle x^2 + 1 \rangle \cong \mathbb{Z}_3(u)$$

where u is a root of $x^2 + 1$. In $\mathbb{Z}_3(u)$, $x^2 + 1 = (x + u)(x - u)$, because $u^2 + 1 = 0$. Since $\mathbb{Z}_3(u)$ contains $\pm u$, it is the root field of $x^2 + 1$ over $\mathbb{Z}_3$. Note that $\mathbb{Z}_3(u)$ has nine elements, and its addition and multiplication tables are easy to construct. (See Chapter 27, Exercise C.)

1 Show that, in any extension of $\mathbb{Z}_3$ which contains a root u of

$$a(x) = x^3 + 2x + 1 \in \mathbb{Z}_3[x]$$

it happens that $u + 1$ and $u + 2$ are the remaining two roots of $a(x)$. Use this fact to find the root field of $x^3 + 2x + 1$ over $\mathbb{Z}_3$. Write its addition and multiplication tables.

2 Find the root field of $x^3 + x^2 + x + 2$ over $\mathbb{Z}_3$, and write its addition and multiplication tables.

3 Determine whether the fields in parts 1 and 2 are isomorphic.

4 Find the root field of $x^3 + x^2 + 1 \in \mathbb{Z}_2[x]$ over $\mathbb{Z}_2$. Write its addition and multiplication tables.

5 Find the root field over $\mathbb{Z}_2$ of $x^3 + x + 1 \in \mathbb{Z}_2[x]$. (CAUTION: This will prove to be a little more difficult than part 4.)

C. Short Questions Relating to Root Fields

Prove each of the following :

1 Every extension of degree 2 is a root field.

2 If $F \subseteq I \subseteq K$ and K is a root field of $a(x)$ over F, then K is a root field of $a(x)$ over I.

3 The root field over $\mathbb{R}$ of any polynomial in $\mathbb{R}[x]$ is $\mathbb{R}$ or $\mathbb{C}$.

4 If c is a complex root of a cubic $a(x) \in \mathbb{Q}[x]$, then $\mathbb{Q}(c)$ is the root field of $a(x)$ over $\mathbb{Q}$.

5 If $p(x) = x^4 + ax^2 + b \in F[x]$, then $F[x]/\langle p(x) \rangle$ is the root field of $p(x)$ over F.

6 If $K = F(a)$ and K is the root field of some polynomial over F, then K is the root field of the minimum polynomial of a over F.

7 Every root field over F is the root field of some *irreducible* polynomial over F. (HINT: Use part 6 and Theorem 2.)

8 Suppose $[K : F] = n$, where K is a root field over F. Then K is the root field over F of *every* irreducible polynomial of degree n in $F[x]$ having a root in K.

9 Let $a(x)$ be a polynomial of degree n in $F[x]$, and let K be the root field of $a(x)$ over F. Prove that $[K : F]$ divides $n!$

D. Reducing Iterated Extensions to Simple Extensions

1 Find c such that $\mathbb{Q}(\sqrt{2}, \sqrt{-3}) = \mathbb{Q}(c)$. Do the same for $\mathbb{Q}(\sqrt{2}, \sqrt[3]{2})$.

2 Let a be a root of $x^3 - x + 1$, and b a root of $x^2 - 2x - 1$. Find c such that $\mathbb{Q}(a, b) = \mathbb{Q}(c)$. (HINT: Use calculus to show that $x^3 - x + 1$ has one real and two complex roots, and explain why no two of these may differ by a real number.)

3 Find c such that $\mathbb{Q}(\sqrt{2}, \sqrt{3}, \sqrt{-5}) = \mathbb{Q}(c)$.

4 Find an *irreducible* polynomial $p(x)$ such that $\mathbb{Q}(\sqrt{2}, \sqrt{3})$ is the root field of $p(x)$ over $\mathbb{Q}$. (HINT: Use Exercise C6.)

5 Do the same as in part 4 for $\mathbb{Q}(\sqrt{2}, \sqrt{3}, \sqrt{-5})$.

† E. Roots of Unity and Radical Extensions

De Moivre's theorem provides an explicit formula to write the n complex nth roots of 1. (See Chapter 16, Exercise H.) By de Moivre's formula, the nth roots of unity consist of $\omega = \cos(2\pi/n) + i \sin(2\pi/n)$ and its first n powers, namely $1, \omega, \omega^2, \ldots, \omega^{n-1}$. We call ω a *primitive* nth root of unity, because all the other nth roots of unity are powers of ω. Clearly, every nth root of unity (except 1) is a root of

$$\frac{x^n - 1}{x - 1} = x^{n-1} + x^{n-2} + \cdots + x + 1$$

By Eisenstein's criterion, this polynomial is irreducible if n is a prime (see Chapter 26, Exercise D). *Prove each of the following, where ω denotes a primitive nth root of unity:*

1 $\mathbb{Q}(\omega)$ is the root field of $x^n - 1$ over $\mathbb{Q}$.

2 If n is a prime, $[\mathbb{Q}(\omega) : \mathbb{Q}] = n - 1$.

3 If n is a prime, ω^{n-1} is equal to a linear combination of $1, \omega, \ldots, \omega^{n-2}$ with rational coefficients.

4 Find $[\mathbb{Q}(\omega) : \mathbb{Q}]$, where ω is a primitive nth root of unity, for $n = 6, 7$, and 8.

5 For any $r \in \{1, 2, \ldots, n - 1\}$, $\sqrt[n]{a}\omega^r$ is an nth root of a. Conclude that $\sqrt[n]{a}, \sqrt[n]{a}\omega$, $\ldots, \sqrt[n]{a}\omega^{n-1}$ are the n complex nth roots of a.

6 $\mathbb{Q}(\omega, \sqrt[n]{a})$ is the root field of $x^n - a$ over $\mathbb{Q}$.

7 Find the degree of $\mathbb{Q}(\omega, \sqrt[3]{2})$ over $\mathbb{Q}$, where ω is a primitive cube root of 1. Also show that $\mathbb{Q}(\omega, \sqrt[3]{2}) = \mathbb{Q}(\sqrt[3]{2}, i\sqrt{3})$. (HINT: Compute ω.)

8 If K is the root field of any polynomial over $\mathbb{Q}$, and K contains an nth root of *any* number a, it contains all the nth roots of unity.

† F. Separable and Inseparable Polynomials

Let F be a field. An irreducible polynomial $p(x)$ in $F[x]$ is said to be *separable* over F if it has no multiple roots in any extension of F. If $p(x)$ does have a multiple root in some extension, it is *inseparable* over F. *Prove the following:*

1 If F has characteristic 0, every irreducible polynomial in $F[x]$ is separable.

Thus, for characteristic 0, there is no question whether an irreducible polynomial is separable or not. However, for characteristic $p \neq 0$, it is different. This case is treated next. In the following problems, let F be a field of characteristic $p \neq 0$.

2 If $a'(x) = 0$, the only nonzero terms of $a(x)$ are of the form $a_{mp} x^{mp}$ for some m. [In other words, $a(x)$ is a polynomial in powers of x^p.]

3 If an irreducible polynomial $a(x)$ is inseparable over F, then $a(x)$ is a polynomial in powers of x^p. (HINT: Use part 2, and reason as in the proof of Theorem 1.)

4 Use Chapter 27, Exercise J (especially the conclusion following J6) to prove the converse of part 3.

Thus, if F is a field of characteristic $p \neq 0$, an irreducible polynomial $a(x) \in F[x]$ is inseparable iff $a(x)$ is a polynomial in powers of x^p. For finite fields, we can say even more:

5 If F is any field of characteristic $p \neq 0$, then in $F[x]$,

$$(a_0 + a_1 x + \cdots + a_n x^n)^p = a_0^p + a_1^p x^p + \cdots + a_n^p x^{np}$$

(HINT: See Chapter 24, Exercise D4.)

6 If F is a finite field of characteristic $p \neq 0$, then in $F[x]$, every polynomial $a(x^p)$ is equal to $[b(x)]^p$ for some $b(x)$. [HINT: Use part 5 and the fact that in a finite field of characteristic p, every element has a pth root (see Chapter 20, Exercise F).]

7 Use parts 3 and 6 to prove: In any finite field, every irreducible polynomial is separable.

Thus, fields of characteristic 0 and finite fields share the property that *irreducible polynomials have no multiple roots*. The only remaining case is that of infinite fields with finite characteristic. It is treated in the next exercise set.

† G. Multiple Roots over Infinite Fields of Nonzero Characteristic

If $\mathbb{Z}_p[y]$ is the domain of polynomials (in the letter y) over $\mathbb{Z}_p$, let $E = \mathbb{Z}_p(y)$ be the field of quotients of $\mathbb{Z}_p[y]$. Let K denote the subfield $\mathbb{Z}_p(y^p)$ of $\mathbb{Z}_p(y)$.

1 Explain why $\mathbb{Z}_p(y)$ and $\mathbb{Z}_p(y^p)$ are infinite fields of characteristic p.

2 Prove that $a(x) = x^p - y^p$ has the factorization $x^p - y^p = (x - y)^p$ in $E[x]$, but is irreducible in $K[x]$. Conclude that there is an irreducible polynomial $a(x)$ in $K[x]$ with a root whose multiplicity is p.

Thus, over an infinite field of nonzero characteristic, an irreducible polynomial may have multiple roots. Even these fields, however, have a remarkable property: *all the roots of any irreducible polynomial have the same multiplicity*. The details follow: Let F be any field, $p(x)$ irreducible in $F[x]$, a and b two distinct roots of $p(x)$, and K the root field of $p(x)$ over F. Let $i: K \to i(K) = K'$ be the isomorphism of Theorem 4, and $\bar{i}: K[x] \to K'[x]$ the isomorphism described immediately preceding Theorem 3.

3 Prove that $\bar{i}$ leaves $p(x)$ fixed.

4 Prove that $\bar{\imath}((x - a)^m) = (x - b)^m$.

5 Prove that a and b have the same multiplicity.

† H. An Isomorphism Extension Theorem (Proof of Theorem 3)

Let F_1, F_2, h, $p(x)$, a, b, and $\bar{h}$ be as in the statement of Theorem 3. To prove that $\bar{h}$ is an isomorphism, it must first be shown that it is properly defined: that is, if $c(a) = d(a)$ in $F_1(a)$, then $\bar{h}(c(a)) = \bar{h}(d(a))$.

1 If $c(a) = d(a)$, prove that $c(x) - d(x)$ is a multiple of $p(x)$. Deduce from this that $hc(x) - hd(x)$ is a multiple of $hp(x)$.

2 Use part 1 to prove that $\bar{h}(c(a)) = \bar{h}(d(a))$.

3 Reversing the steps of the preceding argument, show that $\bar{h}$ is injective.

4 Show that $\bar{h}$ is surjective.

5 Show that $\bar{h}$ is a homomorphism.

† I. Uniqueness of the Root Field

Let $h: F_1 \rightarrow F_2$ be an isomorphism. If $a(x) \in F_1[x]$, let K_1 be the root field of $a(x)$ over F_1, and K_2 the root field of $ha(x)$ over F_2. *Prove the following:*

1 If $p(x)$ is an irreducible factor of $a(x)$, $u \in K_1$ is a root of $p(x)$, and $v \in K_2$ is a root of $hp(x)$, then $F_1(u) \cong F_2(v)$.

2 $F_1(u) = K_1$ iff $F_2(v) = K_2$.

3 If $F_1(u) \neq K_1$ [hence $F_2(v) \neq K_2$], write $F'_1 = F_1(u)$ and $F'_2 = F_2(v)$. Let $q(x) \in F'_1[x]$ be an irreducible factor of $a(x)$, of degree > 1. (Explain why such a factor is certain to exist in $F'_1[x]$.) Let $w \in K_1$ be a root of $q(x)$, and $z \in K_2$ a root of $hq(x)$. Prove that $F'_1(w) \cong F'_2(z)$.

4 Use parts 1 to 3 to form an inductive proof that $K_1 \cong K_2$.

5 Draw the following conclusion: The root field of a polynomial $a(x)$ over a field F is unique up to isomorphism.

† J. Extending Isomorphisms

In the following, let F be a subfield of $\mathbb{C}$. An injective homomorphism $h: F \rightarrow \mathbb{C}$ is called a *monomorphism*; it is obviously an isomorphism $F \rightarrow h(F)$.

1 Let ω be a complex pth root of unity (where p is a prime), and let $h: \mathbb{Q}(\omega) \rightarrow \mathbb{C}$ be a monomorphism fixing $\mathbb{Q}$. Explain why h is completely determined by the value of $h(\omega)$. Then prove that there exist exactly $p - 1$ monomorphisms $\mathbb{Q}(\omega) \rightarrow \mathbb{C}$ which fix $\mathbb{Q}$.

2 Let $p(x)$ be irreducible in $F[x]$, and c a complex root of $p(x)$. Let $h: F \rightarrow \mathbb{C}$ be a monomorphism. If deg $p(x) = n$, prove that there are exactly n monomorphisms $F(c) \to \mathbb{C}$ which are extensions of h.

3 Let $F \subseteq K$, with $[K : F] = n$. If $h: F \to \mathbb{C}$ is a monomorphism, prove that there are exactly n monomorphisms $K \to \mathbb{C}$ which are extensions of h.

4 Prove: The only possible monomorphism $h: \mathbb{Q} \to \mathbb{C}$ is $h(x) = x$. Thus, any monomorphism $h: \mathbb{Q}(a) \to \mathbb{C}$ necessarily fixes $\mathbb{Q}$.

5 Prove: There are exactly three monomorphisms $\mathbb{Q}(\sqrt[3]{2}) \to \mathbb{C}$, and they are determined by the conditions: $\sqrt[3]{2} \to \sqrt[3]{2}$; $\sqrt[3]{2} \to \sqrt[3]{2}\omega$; $\sqrt[3]{2} \to \sqrt[3]{2}\omega^2$, where ω is a primitive cube root of unity.

K. Normal Extensions

If K is the root field of some polynomial $a(x)$ over F, K is also called a *normal extension* of F. There are other possible ways of defining normal extension, which are equivalent to the above. We consider the two most common ones here: they are precisely the properties expressed in Theorems 7 and 6. Let K be a finite extension of F. *Prove the following :*

1 Suppose that for every irreducible polynomial $p(x)$ in $F[x]$, if $p(x)$ has one root in K, then $p(x)$ must have all its roots in K. Prove that K is a normal extension of F.

2 Suppose that, if h is any isomorphism with domain K which fixes F, then $h(K) \subseteq K$. Prove that K is a normal extension of F.

THIRTY-TWO
GALOIS THEORY: THE HEART OF THE MATTER

If K is a field and h is an isomorphism from K to K, we call h an *automorphism* of K (automorphism = "self-isomorphism").

Several of the isomorphisms we looked at in Chapter 31 were, in fact, automorphisms. For example, suppose K is a finite extension of F and h is any isomorphism whose domain is K. If $h(K) \subseteq K$, then necessarily $h(K) = K$, and therefore h is an automorphism of K. The reason is quite simple: $h(K)$ is isomorphic to K, hence K and $h(K)$ are vector spaces of the *same dimension* over F. Therefore, if $h(K) \subseteq K$, $h(K)$ is a subspace of K having the same dimension as K, and so $h(K)$ must be all of K.

This observation leads us to restate a few results of Chapter 31 in a simpler form. (First we restate Theorem 6, then Theorem 4 in a form which uses Theorem 6.)

Let K be the root field of some polynomial over F:

(*) *Any isomorphism with domain K which fixes F is an automorphism of K.*

(**) *If a and b are roots of an irreducible polynomial $p(x)$ in $F[x]$, there is an automorphism of K fixing F and sending a to b.*

(*) is merely a restatement of Theorem 6 of Chapter 31, incorporating the observation that $h(K) = K$.

(**) is a result of combining Theorem 4 of Chapter 31, Theorem 5 of Chapter 31 [applied to $F(a)$ instead of to F), and (*), in that order.

Let K be the root field of a polynomial $a(x)$ in $F[x]$. If $c_1, c_2, \ldots, c_n$ are the roots of $a(x)$, then $K = F(c_1, c_2, \ldots, c_n)$, and, by (*) of page 316, any automorphism h of K which fixes F *permutes* $c_1, c_2, \ldots, c_n$. On the other hand, remember that every element a in $F(c_1, c_2, \ldots, c_n)$ is a sum of terms of the form

$$kc_1^{i_1} c_2^{i_2} \cdots c_n^{i_n}$$

where the coefficient k of each term is in F. If h is an automorphism which fixes F, h does not change the coefficients, so $h(a)$ is completely determined once we know $h(c_1), \ldots, h(c_n)$. Thus, *every automorphism of K fixing F is completely determined by a permutation of the roots of $a(x)$.*

This is *very* important!

What it means is that we may identify the automorphisms of K which fix F with permutations of the roots of $a(x)$.

It must be pointed out here that, just as the symmetries of geometric figures determine their geometric properties, so the symmetries of equations (that is, permutations of their roots) give us all the vital information needed to analyze their solutions. Thus, if K is the root field of our polynomial $a(x)$ over F, we will now pay very close attention to the automorphisms of K which fix F.

To begin with, how many such automorphisms are there? The answer is a classic example of mathematical elegance and simplicity:

Theorem 1 *Let K be the root field of some polynomial over F. The number of automorphisms of K fixing F is equal to the degree of K over F.*

Let $[K:F] = n$, and let us show that K has exactly n automorphisms fixing F. By Theorem 2 of Chapter 31, $K = F(a)$ for some $a \in K$. Let $p(x)$ be the minimum polynomial of a over F; if b is any root of $p(x)$, then by (**), there is an automorphism of K fixing F and sending a to b. Since $p(x)$ has n roots, there are exactly n choices of b, and therefore n automorphisms of K fixing F.

[Remember that every automorphism h which fixes F permutes the roots of $p(x)$ therefore sends a to *some* root of $p(x)$; and h is completely determined once we have chosen $h(a)$.]

For example, we have already seen that $\mathbb{Q}(\sqrt{2})$ is of degree 2 over $\mathbb{Q}$. $\mathbb{Q}(\sqrt{2})$ is the root field of $x^2 - 2$ over $\mathbb{Q}$ because $\mathbb{Q}(\sqrt{2})$ contains both roots of $x^2 - 2$, namely $\pm\sqrt{2}$. By Theorem 1, there are exactly two automorphisms of $\mathbb{Q}(\sqrt{2})$ fixing $\mathbb{Q}$: one sends $\sqrt{2}$ to $\sqrt{2}$; it is the identity function. The other sends $\sqrt{2}$ to $-\sqrt{2}$, and is therefore the function $a + b\sqrt{2} \rightarrow a - b\sqrt{2}$.

Similarly, we saw that $\mathbb{C} = \mathbb{R}(i)$, and $\mathbb{C}$ is of degree 2 over $\mathbb{R}$. The two automorphisms of $\mathbb{C}$ which fix $\mathbb{R}$ are the identity function, and the function $a + bi \rightarrow a - bi$ which sends every complex number to its complex conjugate.

As a final example, we have seen that $\mathbb{Q}(\sqrt{2}, \sqrt{3})$ is an extension of degree 4 over $\mathbb{Q}$, so by Theorem 1, there are four automorphisms of $\mathbb{Q}(\sqrt{2}, \sqrt{3})$ which fix $\mathbb{Q}$: Now, $\mathbb{Q}(\sqrt{2}, \sqrt{3})$ is the root field of $(x^2 - 2)(x^2 - 3)$ over $\mathbb{Q}$, for it contains the roots of this polynomial, and any extension of $\mathbb{Q}$ containing the roots of $(x^2 - 2)(x^2 - 3)$ certainly contains $\sqrt{2}$ and $\sqrt{3}$. Thus, by (*) on page 316, each of the four automorphisms which fix $\mathbb{Q}$ sends roots of $x^2 - 2$ to roots of $x^2 - 2$, and roots of $x^2 - 3$ to roots of $x^2 - 3$. But there are only four possible ways of doing this, namely

$$\left\{ \begin{array}{l} \sqrt{2} \rightarrow \sqrt{2} \\ \sqrt{3} \rightarrow \sqrt{3} \end{array} \right\} \qquad \left\{ \begin{array}{l} \sqrt{2} \rightarrow -\sqrt{2} \\ \sqrt{3} \rightarrow \sqrt{3} \end{array} \right\}$$

$$\left\{ \begin{array}{l} \sqrt{2} \rightarrow \sqrt{2} \\ \sqrt{3} \rightarrow -\sqrt{3} \end{array} \right\} \quad \text{and} \quad \left\{ \begin{array}{l} \sqrt{2} \rightarrow -\sqrt{2} \\ \sqrt{3} \rightarrow -\sqrt{3} \end{array} \right\}$$

Since every element of $\mathbb{Q}(\sqrt{2}, \sqrt{3})$ is of the form $a + b\sqrt{2} + c\sqrt{3} + d\sqrt{6}$, these four automorphisms (we shall call them ε, α, β, and γ) are the following

$$a + b\sqrt{2} + c\sqrt{3} + d\sqrt{6} \xrightarrow{\varepsilon} a + b\sqrt{2} + c\sqrt{3} + d\sqrt{6}$$

$$a + b\sqrt{2} + c\sqrt{3} + d\sqrt{6} \xrightarrow{\alpha} a - b\sqrt{2} + c\sqrt{3} - d\sqrt{6}$$

$$a + b\sqrt{2} + c\sqrt{3} + d\sqrt{6} \xrightarrow{\beta} a + b\sqrt{2} - c\sqrt{3} - d\sqrt{6}$$

$$a + b\sqrt{2} + c\sqrt{3} + d\sqrt{6} \xrightarrow{\gamma} a - b\sqrt{2} - c\sqrt{3} + d\sqrt{6}$$

If K is an extension of F, the automorphisms of K which fix F *form a group.* (The operation, of course, is composition.) This is perfectly obvious: for if g and h fix F, then for every x in F,

$$x \xrightarrow{h} x \qquad \text{and} \qquad x \xrightarrow{g} x \qquad \text{so} \qquad x \xrightarrow{h} x \xrightarrow{g} x$$

that is, $g \circ h$ fixes F. Furthermore, if

$$x \xrightarrow{h} x \qquad \text{then} \qquad x \xleftarrow{h^{-1}} x$$

that is, if h fixes F so does h^{-1}.

This fact is perfectly obvious, but nonetheless of great importance, for it means that we can now use all of our accumulated knowledge about groups to help us analyze the solutions of polynomial equations. And that is precisely what Galois theory is all about.

If K is the root field of a polynomial $a(x)$ in $F[x]$, *the group of all the automorphisms of K which fix F is called the Galois group of $a(x)$*. We also call it the *Galois group of K over F*, and designate it by the symbol

$$Gal(K : F)$$

In our last example we saw that there are four automorphisms of $\mathbb{Q}(\sqrt{2}, \sqrt{3})$ which fix $\mathbb{Q}$. We called them ε, α, β, and γ. Thus, the Galois group of $\mathbb{Q}(\sqrt{2}, \sqrt{3})$ over $\mathbb{Q}$ is $Gal(\mathbb{Q}(\sqrt{2}, \sqrt{3}) : \mathbb{Q}) = \{\varepsilon, \alpha, \beta, \gamma\}$; the operation is composition, giving us the table

$\circ$	ε	α	β	γ
ε	ε	α	β	γ
α	α	ε	γ	β
β	β	γ	ε	α
γ	γ	β	α	ε

As one can see, this is an abelian group in which every element is its own inverse; almost at a glance one can verify that it is isomorphic to $\mathbb{Z}_2 \times \mathbb{Z}_2$.

Let K be the root field of $a(x)$, where $a(x)$ is in $F[x]$. In our earlier discussion we saw that every automorphism of K fixing F [that is, every member of the Galois group of $a(x)$] may be identified with a permutation of the roots of $a(x)$. However, it is important to note that *not every permutation of the roots of $a(x)$ need be in the Galois group of $a(x)$*, even when $a(x)$ is irreducible. For example, we saw that $\mathbb{Q}(\sqrt{2}, \sqrt{3}) = \mathbb{Q}(\sqrt{2} + \sqrt{3})$, where $\sqrt{2} + \sqrt{3}$ is a root of the irreducible polynomial $x^4 - 10x^2 + 1$ over $\mathbb{Q}$. Since $x^4 - 10x^2 + 1$ has four roots, there are $4! = 24$ permutations of its roots, only four of which are in its Galois group. This is because only four of the permutations are genuine symmetries of $x^4 - 10x^2 + 1$, in the sense that they determine automorphisms of the root field.

In the discussion throughout the remainder of this chapter, let F and K remain fixed. F is an arbitrary field and K is the root field of some polynomial $a(x)$ in $F[x]$. The thread of our reasoning will lead us to speak about fields I where $F \subseteq I \subseteq K$, that is, fields "between" F and K. We will

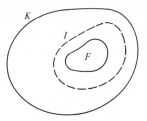

refer to them as *intermediate fields*. Since K is the root field of $a(x)$ over F, it is also the root field of $a(x)$ over I for every intermediate field I.

The letter **G** will denote the Galois group of K over F. With each intermediate field I, we associate the group

$$I^* = Gal(K : I)$$

that is, the group of all the automorphisms of K which fix I. It is obviously a subgroup of **G**. We will call I^* the *fixer* of I.

Conversely, with each subgroup H of **G** we associate the subfield of K containing all the a in K which are not changed by any $\pi \in H$. That is,

$$\{a \in K : \pi(a) = a \text{ for every } \pi \in H\}$$

One verifies in a trice that this is a subfield of K. It obviously contains F, and is therefore one of the intermediate fields. It is called the *fixed field* of H. For brevity and euphony we call it the *fixfield* of H.

Let us recapitulate: Every subgroup H of **G** fixes an intermediate field I, called the fixfield of H. Every intermediate field I is fixed by a subgroup H of **G**, called the fixer of I. This suggests very strongly that there is a one-to-one correspondence between the subgroups of **G** and the fields intermediate between F and K. Indeed, this is correct. This one-to-one correspondence is at the very heart of Galois theory, because it provides the tie-in between properties of field extensions and properties of subgroups.

Just as, in Chapter 29, we were able to use vector algebra to prove new things about field extensions, now we will be able to use group theory to explore field extensions. The vector-space connection was a relative lightweight. The connection with group theory, on the other hand, gives us a tool of tremendous power to study field extensions.

We have not yet proved that the connection between subgroups of **G** and intermediate fields is a one-to-one correspondence. The next two theorems will do that.

Theorem 2 *If H is the fixer of I, then I is the fixfield of H.*

Let H be the fixer of I, and let I' be the fixfield of H. It follows from the definitions of *fixer* and *fixfield* that $I \subseteq I'$, so we must now show that $I' \subseteq I$. We will do this by proving that $a \notin I$ implies $a \notin I'$. Well, if a is an element of K which is not in I, the minimum polynomial $p(x)$ of a over I must have degree ≥ 2 (for otherwise, $a \in I$). Thus, $p(x)$ has another root b. By (**), there is an automorphism of K fixing I and sending a to b. This automorphism moves a, so $a \notin I'$.

Lemma *Let H be a subgroup of* **G**, *and I the fixfield of H. The number of elements in H is equal to* $[K : I]$.

Let H have r elements, namely $h_1, \ldots, h_r$. Let $K = I(a)$. Much of our proof will revolve around the following polynomial:

$$b(x) = (x - h_1(a))(x - h_2(a)) \cdots (x - h_r(a))$$

Since one of the h_i is the identity function, one factor of $b(x)$ is $(x - a)$, and therefore *a is a root of $b(x)$*. In the next paragraph we will see that all the coefficients of $b(x)$ are in I, so $b(x) \in I[x]$. It follows that $b(x)$ is a multiple of the minimum polynomial of a over I, whose degree is exactly $[K : I]$. Since $b(x)$ is of degree r, this means that $r \geq [K : I]$, which is half our theorem.

Well, let us show that all the coefficients of $b(x)$ are in I. We saw on page 314 that every isomorphism $h_i : K \to K$ can be extended to an isomorphism $\bar{h}_i : K[x] \to K[x]$. Because $\bar{h}_i$ is an isomorphism of polynomials, we get

$$\bar{h}_i(b(x)) = \bar{h}_i(x - h_1(a))\bar{h}_i(x - h_2(a)) \cdots \bar{h}_i(x - h_r(a))$$

$$= (x - h_i \circ h_1(a)) \cdots (x - h_i \circ h_r(a))$$

But $h_i \circ h_1, h_i \circ h_2, \ldots, h_i \circ h_r$ are *r distinct* elements of H, and H has exactly r elements, so they are all the elements of H (that is, they are $h_1, \ldots, h_r$, possibly in a different order). So the factors of $\bar{h}_i(b(x))$ are the same as the factors of $b(x)$, merely in a different order, and therefore $\bar{h}_i(b(x)) = b(x)$. Since equal polynomials have equal coefficients, h_i leaves the coefficients of $b(x)$ invariant. Thus, every coefficient of $b(x)$ is in the fixfield of H, that is, in I.

We have just shown that $[K : I] \leq r$. For the opposite inequality, remember that by Theorem 1, $[K : I]$ is equal to the number of I-fixing automorphisms of K. But there are at least r such automorphisms, namely $h_1, \ldots, h_r$. Thus, $[K : I] \geq r$, and we are done.

Theorem 3 *If I is the fixfield of H, then H is the fixer of I.*

Let I be the fixfield of H, and I^* the fixer of I. It follows from the definitions of fixer and fixfield that $H \subseteq I^*$. We will prove equality by showing that there are as many elements in H as in I^*. By the lemma, the order of H is equal to $[K : I]$. By Theorem 2, I is the fixfield of I^*, so by the lemma again, the order of I^* is also equal to $[K : I]$.

It follows immediately from Theorems 2 and 3 that *there is a one-to-one correspondence between the subgroups of* $\text{Gal}(K : F)$ *and the intermediate fields between K and F.* This correspondence, which matches every subgroup with its fixfield (or, equivalently, matches every intermediate field with its fixer) is called a *Galois correspondence.* It is worth observing that larger subfields correspond to smaller subgroups; that is,

$$I_1 \subseteq I_2 \quad \text{iff} \quad I_2^* \subseteq I_1^*$$

As an example, we have seen that the Galois group of $\mathbb{Q}(\sqrt{2}, \sqrt{3})$ over $\mathbb{Q}$ is $\mathbf{G} = \{\varepsilon, \alpha, \beta, \gamma\}$ with the table given on page 326. This group has exactly five subgroups, namely $\{\varepsilon\}$, $\{\varepsilon, \alpha\}$, $\{\varepsilon, \beta\}$, $\{\varepsilon, \gamma\}$, and the whole group $\mathbf{G}$. They may be represented in the "inclusion diagram"

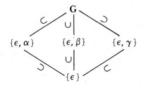

On the other hand, there are exactly five fields intermediate between $\mathbb{Q}$ and $\mathbb{Q}(\sqrt{2}, \sqrt{3})$, which may be represented in the inclusion diagram

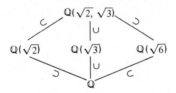

If H is a subgroup of any Galois group, let H° designate the fixfield of H. The subgroups of $\mathbf{G}$ in our example have the following fixfields:

$$\{\varepsilon\}^\circ = \mathbb{Q}(\sqrt{2}, \sqrt{3}) \qquad \{\varepsilon, \alpha\}^\circ = \mathbb{Q}(\sqrt{3}) \qquad \{\varepsilon, \beta\}^\circ = \mathbb{Q}(\sqrt{2})$$

$$\{\varepsilon, \gamma\}^\circ = \mathbb{Q}(\sqrt{6}) \qquad \mathbf{G}^\circ = \mathbb{Q}$$

(This is obvious by inspection of the way ε, α, β, and γ were defined on page 325.) The Galois correspondence, for this example, may therefore be represented as follows:

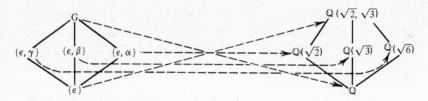

In order to effectively tie in subgroups of **G** with extensions of the field F, we need one more fact, to be presented next.

Suppose $E \subseteq I \subseteq K$, where K is a root field over I and I is a root field over E. (Hence by Theorem 8 on page 316, K is a root field over E.) If $h \in Gal(K : E)$, h is an automorphism of K fixing E. Consider the *restriction of h to I*, that is, h restricted to the smaller domain I. It is an isomorphism with domain I fixing E, so by (*) on page 323, it is an automorphism of I, still fixing E. We have just shown that if $h \in Gal(K : E)$, then the restriction of h to I is in $Gal(I : E)$. This permits us to define a function $\mu : Gal(K : E) \rightarrow Gal(I : E)$ by the rule

$$\mu(h) = \text{the restriction of } h \text{ to } I$$

It is very easy to check that μ is a homomorphism. μ is surjective, because every F-fixing automorphism of I can be extended to an F-fixing automorphism of K, by Theorem 5 in Chapter 31.

Finally, if $h \in Gal(K : E)$, the restriction of h to I is the identity function iff $h(x) = x$ for every $x \in I$, that is, iff h fixes I. This proves that the kernel of μ is $Gal(K : I)$.

To recapitulate: μ is a homomorphism from $Gal(K : E)$ *onto* $Gal(I : E)$ with kernel $Gal(K : I)$. By the FHT, we immediately conclude as follows:

Theorem 4 *Suppose $E \subseteq I \subseteq K$, where I is a root field over E and K is a root field over I. Then*

$$Gal(I : E) \cong \frac{Gal(K : E)}{Gal(K : I)}$$

It follows, in particular, that $Gal(K : I)$ is a normal subgroup of $Gal(K : E)$.

EXERCISES

† **A. Computing a Galois Group**

1 Show that $\mathbb{Q}(i, \sqrt{2})$ is the root field of $(x^2 + 1)(x^2 - 2)$ over $\mathbb{Q}$.

2 Find the degree of $\mathbb{Q}(i, \sqrt{2})$ over $\mathbb{Q}$.

3 List the elements of $Gal(\mathbb{Q}(i, \sqrt{2}) : \mathbb{Q})$ and exhibit its table.

4 Write the inclusion diagram for the subgroups of $Gal(\mathbb{Q}(i, \sqrt{2}) : \mathbb{Q})$, and the inclusion diagram for the fields intermediate between $\mathbb{Q}$ and $\mathbb{Q}(i, \sqrt{2})$. Indicate the Galois correspondence.

† B. Computing a Galois Group of Eight Elements

1 Show that $\mathbb{Q}(\sqrt{2}, \sqrt{3}, \sqrt{5})$ is the root field of $(x^2 - 2)(x^2 - 3)(x^2 - 5)$ over $\mathbb{Q}$.

2 Show that the degree of $\mathbb{Q}(\sqrt{2}, \sqrt{3}, \sqrt{5})$ over $\mathbb{Q}$ is 8.

3 List the eight elements of $\mathbf{G} = Gal(\mathbb{Q}(\sqrt{2}, \sqrt{3}, \sqrt{5}) : \mathbb{Q})$ and write its table.

4 List the subgroups of **G**. (By Lagrange's theorem, any proper subgroup of **G** has either two or four elements.)

5 For each subgroup of **G**, find its fixfield.

6 Indicate the Galois correspondence by means of a diagram like the one on page 330.

† C. A Galois Group Equal to S_3.

1 Show that $\mathbb{Q}(\sqrt[3]{2}, i\sqrt{3})$ is the root field of $x^3 - 2$ over $\mathbb{Q}$, where $\sqrt[3]{2}$ designates the *real* cube root of 2. (HINT: Compute the complex cube roots of unity.)

2 Show that $[\mathbb{Q}(\sqrt[3]{2}) : \mathbb{Q}] = 3$.

3 Explain why $x^2 + 3$ is irreducible over $\mathbb{Q}(\sqrt[3]{2})$, then show that $[\mathbb{Q}(\sqrt[3]{2}, i\sqrt{3}) : \mathbb{Q}(\sqrt[3]{2})] = 2$. Conclude that $[\mathbb{Q}(\sqrt[3]{2}, i\sqrt{3}) : \mathbb{Q}] = 6$.

4 Use part 3 to explain why $Gal(\mathbb{Q}(\sqrt[3]{2}, i\sqrt{3}) : \mathbb{Q})$ has six elements. Then use the discussion following (**) on page 323 to explain why every element of $Gal(\mathbb{Q}(\sqrt[3]{2}, i\sqrt{3}) : \mathbb{Q})$ may be identified with a permutation of the three cube roots of 2.

5 Use part 4 to prove that $Gal(\mathbb{Q}(\sqrt[3]{2}, i\sqrt{3}) : \mathbb{Q}) \cong S_3$.

† D. A Galois Group Equal to D_4

If $\alpha = \sqrt[4]{2}$ is a real fourth root of 2, then the four fourth roots of 2 are $\pm\alpha$ and $\pm i\alpha$. *Explain each of the following, briefly but carefully :*

1 $\mathbb{Q}(\alpha, i)$ is the root field of $x^4 - 2$ over $\mathbb{Q}$.

2 $[\mathbb{Q}(\alpha) : \mathbb{Q}] = 4$.

3 $i \notin \mathbb{Q}(\alpha)$, hence $[\mathbb{Q}(\alpha, i) : \mathbb{Q}(\alpha)] = 2$.

4 $[\mathbb{Q}(\alpha, i) : \mathbb{Q}] = 8$.

5 $\{1, \alpha, \alpha^2, \alpha^3, i, i\alpha, i\alpha^2, i\alpha^3\}$ is a basis for $\mathbb{Q}(\alpha, i)$ over $\mathbb{Q}$.

6 Any $\mathbb{Q}$-fixing automorphism h of $\mathbb{Q}(\alpha, i)$ is determined by its effect on the elements in the basis. These, in turn, are determined by $h(\alpha)$ and $h(i)$.

7 $h(\alpha)$ must be a fourth root of 2 and $h(i)$ must be equal to $\pm i$. Combining the four possibilities for $h(\alpha)$ with the two possibilities for $h(i)$ gives eight possible automorphisms. List them in the format

$$\begin{Bmatrix} \alpha \to \alpha \\ i \to i \end{Bmatrix}, \quad \begin{Bmatrix} \alpha \to -\alpha \\ i \to i \end{Bmatrix}, \quad \cdots$$

8 Compute the table of the group $Gal(\mathbb{Q}(\alpha, i) : \mathbb{Q})$ and show that it is isomorphic to D_4, the group of symmetries of the square.

† E. A Cyclic Galois Group

1 Describe the root field K of $x^7 - 1$ over $\mathbb{Q}$. Explain why $[K : \mathbb{Q}] = 6$.

2 Explain: If α is a primitive seventh root of unity, any $h \in Gal(K : \mathbb{Q})$ must send α to a seventh root of unity. In fact, h is determined by $h(\alpha)$.

3 Use part 2 to list explicitly the six elements of $Gal(K : \mathbb{Q})$. Then write the table of $Gal(K : \mathbb{Q})$ and show that it is cyclic.

4 List all the subgroups of $Gal(K : \mathbb{Q})$, with their fixfields. Exhibit the Galois correspondence.

5 Describe the root field L of $x^6 - 1$ over $\mathbb{Q}$, and show that $[L : \mathbb{Q}] = 3$. Explain why it follows that there are no intermediate fields between $\mathbb{Q}$ and L (except for $\mathbb{Q}$ and L themselves).

6 List the three elements of $Gal(L : \mathbb{Q})$ and write its table. (It will be useful to remember that the sixth roots of unity form a multiplicative *cyclic* group. An automorphism must send any element of order k, in this group, to another element of the same order k.)

† F. A Galois Group Isomorphic to S_5

Let $a(x) = x^5 - 4x^4 + 2x + 2 \in \mathbb{Q}[x]$, and let $r_1, \ldots, r_5$ be the roots of $a(x)$ in $\mathbb{C}$. Let $K = \mathbb{Q}(r_1, \ldots, r_5)$ be the root field of $a(x)$ over $\mathbb{Q}$. *Prove the following* :

1 $a(x)$ is irreducible in $\mathbb{Q}[x]$.

2 $a(x)$ has three real and two complex roots. [HINT: Use calculus to sketch the graph of $y = a(x)$, and show that it crosses the x axis three times.]

3 If r_1 denotes a real root of $a(x)$, $[\mathbb{Q}(r_1) : \mathbb{Q}] = 5$. Use this to prove that $[K : \mathbb{Q}]$ is a multiple of 5.

4 Use part 3 and Cauchy's theorem (Chapter 13, Exercise E) to prove that there is an element α of order 5 in $Gal(K : \mathbb{Q})$. Since α may be identified with a permutation of $\{r_1, \ldots, r_5\}$, explain why it must be a cycle of length 5. (HINT: Any product of disjoint cycles on $\{r_1, \ldots, r_5\}$ has order $\neq 5$.)

5 Explain why there is a transposition in $Gal(K : \mathbb{Q})$. [It permutes the conjugate pair of complex roots of $a(x)$.]

6 Any subgroup of S_5 which contains a cycle of length 5 and a transposition must contain all possible transpositions in S_5, hence all of S_5. Thus, $Gal(K : \mathbb{Q}) = S_5$.

G. Shorter Questions Relating to Automorphisms and Galois Groups

Let F be a field, and K a finite extension of F. *Prove the following :*

1 If an automorphism h of K fixes F and a, then h fixes $F(a)$.

2 $F(a, b)^* = F(a)^* \cap F(b)^*$.

3 Aside from the identity function, there are no $\mathbb{Q}$-fixing automorphisms of $\mathbb{Q}(\sqrt[3]{2})$. [HINT: Note that $\mathbb{Q}(\sqrt[3]{2})$ contains only real numbers.]

4 Explain why the conclusion of part 3 does not contradict Theorem 1.

In the next three parts, let ω be a primitive pth root of unity, where p is a prime.

5 If $h \in Gal(\mathbb{Q}(\omega) : \mathbb{Q})$, then $h(\omega) = \omega^k$ for some k where $1 \leq k \leq p - 1$.

6 Use part 5 to prove that $Gal(\mathbb{Q}(\omega) : \mathbb{Q})$ is an abelian group.

7 Use part 5 to prove that $Gal(\mathbb{Q}(\omega) : \mathbb{Q})$ is a cyclic group.

† H. The Group of Automorphisms of $\mathbb{C}$

Prove the following:

1 The only automorphism of $\mathbb{Q}$ is the identity function. [HINT: If h is an automorphism, $h(1) = 1$, hence $h(2) = 2$, and so on.]

2 Any automorphism of $\mathbb{R}$ sends squares of numbers to squares of numbers, hence positive numbers to positive numbers.

3 Using part 2, prove that if h is any automorphism of $\mathbb{R}$, $a < b$ implies $h(a) < h(b)$.

4 Use parts 1 and 3 to prove that the only automorphism of $\mathbb{R}$ is the identity function.

5 List the elements of $Gal(\mathbb{C} : \mathbb{R})$.

6 Prove that the identity function and the function $a + bi \rightarrow a - bi$ are the only automorphisms of $\mathbb{C}$.

I. Further Questions Relating to Galois Groups

Throughout this set of questions, let K be a root field over F, let $\mathbf{G} = Gal(K : F)$, and let I be any intermediate field. *Prove the following :*

1 $I^* = Gal(K : I)$ is a subgroup of $\mathbf{G}$.

2 If H is a subgroup of $\mathbf{G}$ and $H° = \{a \in K : \pi(a) = a \text{ for every } \pi \in H\}$, then $H°$ is a subfield of K, and $F \subseteq H°$.

3 Let H be the fixer of I, and I' the fixfield of H. Then $I \subseteq I'$. Let I be the fixfield of H, and I^* the fixer of I. Then $H \subseteq I^*$.

4 Let I be a normal extension of F (that is, a root field of some polynomial over F). If $\mathbf{G}$ is abelian, then $Gal(K : I)$ and $Gal(I : F)$ are abelian. (HINT: Use Theorem 4.)

5 Let I be a normal extension of F. If **G** is a cyclic group, then $Gal(K : I)$ and $Gal(I : F)$ are cyclic groups.

6 If **G** is a cyclic group, there exists exactly one intermediate field I of degree k, for each integer k dividing $[K : F]$.

† J. Normal Extensions and Normal Subgroups

Suppose $F \subseteq K$, where K is a normal extension of F. (This means simply that K is the root field of some polynomial in $F[x]$: see Chapter 31, Exercise K.) Let $I_1 \subseteq I_2$ be intermediate fields.

1 Deduce from Theorem 4 that, if I_2 is a normal extension of I_1, then I_2^* is a normal subgroup of I_1^*.

2 Prove the following for any intermediate field I: Let $h \in Gal(K : F)$, $g \in I^*$, $a \in I$, and $b = h(a)$. Then $[h \circ g \circ h^{-1}](b) = b$. Conclude that

$$hI^*h^{-1} \subseteq h(I)^*$$

3 Use part 2 to prove that $hI^*h^{-1} = h(I)^*$.

Two intermediate fields I_1 and I_2 are called *conjugate* iff there is an automorphism [i.e., an element $i \in Gal(K : F)$] such that $i(I_1) = I_2$.

4 Use part 3 to prove: I_1 and I_2 are conjugate iff I_1^* and I_2^* are conjugate subgroups in the Galois group.

5 Use part 4 to prove that for any intermediate fields I_1 and I_2 : if I_2^* is a normal subgroup of I_1^*, then I_2 is a normal extension of I_1.

Combining parts 1 and 5 we have: I_2 is a normal extension of I_1 iff I_2^* is a normal subgroup of I_1^*. (Historically, this result is the origin of the word "normal" in the term "normal subgroup.")

THIRTY-THREE
SOLVING EQUATIONS BY RADICALS

In this final chapter, Galois theory will be used explicitly to answer practical questions about solving equations.

In the introduction to this book we saw that classical algebra was devoted largely to finding methods for solving polynomial equations. The quadratic formula yields the solutions of every equation of degree 2, and similar formulas have been found for polynomials of degrees 3 and 4. But every attempt to find explicit formulas, of the same kind as the quadratic formula, which would solve a general equation of degree 5 or higher ended in failure. The reason for this was finally discovered by the young Galois, who showed that an equation is solvable by the kind of explicit formula we have in mind if and only if its group of symmetries has certain properties. The group of symmetries is, of course, the Galois group which we have already defined, and the required group properties will be formulated in the next few pages.

Galois showed that the groups of symmetries of all equations of degree ≤ 4 have the properties needed for solvability, whereas equations of degree 5 or more do not always have them. Thus, not only is the classical quest for radical formulas to solve all equations of degree > 4 shown to be futile, but a criterion is made available to test any equation and determine if it has solutions given by a radical formula. All this will be made clear in the following pages.

Every quadratic equation $ax^2 + bx + c = 0$ has its roots given by the formula

$$\frac{-b \pm \sqrt{b^2 - 4ac}}{2a}$$

Equations of degree 3 and 4 can be solved by similar formulas. For example, the cubic equation $x^3 + ax + b = 0$ has a solution given by

$$\sqrt[3]{\frac{-b}{2} + \sqrt{D}} + \sqrt[3]{\frac{-b}{2} - \sqrt{D}} \qquad \text{where } D = \frac{a^3}{27} + \frac{b^2}{4} \qquad (*)$$

Such expressions are built up from the coefficients of the given polynomials by repeated addition, subtraction, multiplication, division, *and taking roots*. Because of their use of radicals, they are called *radical expressions* or *radical formulas*. A polynomial $a(x)$ is *solvable by radicals* if there is a radical expression giving its roots in terms of its coefficients.

Let us return to the example of $x^3 + ax + b = 0$, where a and b are rational, and look again at Formula (*). We may interpret this formula to assert that if we start with the field of coefficients $\mathbb{Q}$, adjoin the square root $\sqrt{D}$, then adjoin the cube roots $\sqrt[3]{-b/2 \pm \sqrt{D}}$, we reach a field in which $x^3 + ax + b = 0$ has its roots.

In general, to say that the roots of $a(x)$ are given by a radical expression is the same as saying that we can extend the field of coefficients of $a(x)$ by successively adjoining nth roots (for various n), and in this way obtain a field which contains the roots of $a(x)$. We will express this notion formally now, in the language of field theory.

$F(c_1, \ldots, c_n)$ is called a *radical extension* of F if, for each i, some power of c_i is in $F(c_1, \ldots, c_{i-1})$. In other words, $F(c_1, \ldots, c_n)$ is an iterated extension of F obtained by successively adjoining nth roots, for various n. We say that a polynomial $a(x)$ in $F[x]$ is *solvable by radicals* if there is a radical extension of F containing all the roots of $a(x)$, that is, containing the root field of $a(x)$.

To deal effectively with nth roots we must know a little about them. To begin with, the nth roots of 1, called nth *roots of unity*, are, of course, the solutions of $x^n - 1 = 0$. Thus, for each n, there are exactly n nth roots of unity. As we shall see, everything we need to know about roots will follow from properties of the roots of unity.

In $\mathbb{C}$ the nth roots of unity are obtained by de Moivre's theorem. They consist of a number ω and its first n powers: $1 = \omega^0,\ \omega, \omega^2, \ldots, \omega^{n-1}$. We will not review de Moivre's theorem here because, remarkably, the main facts about roots of unity are true in *every* field of characteristic zero. Everything we need to know emerges from the following theorem:

Theorem 1 *Any finite group of nonzero elements in a field is a cyclic group.* (*The operation in the group is the field's multiplication.*)

If F^* denotes the set of nonzero elements of F, suppose that $G \subseteq F^*$, and that G, with the field's "multiply" operation, is a group of n elements. We will compare G with $\mathbb{Z}_n$ and show that G, like $\mathbb{Z}_n$, has an element of order n and is therefore cyclic.

For every positive integer k which divides n, the equation $x^k = 1$ has *at most* k solutions in F; thus, G contains k *or fewer* elements of order k. On the other hand, in $\mathbb{Z}_n$ there are *exactly* k elements of order k, namely

$$\frac{n}{k}, 2\frac{n}{k}, \ldots, (k-1)\frac{n}{k}$$

Now, every element of G (as well as every element of $\mathbb{Z}_n$) has a well-defined order, which is a divisor of n. Imagine the elements of both groups to be partitioned into classes according to their order, and compare the classes in G with the corresponding classes in $\mathbb{Z}_n$. For each k, G has as many *or fewer* elements of order k than $\mathbb{Z}_n$ does. So if G had no elements of order n (while $\mathbb{Z}_n$ *does* have one), this would mean that G has fewer elements than $\mathbb{Z}_n$, which is false. Thus, G *must* have an element of order n, and therefore G is cyclic.

The nth roots of unity (which are contained in F or a suitable extension of F) obviously form a group with respect to multiplication. By Theorem 1, it is a cyclic group. Any generator of this group is called a *primitive nth root of unity*. Thus, if ω is a primitive nth root of unity, the set of all the nth roots of unity is

$$1, \omega, \omega^2, \ldots, \omega^{n-1}$$

If ω is a primitive nth root of unity, $F(\omega)$ is an *abelian extension of F* in the sense that $g \circ h = h \circ g$ for any two F-fixing automorphisms g and h of $F(\omega)$. Indeed, any automorphism must obviously send nth roots of unity to nth roots of unity. So if $g(\omega) = \omega^r$ and $h(\omega) = \omega^s$, then $g \circ h(\omega) = g(\omega^s) = \omega^{rs}$, and analogously, $h \circ g(\omega) = \omega^{rs}$. Thus, $g \circ h(\omega) = h \circ g(\omega)$. Since g and h fix F, and every element of $F(\omega)$ is a linear combination of powers of ω with coefficients in F, $g \circ h = h \circ g$.

Now, let F contain a primitive nth root of unity, and therefore all the nth roots of unity. Suppose $a \in F$, and a has an nth root b in F. It follows, then, that *all the nth roots of a are in F, for they are* $b, b\omega, b\omega^2, \ldots, b\omega^{n-1}$. Indeed, if c is any other nth root of a, then clearly c/b is an nth root of 1, say

ω^r, hence $c = b\omega^r$. We may infer from the above that *if F contains a primitive nth root of unity, and b is an nth root of a, then $F(b)$ is the root field of $x^n - a$ over F.*

In particular, *$F(b)$ is an abelian extension of F.* Indeed, any F-fixing automorphism of $F(b)$ must send nth roots of a to nth roots of a: for if c is any nth root of a and g is an F-fixing automorphism, then $g(c)^n = g(c^n) = g(a) = a$, hence $g(c)$ is an nth root of a. So if $g(b) = b\omega^r$ and $h(b) = b\omega^s$, then

$$g \circ h(b) = g(b\omega^s) = b\omega^r\omega^s = b\omega^{r+s}$$

and
$$h \circ g(b) = h(b\omega^r) = b\omega^s\omega^r = b\omega^{r+s}$$

hence $g \circ h(b) = h \circ g(b)$. Since g and h fix F, and every element in $F(b)$ is a linear combination of powers of b with coefficients in F, it follows that $g \circ h = h \circ g$.

If $a(x)$ is in $F[x]$, remember that $a(x)$ is solvable by radicals just as long as there exists *some* radical extension of F containing the roots of $a(x)$. [*Any* radical extension of F containing the roots of $a(x)$ will do.] Thus, we may as well assume that any radical extension used here begins by adjoining to F the appropriate roots of unity; *henceforth we will make this assumption.* Thus, if $K = F(c_1, \ldots, c_n)$ is a radical extension of F, then

$$\underbrace{F}_{I_0} \subseteq \underbrace{F(c_1)}_{I_1} \subseteq \underbrace{F(c_1, c_2)}_{I_2} \subseteq \cdots \subseteq \underbrace{F(c_1. \ldots, c_n)}_{I_m = K}$$

is a sequence of simple *abelian* extensions. (The extensions are all *abelian* by the comments in the preceding three paragraphs.) If G denotes the Galois group of K over F, each of these fields I_k has a fixer which is a subgroup of G. These fixers form a sequence

$$\underbrace{K^*}_{= \{\varepsilon\}} \subseteq I^*_{m-1} \subseteq \ldots \subseteq I^*_1 \subseteq \underbrace{F^*}_{= G}$$

For each k, by Theorem 4 on page 330, I^*_k is a normal subgroup of I^*_{k+1}, and $I^*_{k+1}/I^*_k \cong \text{Gal}(I_{k+1} : I_k)$ which is abelian because I_{k+1} is an abelian extension of I_k. The following definition was invented precisely to account for this situation:

*A group G is called **solvable** if it has a sequence of subgroups $\{e\} = H_0 \subseteq H_1 \subseteq \ldots \subseteq H_m = G$ such that for each k, G_k is a normal subgroup of G_{k+1} and G_{k+1}/G_k is abelian.*

We have shown that *if K is a radical extension of F, then Gal(K : F) is a solvable group.* We wish to go further and prove that if $a(x)$ is any polynomial which is solvable by radicals, its Galois group is solvable. To do so, we must first prove that any homomorphic image of a solvable group is solvable. A key ingredient of our proof is the following simple fact, which was explained on page 148: G/H is abelian iff H contains all the products $xyx^{-1}y^{-1}$, for all x and y in G. (The products $xyx^{-1}y^{-1}$ are called "commutators" of G.)

Theorem 2 *Any homomorphic image of a solvable group is a solvable group.*

Let G be a solvable group, with a sequence of subgroups

$$\{e\} \subseteq H_1 \subseteq \cdots \subseteq H_m = G$$

as specified in the definition. Let $f : G \to X$ be a homomorphism from G onto a group X. Then $f(H_0), f(H_1), \ldots, f(H_m)$ are subgroups of X, and clearly $\{e\} \subseteq f(H_0) \subseteq f(H_1) \subseteq \cdots \subseteq f(H_m) = X$. For each i we have the following: if $f(a) \in f(H_i)$ and $f(x) \in f(H_{i+1})$, then $a \in H_i$ and $x \in H_{i+1}$, hence $xax^{-1} \in H_i$ and therefore $f(x)f(a)f(x)^{-1} \in f(H_i)$. So $f(H_i)$ is a normal subgroup of $f(H_{i+1})$. Finally, since H_{i+1}/H_i is abelian, every commutator $xyx^{-1}y^{-1}$ (for all x and y in H_{i+1}) is in H_i, hence every $f(x)f(y)f(x)^{-1}f(y)^{-1}$ is in $f(H_i)$. Thus, $f(H_{i+1})/f(H_i)$ is abelian.

Now we can prove the main result of this chapter:

Theorem 3 *Let $a(x)$ be a polynomial over a field F. If $a(x)$ is solvable by radicals, its Galois group is a solvable group.*

By definition, if K is the root field of $a(x)$, there is an extension by radicals $F(c_1, \ldots, c_n)$ such that $F \subseteq K \subseteq F(c_1, \ldots, c_n)$. It follows by Theorem 4 on page 330 that $Gal(F(c_1, \ldots, c_n) : F)/Gal(F(c_1, \ldots, c_n) : K) \cong Gal(K : F)$, hence by Theorem 4 on page 330, $Gal(K : F)$ is a homomorphic image of $Gal(F(c_1, \ldots, c_n) : F)$ which we know to be solvable. Thus, by Theorem 2 $Gal(K : F)$ is solvable.

Actually, the converse of Theorem 3 is true also. All we need to show is that, if K is an extension of F whose Galois group over F is solvable, then K may be further extended to a radical extension of F. The details are not too difficult and are assigned as Exercise E at the end of this chapter.

Theorem 3 together with its converse say that *a polynomial $a(x)$ is solvable by radicals iff its Galois group is solvable.*

We bring this chapter to a close by showing that there exist groups which are not solvable, and there exist polynomials having such groups as their Galois group. In other words, *there are unsolvable polynomials.* First, here is an unsolvable group:

Theorem 4 *The symmetric group S_5 is not a solvable group.*

Suppose S_5 has a sequence of subgroups

$$\{e\} = H_0 \subseteq H_1 \subseteq \cdots \subseteq H_m = S_5$$

as in the definition of solvable group. Consider the subset of S_5 containing all the cycles (ijk) of length 3. We will show that if H_i contains all the cycles of length 3, so does the next smaller group H_{i-1}. It would follow in m steps that $H_0 = \{e\}$ contains all the cycles of length 3, which is absurd.

So let H_i contain all the cycles of length 3 in S_5. Remember that if α and β are in H_i, then their commutator $\alpha\beta\alpha^{-1}\beta^{-1}$ is in H_{i-1}. But any cycle (ijk) is equal to the commutator

$$(ilj)(jkm)(ilj)^{-1}(jkm)^{-1} = (ilj)(jkm)(jli)(mkj) = (ijk)$$

hence every (ijk) is in H_{i-1}, as claimed.

Before drawing our argument toward a close, we need to know one more fact about groups; it is contained in the following classical result of group theory:

Cauchy's theorem *Let G be a finite group of n elements. If p is any prime number which divides n, then G has an element of order p.*

For example, if G is a group of 30 elements, it has elements of orders 2, 3, and 5. To give our proof a trimmer appearance, we will prove Cauchy's theorem specifically for $p = 5$ (the only case we will use here, anyway). However, the same argument works for any value of p.

Consider all possible 5-tuples (a, b, c, d, k) of elements of G whose product $abcdk = e$. How many distinct 5-tuples of this kind are there? Well, if we select a, b, c, and d at random, there is a unique $k = d^{-1}c^{-1}b^{-1}a^{-1}$ in G making $abcdk = e$. Thus, there are n^4 such 5-tuples.

Call two 5-tuples *equivalent* if one is merely a cyclic permutation of the other. Thus, (a, b, c, d, k) is equivalent to exactly five distinct 5-tuples, namely (a, b, c, d, k), (b, c, d, k, a), (c, d, k, a, b), (d, k, a, b, c) and (k, a, b, c, d). The only exception occurs when a 5-tuple is of the form (a, a, a, a, a) with all its components equal; it is equivalent only to itself. Thus, the equivalence class of any 5-tuple of the form (a, a, a, a, a) has a single member, while all the other equivalence classes have five members.

Are there any equivalence classes, other than $\{(e, e, e, e, e)\}$, with a single member? *If not* then $5\,|\,(n^4 - 1)$ (for there are n^4 5-tuples under consideration, less (e, e, e, e, e)), hence $n^4 \equiv 1 \pmod 5$. But we are assuming that $5\,|\,n$, hence $n^4 \equiv 0 \pmod 5$, which is a contradiction.

This contradiction shows that there must be a 5-tuple $(a, a, a, a, a) \neq (e, e, e, e, e)$ such that $aaaaa = a^5 = e$. Thus, there is an element $a \in G$ of order 5.

We will now exhibit a polynomial in $\mathbb{Q}[x]$ having S_5 as its Galois group (remember that S_5 is *not* a solvable group).

Let $a(x) = x^5 - 5x - 2$. By Eisenstein's criterion, $a(x + 2)$ is irreducible over $\mathbb{Q}$, hence $a(x)$ also is irreducible over $\mathbb{Q}$. By elementary calculus, $a(x)$ has a single maximum at $(-1, 2)$, a single minimum at $(1, -6)$, and a single point of inflection at $(0, -2)$. Thus (see figure), its graph intersects the x axis exactly three times. This means that $a(x)$ has three real roots, r_1, r_2, and r_3, and therefore two complex roots, r_4 and r_5, which must be complex conjugates of each other.

Let K denote the root field of $a(x)$ over $\mathbb{Q}$, and **G** the Galois group of $a(x)$. As we have already noted, every element of **G** may be identified with a

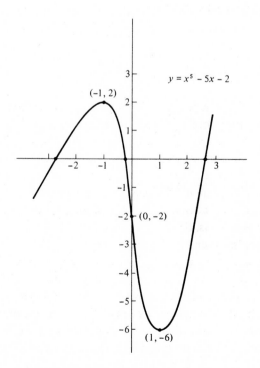

permutation of the roots r_1, r_2, r_3, r_4, r_5 of $a(x)$, so **G** may be viewed as a subgroup of S_5. We will show that **G** is all of S_5.

Now, $[\mathbb{Q}(r_1) : \mathbb{Q}] = 5$ because r_1 is a root of an irreducible polynomial of degree 5 over $\mathbb{Q}$. Since $[K : \mathbb{Q}] = [K : \mathbb{Q}(r_1)][\mathbb{Q}(r_1) : \mathbb{Q}]$, it follows that 5 is a factor of $[K : \mathbb{Q}]$. Then, by Cauchy's theorem, **G** contains an element of order 5. This element must be a cycle of length 5: for by Chapter 8, every other element of S_5 is a product of two or more disjoint cycles of length < 5, and such products cannot have order 5. (Try the typical cases.) Thus, **G** *contains a cycle of length 5.*

Furthermore, **G** *contains a transposition* because complex conjugation $a + bi \to a - bi$ is obviously a $\mathbb{Q}$-fixing automorphism of K; it interchanges the complex roots r_4 and r_5 while leaving r_1, r_2, and r_3 fixed.

Any subgroup $G \subseteq S_5$ which contains a transposition τ and a cycle σ of length 5 necessarily contains all the transpositions. (They are $\sigma\tau\sigma^{-1}$, $\sigma^2\tau\sigma^{-2}$, $\sigma^3\tau\sigma^{-3}$, $\sigma^4\tau\sigma^{-4}$, and their products; check this by direct computation!) Finally, if G contains all the transpositions, it contains everything else: for every member of S_5 is a product of transpositions. Thus, $\mathbf{G} = S_5$.

We have just given an example of a polynomial $a(x)$ of degree 5 over $\mathbb{Q}$ whose Galois group S_5 is not solvable. Thus, $a(x)$ is an example of a polynomial of degree 5 which cannot be solved by radicals. In particular, there cannot be a radical formula (on the model of the quadratic formula) to solve all polynomial equations of degree 5, since this would imply that every polynomial equation of degree 5 has a radical solution, and we have just exhibited one which does not have such a solution.

In Exercise B it is shown that S_3, S_4, and all their subgroups are solvable, hence every polynomial of degree ≤ 4 is solvable by radicals.

EXERCISES

A. Finding Radical Extensions

1 Find radical extensions of $\mathbb{Q}$ containing the following complex numbers:

(a) $(\sqrt{5} - \sqrt[3]{2})/(\sqrt[4]{3} + \sqrt[3]{4})$ (b) $\sqrt{(1 - \sqrt[9]{2})/\sqrt[3]{1 - \sqrt{5}}}$

(c) $\sqrt[3]{(\sqrt{3} - 2i)^3/(i - \sqrt{11})}$

2 Show that the following polynomials in $\mathbb{Q}[x]$ are not solvable by radicals:

(a) $2x^5 - 5x^4 + 5$ (b) $x^5 - 4x^2 + 2$ (c) $x^5 - 4x^4 + 2x + 2$

3 Show that $a(x) = x^5 - 10x^4 + 40x^3 - 80x^2 + 79x - 30$ is solvable by radicals over $\mathbb{Q}$, and give its root field. [HINT: Compute $(x - 2)^5 - (x - 2)$.]

4 Show that $ax^8 + bx^6 + cx^4 + dx^2 + e$ is solvable by radicals over any field.

(HINT: Let $y = x^2$; use the fact that every fourth-degree polynomial is solvable by radicals.)

5 Explain why parts 3 and 4 do not contradict the principal finding of this chapter: that polynomial equations of degree $n \geq 5$ do not have a general solution by radicals.

† B. Solvable Groups

Let G be a group. The symbol $H \lhd G$ is commonly used as an abbreviation of "H is a *normal* subgroup of G." A *normal series* of G is a finite sequence $H_0, H_1, \ldots, H_n$ of subgroups of G such that

$$\{e\} = H_0 \lhd H_1 \lhd \cdots \lhd H_n = G$$

Such a series is called a *solvable series* if each quotient group H_{i+1}/H_i is abelian. G is called a *solvable group* if it has a solvable series.

1 Explain why every abelian group is, trivially, a solvable group.

2 Let G be a solvable group, with a solvable series $H_0, \ldots, H_n$. Let K be a subgroup of G. Show that $J_0 = K \cap H_0, \ldots, J_n = K \cap H_n$ is a normal series of K.

3 Use the remark immediately preceding Theorem 2 to prove that $J_0, \ldots, J_n$ is a solvable series of K.

4 Use parts 2 and 3 to prove: Every subgroup of a solvable group is solvable.

5 Verify that $\{\varepsilon\} \subseteq \{\varepsilon, \beta, \delta\} \subseteq S_3$ is a solvable series for S_3. Conclude that S_3, and all of its subgroups, are solvable.

6 In S_4, let A_4 be the group of all the even permutations, and let

$$B = \{\varepsilon, (12)(34), (13)(24), (14)(23)\}$$

Show that $\{\varepsilon\} \subseteq B \subseteq A_4 \subseteq S_4$ is a solvable series for S_4. Conclude that S_4 and all its subgroups are solvable.

The next three sets of exercises are devoted to proving the converse of Theorem 3 : If the Galois group of $a(x)$ is solvable, then $a(x)$ is solvable by radicals.

† C. pth Roots of Elements in a Field

Let p be a prime number, and ω a primitive pth root of unity in the field F.

1 If d is any root of $x^p - a \in F[x]$, show that $F(\omega, d)$ is a root field of $x^p - a$.

Suppose $x^p - a$ is *not* irreducible in $F[x]$.

2 Explain why $x^p - a$ factors as $x^p - a = p(x)f(x)$ where both factors have degree ≥ 2.

3 If deg $p(x) = m$, explain why the constant term of $p(x)$ (let us call it b) is equal to the product of m pth roots of a. Conclude that $b = \omega^k d^m$ for some k.

4 Use part 3 to prove that $b^p = a^m$.

5 Explain why m and p are relatively prime. Explain why it follows that there are integers s and t such that $sm + tp = 1$.

6 Explain why $b^{sp} = a^{sm}$. Use this to show that $(b^s a^t)^p = a$.

7 Conclude: If $x^p - a$ is *not* irreducible in $F[x]$, it has a root (namely $b^s a^t$) in F.

We have proved: $x^p - a$ *either has a root in F or is irreducible over F.*

† D. Another Way of Defining Solvable Groups

Let G be a group. The symbol $H \lhd G$ should be read, "H is a normal subgroup of G." A *maximal* normal subgroup of G is an $H \lhd G$ such that, if $H \lhd J \lhd G$, then necessarily $J = H$ or $J = G$. *Prove the following*:

1 If G is a *finite* group, every normal subgroup of G is contained in a maximal normal subgroup.

2 Let $f : G \to H$ be a homomorphism. If $J \lhd H$, then $f^{-1}(J) \lhd G$.

3 Let $K \lhd G$. If $\mathscr{J}$ is a subgroup of G/K, let $\hat{\mathscr{J}}$ denote the union of all the cosets which are members of $\mathscr{J}$. If $\mathscr{J} \lhd G/K$, then $\hat{\mathscr{J}} \lhd G$. (Use part 2.)

4 If K is a maximal normal subgroup of G, then G/K has no nontrivial normal subgroups. (Use part 3.)

5 If an abelian group G has no nontrivial subgroups, G must be a cyclic group of prime order. (Otherwise, choose some $a \in G$ such that $\langle a \rangle$ is a proper subgroup of G.)

6 If $H \lhd K \lhd G$, then G/K is a homomorphic image of G/H.

7 Let $H \lhd G$, where G/H is abelian. Then G has subgroups $H_0, \ldots, H_q$ such that $H = H_0 \lhd H_1 \lhd \cdots \lhd H_q = G$, where each quotient group H_{i+1}/H_i is cyclic of prime order.

It follows from part 7 that if G is a solvable group, then, by "filling in gaps," G has a normal series in which all the quotient groups are cyclic of prime order. Solvable groups are often defined in this fashion.

E. If *Gal*(*K* : *F*) Is Solvable, *K* Is a Radical Extension of *F*

Let K be a finite extension of F, with $G = Gal(K : F)$ a solvable group. As remarked in the text, we will assume that F contains the required roots of unity. By Exercise D, let $H_0, \ldots, H_n$ be a solvable series for G in which every quotient H_{i+1}/H_i is cyclic of prime order. For any $i = 1, \ldots, n$, let F_i and F_{i+1} be the fixfields of H_i and H_{i+1}. *Prove the following*:

1 F_i is a normal extension of F_{i+1}, and $[F_i : F_{i+1}]$ is a prime p.

2 Let π be a generator of $Gal(F_i : F_{i+1})$, ω a pth root of unity in F_{i+1}, and $b \in F_i$. Set

$$c = b + \omega\pi^{-1}(b) + \omega^2\pi^{-2}(b) + \cdots + \omega^{p-1}\pi^{-(p-1)}(b)$$

Show that $\pi(c) = \omega c$.

3 Use part 2 to prove that $\pi^k(c^p) = c^p$ for every k, and deduce from this that $c^p \in F_{i+1}$.

4 Use Exercise C to prove that $x^p - c^p$ is irreducible in $F_{i+1}[x]$.

5 Prove that F_i is the root field of $x^p - c^p$ over F_{i+1}.

6 Conclude that K is a radical extension of F.

INDEX